Germ Theory

Germ Theory
Medical Pioneers in Infectious Diseases

SECOND EDITION

ROBERT P. GAYNES, MD

ASM
PRESS

Washington, DC

WILEY

Editorial Correspondence:
ASM Press, 1752 N Street, NW, Washington, DC 20036-2904, USA

Registered Offices:
John Wiley & Sons, Inc., 111 River Street, Hoboken, NJ 07030, USA

For details of our global editorial offices, customer services, and more information about Wiley products, visit us at www.wiley.com.

Wiley also publishes its books in a variety of electronic formats and by print-on-demand.

Some content that appears in standard print versions of this book may not be available in other formats.

Library of Congress Cataloging-in-Publication Data Applied for

ISBN 9781683673767 (Paperback);
ISBN 9781683673774 (Adobe PDF);
ISBN 9781683673781 (e-Pub)

Cover image: Front, top row, left to right: Antony van Leeuwenhoek, Anthony Fauci, Louis Pasteur, Françoise Barré-Sinoussi (courtesy of Institut Pasteur, credit: François Gardy); bottom row, left to right: Robert Koch, Lillian Wald, Barry Marshall, Alexander Fleming (courtesy of the Wellcome Collection, CC BY 4.0); back cover, top row, left to right: Ignaz Semmelweis, Avicenna, Paul Ehrlich, Edward Jenner; bottom row, left to right: Edwin Chadwick, Girolamo Fracastoro, Hippocrates, Joseph Lister. Background and spine image: electron microscope schematic (courtesy of David Eccles [Gringer] CC BY-SA 3.0).
Cover design: Debra Naylor, Naylor Design, Inc

Set in 10.5/13pt ArnoPro by Straive, Pondicherry, India

SKY10068098_022324

Contents

List of Illustrations

Foreword

This second and considerably expanded edition of *Germ Theory: Medical Pioneers in Infectious Diseases* by Robert Gaynes is an important and timely contribution. The germ theory is foundational to the scientific understanding of the microbial world and the threats it poses to our society and even our species. In this work the author has undertaken the ambitious task of tracing the origins, development, and impact of this idea across two and a half millennia. His sweeping, panoramic study ranges from ancient Greece to the present. Today we find ourselves in the precarious position of being subject to increasingly numerous zoonotic pathogens, as the ever-lengthening list of emerging and reemerging diseases from Avian flu to Zika to Ebola and COVID-19 reminds us. Especially in this context, *Germ Theory* provides a compelling account of how medicine has evolved—and is still evolving—in its understanding of infectious diseases and the tools it deploys to control them.

The organizational approach of the study is biographical. Each of the book's eighteen chapters explores the role of a key figure in understanding the role of microorganisms as causes of disease and promoting an array of strategies to combat them—vaccination, sanitation, public health, antibiotics. The figures, including Edward Jenner, Louis Pasteur, and Anthony Fauci, are well known. But in each case Gaynes has skillfully deepened our understanding of the process of scientific discovery and of the difficulties confronting those who propose innovative and unorthodox ideas. Scientists, physicians, historians, and other specialists will find original and illuminating material here, including the author's interviews with a number of the recent protagonists of the narrative. At the same time, however, the biographical approach humanizes the subject matter and helps to make complex scientific concepts accessible to the general reader new to the field. The author succeeds throughout in conveying the excitement and importance of the topic while avoiding jargon and obscurity. He also eschews any sense of linear triumphalism, carefully choosing instead to present the challenges and the dangers of the present with sobering clarity. Extremely well written and presented with the authority of an author who is himself a leading medical scientist, this work deserves a wide readership.

FRANK M. SNOWDEN, PHD
Andrew Downey Orrick Professor Emeritus of History,
Yale University;
former Chair, Program in History of Science and History of Medicine,
Yale University

Preface

History is simply the biography of the mind of man; and our interest in history, and its educational value to us, is directly proportionate to the completeness of our study of the individuals through whom this mind has been manifested. To understand clearly our position in any science today, we must go back to its beginnings, and trace its gradual development.

Sir William Osler

The first edition of this text was inspired by the reception my 2008 seminar received from members of the Emory Infectious Disease Division. I presented the history of our field by way of short biographies of some of the people who changed it. The seminar began at the initial stages of Western medicine in ancient Greece and ended with the discovery of penicillin and the beginnings of modern antimicrobial therapy. Faculty, fellows, postdoctoral students, residents, and medical students all appreciated the seminar and commented on their lack of acquaintance with the historical roots of their chosen discipline. A presentation at the Centers for Disease Control and Prevention (CDC) some months later yielded similar comments from those in public health.

In the midst of the COVID-19 pandemic, few would question the importance of the germ theory of disease and its effects on our world. Building on the first edition, this second edition includes a more complete discussion of the origin of the germ theory, including the research of Codell Carter, Margaret Pelling, Michael Worboys, and others who contend that the germ theory notion had its more probable beginnings in the practices of quarantine, first for leprosy and later for bubonic plague. Chapter 6 now includes a discussion of the sanitary movement and contagion in 19th-century medicine. Chapter 14 still highlights the work of Fleming, Florey, Chain, Heatley, and others in England and America on the discovery of penicillin; it adds compelling stories from continental Europe—in France, the

Netherlands, and Denmark during World War II. Scientists in these countries attempted to produce penicillin while keeping their efforts secret from the Nazis. The Danish story is told for the first time in English thanks to Dr. Peder Worning's help with the translation of Danish documents recently declassified.

The first edition ended with the story of penicillin and its wide-scale distribution in the mid-1940s, ushering in the modern antibiotic era. The second edition attempts to remedy the truncation of the first edition in the mid-20th century noted by one reviewer (1). I have added three new chapters to recognize the developments relating to the germ theory of disease that have unfolded in the last 75 years. Notably, I have interviewed Dr. Françoise Barré-Sinoussi, 2008 Nobel laureate for the discovery of HIV, the most important new pathogen discovered in the last 75 years; Dr. Barry Marshall, 2005 Nobel laureate for discovering the link between *Helicobacter pylori* and peptic ulcer disease, opening the concept that microorganisms can trigger diseases that were long considered chronic; and Dr. Anthony Fauci, Director of the U.S. National Institute of Allergy and Infectious Diseases, dubbed America's Top Infectious Disease Doctor. Fauci is well known in the U.S. for his public information on COVID-19 but has made numerous other contributions to the field of infectious diseases and immunology. All three of these noteworthy individuals are quoted from my interviews with them and have reviewed their chapters for accuracy. It was an honor to have spoken to them. I hope I have been able to place their contributions to the germ theory of disease in context to the exciting and challenging times that we face.

I chose to weave the narrative of the origins of the germ theory of disease through short biographies of 13 men and 2 women who changed the very fabric of our knowledge. Guided by others who followed a similar path—notably Sherwin Nuland, author of *Doctors: The Biography of Medicine*—I selected the biographical approach to humanize further the persons who made the significant discoveries. I also chose this style to enhance accessibility, as this book is intended not just for physicians or students of medicine but to be accessible to anyone with an interest in microbiology, infectious diseases, medical history, and, to a degree, biography. The stories of these medical pioneers demonstrate both the impact of their early life influences on their innovations and their frustrations with their societies' inability to accept some of the greatest discoveries in the history of medicine. The biographical approach illustrates how change in medical thought has occurred. Since paradigm shifts in our scientific thinking will continue, the study of historical transformations functions to encourage a requisite open-mindedness to new shifts in medical thinking.

ROBERT P. GAYNES, MD

1. **Ewald PW.** 2013. *Q Rev Biol* **88**:151.

Special Note

An essential part of the American Society for Microbiology's (ASM) mission is to embrace inclusive diversity in the STEM community, including in its book publications. Being inclusive enhances innovation, broadens the health research agenda, and furthers scientific advancement. ASM and this book's author are committed to promoting and advancing the microbial sciences through the elevation, embodiment, and sustainability of inclusive diversity with equity, access, and accountability (IDEAA).

ASM and this book's author also acknowledge that the history of science, including the history of germ theory, has been focused through a narrow lens and as a result historically excluded and underrepresented individuals were either barred entry to the field, allowed to participate in limited or special capacity, and/or had their contributions completely overlooked, in order to perpetuate the status quo. This approach and practice across history has resulted in low diversity in the field, which is reflected in the profiles contained herein. ASM's goal is that we continue advancing the microbial sciences through IDEAA. By doing so, we will begin to mitigate the lack of diversity in the field and thus ensure future historical profiles will reflect the large diversity of scientists making impacts at all levels, including those rare, paradigm-shifting contributors such as Lillian Wald and Françoise Barré-Sinoussi who are focused on in this text.

ASM and this book's author recognize that there are many more stories across the field that we might not have represented within this text. We invite the reader to share stories and feedback on individuals whom we should include as integral players and add to the historical record in the next edition. In the interest of ASM's

IDEAA mission we also encourage the reader to view the below resources to help foster the next generation of microbiologists:

Article: Inclusive Approaches to Mentoring Historically Underrepresented Groups, https://asm.org/Articles/2022/June/Inclusive-Approaches-to-Mentoring-Historically-Und

Webinar: Strengthening Career Pathways in Science for Underrepresented Groups, https://asm.org/Webinars/Strengthening-Career-Pathways-in-Science-for-Under

Meeting: Annual Biomedical Research Conference for Minoritized Scientists (ABRCMS), https://abrcms.org/

Acknowledgments

Having benefited from the feedback of a number of generous individuals, I want to thank Kirvin Gilbert, Lisa Macklin, James Curran, Alicia Hidron, Elissa Meites, Abeer Moanna, Mark Mulligan, David Rimland, Michael Schlossberg, Robert Rosman, Peter Rogers, and Harold Jaffe for their patience, sage advice, and encouragement.

I am particularly indebted to Drs. Françoise Barré-Sinoussi, Barry Marshall, and Anthony Fauci for their willingness to be interviewed and their subsequent review of their respective chapters for accuracy. I also want to recognize the love and support from my children, Sara and Matthew, and most of all, my wife, Sherry.

I hope that this final product will impart to the reader the knowledge, understanding, and passion that I discovered in writing it.

About the Author

Robert P. Gaynes, MD, is a Professor Emeritus of Medicine (Infectious Diseases) at Emory University School of Medicine, where he continues to teach the history of medicine.

After graduating *magna cum laude* from the University of Illinois in Urbana, Dr. Gaynes earned his medical degree from the University of Chicago Pritzker School of Medicine with honors in 1979. He completed a residency in internal medicine at Michael Reese Hospital in Chicago.

After serving for 2 years in the Epidemic Intelligence Service at the Centers for Disease Control and Prevention (CDC), he returned to complete a fellowship in infectious diseases at the University of Chicago Hospitals and Clinics. In the 1980s, he served as an Assistant Professor of Medicine and Hospital Epidemiologist at the University of Michigan Hospitals in Ann Arbor, MI.

From 1989 to 2009, Dr. Gaynes worked at the CDC in several positions in the Division of Healthcare Quality Promotion and for over a decade as Chief of the Surveillance Activity in the Hospital Infections Program and as the Director of CDC's National Nosocomial Infection Surveillance System.

From 2009 to 2022, he served as an attending physician and the Chair of the Infection Control Committee, Antimicrobial Stewardship Committee, and COVID-19 Vaccine Planning Committee at the Atlanta VA Hospital. As Professor of Medicine at Emory University School of Medicine during this period, Dr. Gaynes lectured on various infectious disease topics and taught courses on the history of medicine. He has authored or coauthored more than 150 papers and book chapters on infectious disease topics.

Board certified in Internal Medicine and in Infectious Diseases, Dr. Gaynes is a Fellow of the Infectious Diseases Society of America. In addition, he is a reviewer for numerous scientific journals and served on the Editorial Board of *Infection Control & Hospital Epidemiology.*

He is a husband, father, and grandfather who enjoys history, racquetball, gourmet cooking, and travel.

1 Introduction

What is the greatest contribution that medicine has made to humanity? In 2007, more than 11,000 readers of the *British Medical Journal* were surveyed to answer that question. The answers were (in order): (i) public health sanitation; (ii) discovery of antibiotics; (iii) discovery of anesthesia; (iv) discovery of vaccination; (v) discovery of the structure of DNA; and (vi) discovery of the germ theory of disease (1). Of the greatest contributions ever made to medicine, four of the top six were either the discovery of the germ theory of disease or innovations that were a direct result of that discovery. While the answers to this provocative question can be argued, the germ theory remains one of the most important contributions in the 2,500-year history of Western medicine and a relatively recent one—only 150 years old.

Modern society has taken for granted the value of public health sanitation; the vanishing of vaccine-preventable diseases such as smallpox, measles, and polio; and the existence of antibiotics to treat infectious diseases. In the 1970s, we even had the hubris to declare infectious diseases "conquered." However, infectious diseases have emerged or reemerged to devastate our modern world. I didn't realize when I made my decision to go into the specialty of infectious diseases in 1978, my last year of medical school, that I would be witness to this emergence. I thought that antibiotic treatments could actually cure people, not just treat them. Unlike chronic illnesses such as diabetes, bacterial pneumonia or a urinary tract infection, once treated with antibiotics, could be cured. I was not alone in this thinking.

I entered the specialty at a time when it was believed that medical science had nearly done it all—that there would be little left to do since we had such powerful agents for treating and curing infectious diseases. I recall my first meeting of the Infectious Diseases Society of America in 1981, when one of the foremost authorities in the field told the audience that "all infectious disease doctors would be doing in the next decade would be culturing each other." This complacency would be short-lived; such complacency has always been short-lived in medicine. During the same meeting, James Curran, at the time working at the Centers for Disease Control (CDC) and now recently retired as dean of the Rollins School of Public Health at Emory University, was scheduled to describe the first cases of what would soon be called AIDS (acquired immunodeficiency syndrome). Curran was the last of four speakers in a session that began with 150 people in the audience. But word of these cases had spread through the conference. By the time Curran spoke, over 1,000 people had crammed into a room at a downtown hotel in Chicago, IL. Even though the attendees at the meeting were told that infectious diseases were "conquered," the infectious disease community of doctors, microbiologists, and public health officials had clearly recognized that something new was happening.

Many new diseases, such as AIDS, hepatitis C virus, hantavirus, SARS (severe acute respiratory syndrome), MERS (Middle East respiratory syndrome), Zika, and COVID-19 infection, have been described in the last 40 years. The microorganisms that cause these diseases were not actually new—they clearly existed before medical science became aware of them. In some cases, new techniques were developed to identify organisms that had always been there but that we could not detect. More often, the novelty for many of the new diseases is not the emergence of a new microorganism but the novel way the pathogen found its way into humans. Microorganisms are now thought to trigger diseases long considered to be chronic in nature, including peptic ulcers, numerous cancers, arthritis, type 1 diabetes, and others. In 1981, no one imagined how one of these newly emerged diseases, AIDS, which is caused by human immunodeficiency virus (HIV), would change the human landscape of entire continents, alter our conceptual framework of how we treat or even think about an infectious disease, and completely confound our understanding of the human immune system and vaccine development. Additionally, the upheaval caused by a virus causing respiratory disease that began in China at the end of 2019 was almost unimaginable until it spread rapidly through the world, resulting in the worldwide COVID-19 pandemic, a story that is still being scripted at the time of this writing.

Consider the vast changes wrought by HIV. In sub-Saharan Africa, HIV has completely altered the human demographics. For many decades, even centuries, before HIV, infectious diseases were the leading cause of death in Africa, but the deaths were childhood deaths. Malaria, infectious diarrhea, measles, and other childhood infectious diseases took their toll on the youngest inhabitants. As tragic as it sounds, most children in Africa did not see their first birthday, a fact that was simply accepted. The dynamic was high birth rates and high childhood death rates. In the course of a generation, HIV became the leading cause of death in Africa, but the deaths were not childhood deaths. The mortality rates among adults from HIV exceeded even the highest childhood mortality rates. In 2007, according to the Joint United Nations Program on HIV/AIDS, 60% of all HIV infections in the world were in Africa, even though Africa has only 12% of the world's population. In 2007, the life expectancy in Africa was 47 years with HIV and 62 years without HIV. In 2022, Africa remains the most severely affected region, with nearly 1 in every 25 adults (3.4%) living with HIV and accounting for more than two-thirds of people living with HIV worldwide (https://www.who.int/data/gho/data/themes/hiv-aids#cms). The medical, economic, social, and political effects of this one infectious disease are so great that the capacity to cope with its burden is stretched thin and, in many cases, has nearly collapsed. Curran was a witness to AIDS during its first 40 years in our modern world and described it this way:

> When you walked in the Castro district [of San Francisco] in the early 1990s, AIDS was palpable. The same thing is still true in some African cities. The hospital wards of African hospitals are filled with AIDS patients, sometimes two to three to a bed with young adult people dying. An AIDS death prior to antiretroviral therapy was a horrible process for most people. They lost weight, became demented, had unremitting diarrhea, Kaposi's lesions all over themselves, and people did not want to be near them. (J. Curran, unpublished data)

In Africa, while mortality for HIV-positive patients has improved since 2007 (see chapter 17, PEPFAR Successes), this kind of lonely, horrific death still occurred more than 420,000 times in 2021. HIV has forced a change in our social attitudes. Consider the sexual attitudes before HIV appeared. For centuries, syphilis was the most feared sexually transmitted disease since it could not be effectively treated until the 20th century. During the early 1900s, this disease became treatable with arsenic-based compounds. The concern about syphilis virtually disappeared with the introduction of penicillin. In the 1960s and 1970s, people had the perception that there were no incurable sexually transmitted pathogens. True, genital herpes existed and was a concern. While genital herpes was incurable, it did not cause death and could be treated with medicines. Sexual health could simply be managed by a visit to the doctor. Along came HIV. Once its sexual transmissibility was established, we were faced with a sexually transmitted pathogen that not only could not be handled by a trip to the doctor but also caused death. Sexual practices and attitudes changed. Among adolescents, for example, the rate of unprotected sex has decreased (2). But relentless efforts are needed for each new generation to implement these behavioral interventions.

The medical community itself has been forced to change its thinking because of HIV. Biologically, the silent nature of the infection, i.e., the long incubation period of years, even decades, leaves us with tens of millions of human incubators. Many HIV-positive individuals do not even know that they are infected. When the CDC reported the first five AIDS cases in 1981, an estimated 250,000 people were already infected with HIV in the United States alone. HIV has a silent, long incubation period followed by a lingering illness. Infectious disease physicians were just not equipped to face a chronically ill patient population in such vast numbers. The appearance of HIV/AIDS forced a complete reexamination of what we thought we knew about treating infectious diseases. Contrast HIV to influenza, which crashes into a community and leaves in a few short weeks. With HIV, an infectious disease becomes a chronic illness.

As the established paradigms for treating vast numbers of AIDS patients broke down for infectious disease doctors, public health officials struggled with the time-honored models they had used for monitoring infectious diseases. When a new infectious disease enters a community, it is often called an epidemic. The Merriam-Webster dictionary defines an epidemic as the occurrence of more cases of a disease than would be expected in a community or region during a given period of time. With some infectious diseases, an epidemic is easy to spot. Often, making the determination that more cases of an infectious disease have occurred than would be expected in a region can be challenging. One initial and basic approach used in public health to make this determination is development of an epidemic curve, a technique I learned on my first day of work at the CDC. After a known exposure, the number of cases of an infectious disease is plotted against the time of onset of illness among individuals with the disease to graph an epidemic curve. The shape of the epidemic curve might suggest the mode of transmission of the infectious agent. An epidemic curve where all the cases show signs of infection at nearly the same time and the number of cases quickly tapers off suggests that an infectious agent was transmitted from a common source, e.g., contaminated food at a picnic. With a different type of infectious disease, e.g., influenza, the curve may show an initial or index case, followed by two cases, then four, then eight, etc., until all individuals who are susceptible in a community have acquired the infection and the number of new cases decreases. This shape of an epidemic curve suggests person-to-person transmission of an agent. There are combinations of shapes in epidemic curves, but the overall shapes of these curves are the same. The number of cases goes up and then comes back down. In public health, the goals are to shorten the epidemic or work to prevent the problem or a similar problem from occurring, i.e., decrease the width and/or height of the epidemic curve and prevent the occurrence of any epidemic curves in other communities. This is the infectious disease epidemic paradigm. Doctors and public health officials were accustomed to this approach. In fact, we thought we were pretty good at it. HIV was different. HIV produced a chronic infectious disease, one in which the virus entered the body, remained silent (although it could be transmitted) for years, and then produced a lingering disease in which the virus did not disappear in an individual while he or she remained alive, even during treatment. The curve measuring HIV cases went up but did not come back down. Other viruses, such as hepatitis B or C virus, had long incubation periods before illness developed and were known before HIV. However, only a portion of patients initially infected with hepatitis B or C developed chronic illness, although we continue to grapple with people with chronic liver disease from hepatitis B and C. The rapid spread of HIV was coupled with an almost completely ineffective immune response from a person who had the virus. The virus was not eliminated by antibody production. In fact, HIV went on to damage the very immune system we had always thought would eliminate a pathogen. HIV's ravage of the immune system left the person susceptible to other infectious agents that he or she would normally not be susceptible to, so-called opportunistic infections. In the first few years of AIDS, it was invariably fatal within 4 years of an AIDS diagnosis. No one was prepared to deal with an infectious agent that could produce disease like this. The mentality of the medical community had to change to deal with a chronic infectious disease. Public health turned to the chronic disease experts for guidance on how to predict, moderate, and control it. We could not cure HIV and move on to something else. In the United States, we have been somewhat comforted by the success of treatment of HIV, which has improved dramatically since the mid-1990s. Life expectancy of adults with HIV infection who take combination antiretroviral drug therapy may now be near that of life expectancy of individuals without HIV infection (3). Still, treatment does not cure the individual. Indeed, intermittent treatment often leads to changes, i.e., mutations, in HIV that result in treatments that are ineffective or more difficult to administer, a problem that has troubled infectious disease management since the advent of modern antimicrobial therapy. Success in HIV treatment in the United States and elsewhere remains challenging and expensive. In Africa and other less developed regions, the prospect of treatment has improved since 2010 but there are still more than 27 million people in Africa with HIV. The cost of drugs and the infrastructure to deliver and monitor drug treatment will remain substantial. The "what goes up must come down" or epidemic thinking will not solve HIV or other endemic infectious diseases, such as tuberculosis or malaria, in Africa. We have had to reconsider our infectious disease treatment/prevention model to one that effectively approaches chronic or endemic infectious diseases. According to Curran, "this model is far less heroic, but arguably more important. And that's why there needs to be a change in thinking" (Curran, unpublished).

Nowhere has HIV challenged our conceptual framework more than in the development of an HIV vaccine. The world desperately needs an effective immunization for HIV as the means to control or even eliminate AIDS. Immunizations have been one of the most effective methods for control of disease, even eradicating the scourge of smallpox. The traditional approach to immunization formulation is to kill or inactivate a pathogen and inject it into people. Vaccines formulated in this way rely upon duplicating a successful human immune response to the natural exposure of a pathogen. Following the discovery of HIV as a causative agent of AIDS in 1983, the expectation was to rapidly develop a vaccine but, as of 2022, we still do not have a licensed vaccine. Progress has been hindered by the extensive genetic variability of HIV. We still do not have a complete understanding of immune responses required to protect against HIV acquisition (4). The traditional methodology of vaccine development has not worked for HIV.

HIV has been the most visible pathogen that has changed the concept that infectious diseases could be conquered, a concept that, according to Curran, "is now not credible" (Curran, unpublished). We have every reason to believe that other microorganisms might find their way into humans as we carelessly expand our way into previously undisturbed ecosystems. A new pathogen emerging anywhere in the world is cause for worldwide concern. The most obvious example is SARS coronavirus-2 (SARS-CoV-2). The rapid geographic dispersal of this virus causing COVID-19 infection is a clear reminder of how small our world has become. But there are other worries that shake our notion that infectious diseases have been subjugated.

The complacency that led us to the notion that we have conquered infectious diseases was due to our apparent successes—immunizations and antibiotics. Until HIV, development of vaccines was a major success of our modern medical world. The chief difficulty was administering the vaccines to the people who needed them. Vaccines were and still are underutilized. But immunizations can only prevent illness. Antibiotics treat illness. From the discovery of the first antibiotic, medical science and microorganisms have been in a constant struggle. Since the beginning of modern antibiotic therapy, resistance to antibiotics has steadily increased, and it is increasing faster than we can discover new therapies. Recent discoveries of vancomycin-resistant *Staphylococcus aureus* and bacteria containing enzymes called carbapenemases that degrade some of the most potent antibiotics, the carbapenems, are but two examples of the difficulties facing contemporary treatment of infectious diseases. But we cannot simply blame the appearance of antibiotic-resistant pathogens on clever, evolutionary tricks of microorganisms.

The widespread resistance among bacteria is a problem largely of our own making. Antibiotic use leads to antibiotic resistance. While it is blatantly obvious, most health care providers generally ignore this statement. Our infectious disease treatment paradigm involves use of broad-spectrum antibiotics, usually without specific knowledge of the offending pathogens. There is hidden harm that occurs as a result of this antibiotic exposure. Sometimes a pathogen causing an infection will begin susceptible to the agent but become resistant to it during therapy. But often, antibiotic resistance is actually the result of collateral damage. The resistance develops among pathogens that are not responsible for the infectious syndrome manifested by the patient. Humans normally harbor bacteria in the mouth, the colon, the skin, and other sites. The normal bacterial flora in humans is immense and often protects the body when harmful bacteria invade. When powerful antibiotics destroy much of the normal, protective bacteria, the void is filled by small numbers of resistant bacteria that we all may harbor. These resistant subpopulations of bacteria are given a chance to proliferate after antibiotic therapy. The resistant organisms may not even cause damage to the person treated with broad-spectrum drugs, but this person may become a reservoir or a human factory of a silent source of resistant organisms to those around him. Infection control efforts often fail since no one may be aware that the recently treated individual harbors antibiotic-resistant organisms. The number of people colonized but not infected with a resistant organism may be 50 times greater than the number of people who have an infection with a resistant organism. This so-called iceberg effect, where most of the resistance problem is hidden below the surface where we cannot see it, has been best described for hospitals, but resistance has become a problem outside hospitals. Community-acquired methicillin-resistant *S. aureus* (MRSA) infecting large numbers of patients who had not been in hospitals has been described in recent years. MRSA can be spread from one person to another. Our attempts to limit spread of this organism through infection control efforts appear to have been largely ineffective. Current attempts at hospital-based efforts to control MRSA are controversial and may be too little, too late, since so many MRSA infections now occur outside hospitals. Indeed, determining the effectiveness of control measures for MRSA or other resistant microorganisms is complex. Ultimately, the reservoir of the resistance will be found to involve exposure to antibiotics, perhaps in a person's distant past or in another person who served as the source of the MRSA.

The proliferation of antibiotic-resistant organisms is challenging the current tenet of infectious disease practice—use broad-spectrum antibiotics to "kill the bugs" with little or no concern for the collateral damage of antibiotic resistance. We have abused one of our greatest medical treasures, antibiotics. Abuse of antibiotics has opened the door to a new era with untreatable antibiotic-resistant infections, a post-antibiotic era. Pharmaceutical companies are not going to bail us out of this problem, since most have limited or even eliminated antibiotic drug discovery programs, largely because of economics. For example, from 2019–2022 there were no new antibacterial antibiotics approved by the Food and Drug Administration in the United States (https://www.fda.gov/drugs/new-drugs-fda-cders-new-molecular-entities-and-new-therapeutic-biological-products/novel-drug-approvals-2022). While some progress has been made in the last 25 years, finding new antibiotics has been time-consuming, difficult, and expensive, with no guarantee of success.

The dilemmas from newly emerging infectious diseases—HIV, an often fatal, chronic infectious disease, and SARS-CoV-2, which have stressed, to the breaking point in some places, our ability to prevent infection, find treatment, and provide it to millions; HIV vaccine failures that require a complete reassessment of the concepts of the human immune response; untreatable, antibiotic-resistant infections that portend a post-antibiotic era; and the unsettling notions that new pathogens like SARS-CoV-2 or a microorganism that causes disease even worse than COVID-19 infection will likely emerge, and that a problem that emerges in one corner of the globe can be rapidly transported anywhere in the world—have forced the medical and public health community to scrutinize itself. Medical science must reconsider all approaches to detecting and diagnosing emerging infectious diseases, to developing vaccines, to discovering new antibiotics, and to using wisely the precious few antibiotics that we have.

Where do we start? Before we can figure out where we should go, we should examine how we got here. We should look back to our initial advances in understanding infectious diseases. Appreciating how major contributions were made by some of the greatest doctors in history will help us reconsider our approaches to the practice of infectious diseases. Just as economists and policymakers look back to the Depression to develop policies to relieve us from economic woes, we should examine the history of infectious diseases. Historians debate the best approach to study the past. Thomas Carlyle, a well-known historian in the mid-1800s, once wrote, "The history of the world is but the biography of great men" (5). As an amateur historian, I have chosen to view history this way in order to detail the lives of 13 men and 2 women who were pioneers of our current conceptual framework of infectious diseases. Here, I must make a few qualifying remarks. First, the conceptual framework for the practice of infectious diseases is from Western medicine. This book is largely limited to Western medicine. Some readers may have a limited appreciation for the Eurocentric approach. Yet while my purpose is not to glorify Eurocentric concepts, Europe is where our conceptual framework for the practice of infectious diseases medicine was conceived. Contributions from other cultures, as germane as they are to a holistic perspective on the method of discovery, were largely ignored by Western medicine—often to its detriment. Second, there will be those who question the choice of the individuals who were included. While I doubt that one can argue against those included in this book, the persons excluded from any discussion will no doubt generate controversy. I have chosen individuals whose contributions were essential to understanding the origins of our approach to the practice of infectious diseases medicine and whose stories were compelling to me. I will admit to small detours into anecdote to better understand the person. The influences on the lives of these discoverers, often early influences, the era in which they lived, their personalities, and, in many cases, their daring, permitted these individuals to make extraordinary contributions even when those contributions were not initially well received. Discoveries made by an individual need to be accepted into the practices of the community before they can actually become advances. As we shall see, at times, the discovery had little impact until the practices of the community were reevaluated. These reexaminations provide lessons and, perhaps, some hope as we are forced to look introspectively toward our own practice of infectious diseases medicine.

REFERENCES

1. **Ferriman A.** 2007. *BMJ* readers choose the "sanitary revolution" as greatest medical advance since 1840. *BMJ* **334**:111. http://dx.doi.org/10.1136/bmj.39097.611806.DB.
2. **Eaton DK, Kann L, Kinchen S, Shanklin S, Ross J, Hawkins J, Harris WA, Lowry R, McManus T, Chyen D, Lim C, Whittle L, Brener ND, Wechsler H, Centers for Disease Control and Prevention (CDC).** 2010. Youth risk behavior surveillance—United States, 2009. *MMWR Surveill Summ* **59**:1–142.
3. **Marcus JL, Leyden WA, Alexeeff SE, Anderson AN, Hechter RC, Hu H, Lam JO, Towner WJ, Yuan Q, Horberg MA, Silverberg MJ.** 2020. Comparison of overall and comorbidity-free life expectancy between insured adults with and without HIV infection, 2000–2016. *JAMA Netw Open* **3**:e207954.
4. **Ng'uni T, Chasara C, Ndhlovu ZM.** 2020. Major scientific hurdles in HIV vaccine development: historical perspective and future directions. *Front Immunol* **11**:590–780.
5. **Hirsch ED.** 2002. *The New Dictionary of Cultural Literacy*, 3rd ed. Houghton Mifflin Harcourt, Boston, MA.

2 Hippocrates, the Father of Modern Medicine

The history of infectious diseases and the practices that attempted to treat them is largely a history of medicine itself. So, we must begin with the foundation of Western medicine. Physicians command a high esteem compared with those engaged in nearly any other occupation around the world. This distinguished societal view of physicians originated in ancient Greece 2,500 years ago, when the medical profession assumed a new and distinctive character. To understand this transformation, consider how illness was treated before the time of Hippocrates and his colleagues.

MEDICINE BEFORE HIPPOCRATES

In the centuries before Hippocrates (circa 460 to 370 BCE [Fig. 2.1]), ancient Greeks believed that divine interventions were required for curing illness. This belief was perhaps understandable when one considers how rapidly certain illnesses can afflict someone. Infections can have the swiftest onset of all disease entities—one moment someone is healthy, and the next moment, he or she is seriously ill. I have had patients describe the precise moment, down to the hour and minute, when symptoms of one of the most common bacterial causes of pneumonia, pneumococcal pneumonia, began. It is understandable that humans would have turned to a higher power to explain this rapid, otherwise inexplicable event. Around 800 BCE, temples/hospitals sprang up throughout the ancient Greek world. These shrines were called asclepieia, after Asclepius, Greek god of

Germ Theory: Medical Pioneers in Infectious Diseases, Second Edition. Robert P. Gaynes.
© 2023 American Society for Microbiology.

FIGURE 2.1 *Hippocrates. Courtesy of the National Library of Medicine (NLM image ID B014575).*

medicine. The origins of the principles behind the asclepieia and the practices found inside these temples are murky. If you were ill during that time, you would go to the temple/hospital to ask a priest/physician what you might do to effect a cure. Asclepieia were more like spas than hospitals. They were often located in beautiful country settings and placed an emphasis on pure air, clear water from springs which probably had high mineral content, exercise, healthy diets, and rest. There was much fanfare from spiritual priests who cultivated a restful atmosphere. In the resort-like setting, the mystical priests were thought to have a direct communication with divinities. The straightforward, sensible approach to diet, rest, and exercise would be viewed as conducive to restoring health even today. However, any similarity to modern restorative health spas disappears when one considers what the ancient Greeks accepted for improvement of their health. Aesculapian snakes, endemic to areas in Greece, "anointed" the temples and were encouraged to slither over the ill. Snakes were considered sacred to the god Asclepius, who is usually pictured with a serpent wound around his long staff, a symbol for medicine still seen today. This symbol, common among many of today's medical organizations, such as the World Health Organization and the American Medical Association, may seem odd to some. If a modern advertising agency were to ask a focus group for a symbol for medicine—one that represented trust, admiration, and respect, it seems unlikely that the symbol chosen would be a snake! Yet ancient societies often worshipped serpents and regarded snakes as creatures with the gift of life and renewal, probably related to the snakes' shedding of their skin. Still, the staff of Asclepius is the first purely medical association of snakes (1). In addition to the physical presence of snakes in the facility, the imagery of snakes was considered critically important. The ill would sleep in the facility, sometimes for several days. Each morning, a person would relay his or her dreams to the priest, who would interpret them. The images from the dreams would dictate what was required for cure. If a snake appeared in the dream of someone sleeping in the temple, it was believed that the god of healing, Asclepius, himself had appeared to the person to effect a cure. I suspect that Sigmund Freud might have had a different interpretation of that vision. During this time in ancient Greece, disease was the result of divine influence, so treatment required similar influence. If the patient improved, the gods got the credit. If the person did not improve, the individual was to blame, for he or she must have done something to anger the gods.

Every physician can recall a patient who placed enormous faith in the divine power of healing. There is nothing intrinsically adversarial in combining medical science with divine faith. But in its infancy, medicine could not move forward without a separation from the supernatural. A field that believes that the causes and treatment of illness are based solely on invisible and unknowable forces is contrary to a discipline in which the causes of disease can be discovered, and treatment can be aimed at those causes. The move away from divine intervention was a crucial step for medicine. One of the greatest contributions of the ancient Greeks, perhaps its greatest, was the conceptual shift to the belief that disease has a natural, knowable cause. Once Western medicine declared its independence from religious practices, it could advance into a truly scientific discipline.

HIPPOCRATES

The first individual in Western civilization to be known for his contributions to medicine from the ancient Greek civilization is Hippocrates. While over 70 surviving volumes of writings are attributed to him, historians have determined that different authors actually wrote these texts. Because of the authorial uncertainty, the collective works have been called the Hippocratic Corpus or body of work, and the physicians who practiced the art, the Hippocratic physicians. For unassailable fact, we know almost nothing of Hippocrates' life. We do know that Hippocrates was born in 460 BCE on the island of Cos, off the coast of present-day Turkey, but that's about it. Every other description of Hippocrates' life is impossible to verify and has been combined with or distorted by legend. The only verification of the tiny biographical fact about his birth comes from two dialogues of Plato, the *Protagoras* and the *Phaedrus*. Historians in the 1800s fruitlessly labored to decipher which of the surviving ancient Greek texts were written by Hippocrates (2). It is said that Hippocrates lived for a long period, somewhere between 85 and 110 years. We do not know what he looked like, although there are multiple statues said to convey his likeness. These statues differ in appearance, but all portray Hippocrates as a sensitive, wise, and distinguished man. His contemporaries, from a period bursting with intellectual activity, included Socrates, Plato, Aristotle, Pericles, Euripides, Aeschylus, Sophocles, and Aristophanes. In the 60-year period when Hippocrates was at the peak of his career, Western civilization witnessed the foundations of art, architecture, political science, philosophy, drama, and, of course, medicine.

NATURAL CAUSE FOR DISEASE

Medicine's move away from the supernatural to natural causes of illness was, like most paradigm shifts, a product of its time and sprang from revolutionary ancient Greek philosophies that seemingly had little to do with medicine. We do not precisely know the time when the shift occurred. The change in thinking seemed to take place not in one of the great city-states such as Athens but on the edges of the Greek peninsula (3). The century before Hippocrates was the time of Thales, circa 640 to 546 BCE, perhaps the first true scientist of Western civilization. Like Hippocrates, he lived on an island in the Aegean Sea, Miletus. Sea trade was the foundation of the culture, whether it was trade of goods or ideas. Philosophies were eventually transported around ancient Greece like assorted materials. The philosophy that Thales developed influenced his students, whose writings serve as the basis for what we know about him. He made contributions to mathematics, astronomy, navigation, and geometry. Thales was perhaps the first person in Western civilization to believe that the physical world was comprehensible, not a result of divine deeds or gifts. He seemed to accept the existence of gods but did not rely on divine actions to explain the physical world. Thales theorized that the basic elements in all life were water (the primary element), earth, and air. Anaximander, from 560 BCE, extended Thales' thoughts and espoused the philosophy that the universe was composed of opposite forces in balance. Natural, understandable, and universal laws governed these forces. The philosopher Heraclitus, from about 500 BCE, living north of Miletus on Ephesus, contended that fire, rather than water, was the principal element but embraced the idea that the universe was composed of opposing forces that required balance. By the sixth century BCE, four basic elements were accepted as the building blocks of the universe: water, earth, fire, and air. Each element had its corresponding characteristic: wet, dry, hot, and cold. These elements were physical and observable, and they normally existed in a type of oppositional balance. This doctrine of four basic elements and their physical characteristics found in a balance would profoundly influence the medical theory of the Hippocratic Corpus, and medicine in general, for centuries.

Alcmaeon, a member of the Crotona school in the fifth century BCE, was perhaps the first person to move the cosmic philosophy of the basic four elements of the universe to a focus on humans and their health. Works by Aristotle provide the principal sources for what we know about Alcmaeon's views. Alcmaeon held that like the universe, health is a harmony of forces, and disease is a disturbance of that harmony. Empedocles (493 to 433 BCE) furthered this concept and has sometimes been credited with formation of the theory that health is a balance of forces. According to Empedocles, the four elements are joined together during life and are separated upon a person's death. He advanced the importance of air as having substance and the ability to exert pressure. The flow of air was connected to the flow of blood in the body. During this period, Democritus influenced medical and scientific thought with a theory that all objects, animate and inanimate, were composed of invisible particles, or atoms. At this time of intellectual ferment in Greek history, Hippocrates was born.

THE FOUR HUMORS AND DISEASE

Although we remain unsure as to what Hippocrates, the man, actually contributed, the Hippocratic Corpus provides fairly clear principles of health and disease. These principles brought together the previous theories that the universe was composed of four elements (water, air, fire, and earth) with the specific qualities associated with each element (wet, dry, hot, and cold) and the concept of harmony or balance. Additionally, certain of these elements were in relative abundance during each of the four seasons of the year. A key principle in Hippocratic theory was that four visible secretions of the body, blood, phlegm, yellow bile, and black bile, could be related to the four basic elements of the universe. These four secretions, or humors, shared essentially the same physical characteristics—hot (blood), cold (phlegm), moist (yellow bile), and dry (black bile)—as fire, air, water, and earth.

The human body was a microcosm of the universe, a concept that must have been pleasing to the Greek intellect of the time. Most importantly, the causes of illness were not divine but knowable and followed logically from understanding the balance inherent in the universe.

Hippocrates is said to have learned medicine from his father, Heraclides. The sketchy biographical information available on Hippocrates states that he traveled extensively and practiced medicine in cities and islands throughout ancient Greece, where he was probably exposed to a variety of philosophies. He is credited with setting up a school on his home island of Cos, the so-called Coan school, which had a holistic approach to health and disease. Hippocratic physicians believed that illness must be interpreted as an event that occurred in the circumstances of the patient as a whole. The concept of harmony or balance was central to this philosophy. The specific diagnosis was less important to the Hippocratic physician than helping the body restore a natural balance to its native state. Prognosis was much favored over diagnosis. Given the limited medical knowledge of the day, this preference served them well. The focus for the Hippocratic physician was to attempt to determine which symptoms represented serious disease or illness that would improve with their (limited) therapy.

THE COAN SCHOOL OF MEDICINE

The Coan school exerted the main influence on medicine in ancient Greece but was not the only medical influence of this age. The most significant was a separate school in the town of Cnidus, facing Cos on the mainland of Asia Minor, called the Cnidian school. Like the Coan school, the Cnidians held that the causes of disease were knowable. However, their approach to their patients more closely resembled that of modern medicine. The Cnidians took great efforts to distinguish various diseases in each organ. For example, they categorized 12 diseases of the bladder and 4 in the kidneys. Extensive categorization, a common method used by the Greeks at the time, was largely based upon the principle that diseases were caused by nutrients that were not discharged from the belly. The leftover residues were responsible for a variety of illnesses. The Cnidian approach to medicine was one of diagnosis, not prognosis. Given the rudimentary knowledge of anatomy and physiology of the day—human dissection was prohibited in ancient Greece—the Cnidian approach suffered greatly since one could not have offered much after the effort in diagnosis was undertaken. Diagnosis is of little use unless it can be followed by a specific treatment. The usefulness of the Cnidian view would have to wait until medical knowledge could progress in anatomy, physiology, and even cellular function. The wait would be a long one—over 2,000 years. Although medical knowledge was extended during the next 20 centuries, progress in Western medicine was hampered by the steadfast adherence to humoral theory and the Coan approach to disease. Elements of the germ theory of infectious diseases were discovered during these 2,000 years of waiting but could not be incorporated into medicine until the holistic approach of the humoral theory of disease was fully abandoned. In its time, the Coan school fared better than the Cnidian school, gaining greater fame and influence than the Cnidian methods, although these two schools probably shared knowledge.

The Coan school maintained that personal health was a natural mixture of the four humors: blood, phlegm, yellow bile, and black bile. Particular temperaments or personalities were associated with a relative abundance of a particular humor. The person with a larger quantity of black bile (from the Greek, *melan* = black and *cholic* = bile) would be said to be depressed, introspective, and sentimental. Our modern-day words for depictions of certain personality types—melancholy, sanguine (from the Latin, *sanguis* = blood), choleric (from the Latin, *cholericus*), and phlegmatic (from the Latin, *phlegmaticus*)—can be traced directly from the humoral theory, with the same connotations of their human personality descriptions as in Hippocratic times.

The Hippocratic physicians had a solid grasp of the exterior anatomy of a human but were essentially clueless about what happened beneath the surface. The relationship between health and disease had to be deduced from what went into the body and what could be observed to come back out. The fundamental principle of the humoral theory considered disease to be a disruption of the balance of humors or a dyscrasia, from the Greek, meaning bad mixture of humors. Surprisingly, this term is still found in modern medical lexicon, often used by hematologists. Unsure of its precise nature, hematologists sometime will refer to a patient's undiagnosed blood disease as a blood dyscrasia. Disease, caused by the excess or deficiency of one or more of these humors, could be from external or internal forces. The seasons of the year or climate could affect this balance. For example, phlegm, which the ancient Greeks found to be colder than the other humors, increases in winter. Winter's natural attribute, cold, could explain the tendency for certain diseases that had an excess of phlegm, the humor with the cold attribute, to occur. Alternatively, the lifestyle of the patient, particularly the food he ate, could also affect the balance. The final element that affected the humoral balance was pneuma, or vital air. The doctrine of pneuma is never well explained in the Hippocratic Corpus. Further explanations of its "physiology" would occur a century later with the practice of Alexandrian medicine in Egypt. Inhaled air was thought to go to the brain first, a logical consideration since the nose and mouth are near the brain. From the brain, pneuma spreads downwards from the great vessels of the body, through minor vessels to the rest of the body (4). Changes in the humidity or winds could affect pneuma. Changes in pneuma could disrupt any of the four humors.

Three of the four humors, yellow bile, phlegm, and black bile, tended to be observed only when illness occurred. Blood seemed to hold a different position in the humoral theory, being an essential element of life. Still, there were times when it seemed to be naturally discharged from the body, such as during menstruation or nosebleeds, leading Hippocratic physicians to devise the first rationale for the practice of bloodletting to remove any excess.

FEVER AND THE FOUR HUMORS

The heart's function as a pump was unknown to the ancient Greeks, but they recognized its essential importance to the body. The heart's main function was thought to be involved in the production of innate heat. The heart attracted pneuma, much as a flame requires air, which was spread through the body through the arteries. When pneuma combined with blood, innate heat was generated. To balance the heating action of the beating heart, the ancient Greek physicians ascribed functions to several anatomic organs. The sac that surrounds the heart, known as the pericardium, was observed to normally contain a small amount of fluid. The Greeks believed that the pericardium was involved in cooling the heat of the heart from the vigorous beating. The lungs were also thought to help cool the hot heart. Treatment of febrile illnesses, what we would now characterize as infectious diseases, involved removing innate heat, often by removing blood, as we shall see. But the Hippocratic approach to disease and to the patient must first be considered.

Disease was thought to have three stages: first, the disruption of the humoral proportions; second, the body's reaction manifested by a symptom, e.g., fever; and third, a "crisis," when the illness ended by discharge of the excessive fluid or death. It is clear in the writings that the Hippocratic physician's role in treating disease processes was to help nature restore the balance of humors. Nowhere in the Hippocratic Corpus was there any description of a divine influence on these humors. Instead, the Hippocratic doctrine placed enormous conviction in nature's tendency to revert the humoral imbalance to a balance. Still, doctors could help the natural process.

THE FOUNDATION OF EVIDENCE-BASED MEDICINE

The approach Hippocratic physicians used to help the natural process cultivated the standard of behavior for physicians in Western civilization. Since they believed that disease had natural, not supernatural, causes, and these causes were knowable, the Hippocratic physicians became expert observers and careful record keepers. The shared experience of the other physicians was considered to be a critical part of training a new physician. One's own experience was thought to be insufficient and even delusive compared with collective wisdom of others in the profession.

Today, we try to practice by similar principles using "evidence-based" medicine. Contemporary principles such as adopting medical recommendations only from well-performed clinical trials are actually extensions of Hippocratic principles. Ideally, all medical decisions would be based upon such trials. However, patients and some physicians are constantly amazed at the frequency with which decisions are made where no such evidence exists. At times, the physician cannot project the findings of a clinical trial to a particular patient because he or she does not fit the precise diagnosis of the patients in the clinical trial. More often, no clinical trial appropriate to the clinical setting exists. As physicians, we must fall back on experience, both our own and that of other physicians. The Hippocratic physicians recognized the importance of this collective knowledge, with some caveats, in one of the most famous of all aphorisms handed down from this period.

Life is short, art is long, opportunity fleeting, experience delusive, and judgment difficult (2).

The insight into the practice of medicine from this 2,500-year-old statement is extraordinary. That medicine requires more knowledge than any one physician can amass in a lifetime is clear to anyone who has tried to practice it. The choice of the word "art" is also noteworthy. Despite our best efforts to develop evidence-based recommendations, medicine is and always will be an inexact science. Patients vary in their manifestations of the same diagnosis and their responses to the same drugs. While physicians must often rely on their own experience, that singular experience may be misleading. One physician's experience may be the result of only a small number of patients with a condition or patients who had unusual manifestations of a diagnosis. The physicians of ancient Greece knew that combined experience was superior. The Hippocratic physicians also recognized that the opportunity for a physician to truly make an impact is deceptively small. Timing is everything. Physicians today can cure a bacterial infection with antibiotics, provided the infection has not spread beyond the initial infected organ, say, a kidney. If the bacteria have spread to distant organs, the resulting condition, known as sepsis, can cause multiple-organ failure and death. The window between curable and fatal can be tiny. Finally, even with all that physicians have learned and accepted about the imprecise nature of the art, deciding what to do for a given patient is always demanding. This aphorism demonstrates just how well the Hippocratic physicians understood the art of medicine. It also demonstrates their value for humility, as it forswears arrogance and would seem to guard against unwavering adherence to any one doctrine. While physicians are indebted to meticulous observation and record keeping from the Hippocratic tradition, their virtuous character provides doctors with the most cherished Hippocratic legacy.

THE HIPPOCRATIC OATH

Hippocrates' practices were marked by the adherence to high moral and ethical standards of behavior. These values are best exemplified by the most enduring symbol from the time—the Hippocratic Oath.

I swear by Apollo the physician, and Asclepius, and Hygeia, and Panacea, and all the gods and goddesses that, according to my ability and judgment, I will keep this Oath and this stipulation—to reckon him who taught me this Art equally dear to me as my parents, to share my substance with him, and relieve his necessities if required; to look upon his offspring in the same light as my own brothers, and to teach them this Art, if they should wish to learn it, without fee or stipulation; and that by precept, discourse, and every other mode of instruction, I will impart a knowledge of the Art to my own sons, and those of my teachers, and to disciples bound by a stipulation and oath according to the law of medicine, but to none others. I will follow that system of regimen which, according to my ability and judgment, I consider for the benefit of my patients, and abstain from whatever is deleterious and mischievous. I will give no deadly medicine to any one if asked, nor suggest any such counsel; and in like manner I will not give to a woman a pessary to produce abortion. With purity and with holiness I will pass my life and practice my Art. I will not cut persons laboring under the stone but will leave this to be done by men who are practitioners of this work. Into whatever houses I enter, I will go into them for the benefit of the sick and will abstain from every voluntary act of mischief and corruption; and further, from the seduction of females or males, of freemen and slaves. Whatever, in connection with my professional practice, or not in connection with it, I see or hear, in the life of men, which ought not to be spoken of abroad, I will not divulge, as reckoning that all such should be kept secret. While I continue to keep this Oath unviolated, may it be granted to me to enjoy life and the practice of the Art, respected by all men, in all times! But should I trespass and violate this Oath, may the reverse be my lot (5)!

No date can be established as to the origin of the oath, but some form appears to have existed during Hippocrates' lifetime. An oath was a solemn and serious undertaking by anyone who took it in ancient Greece. In Greek mythology, even the Greek gods were bound by oaths. The first portion of the oath contains a covenant, sworn to Apollo, Asclepius, Hygeia, Panacea, and all the gods and goddesses. The list order of these deities is no accident. Prior to the time of Hippocrates, medicine was taught father to son. Although the oath reflects this practice—Apollo was the father of Asclepius; Asclepius was father to Hygeia and Panacea—Hippocrates himself did not adhere to this tradition. In fact, he is credited with being the first physician to accept students who were not his own children. The Hippocratic Oath includes statements regarding the nature of the relationships between teacher (note that even the word doctor is from the Latin, meaning teacher) and student. Some historians believe that the Oath was a means to ensure the commitment and ethical behavior of these nonrelated students (6). The high moral and ethical standard of behavior remains one of the most important and durable traditions from the Hippocratic physicians. Without such ethical standards, physicians would be far less effective. As a physician, I am given the privilege to ask of my patients some of the most intimate, personal details of their lives within moments of meeting them. These questions include details of a patient's illicit drug use, alcohol consumption, and sexual practices, all details that would have a direct bearing on a person's risk for an infectious disease. I have this privilege because patients expect that as a physician, I am going to maintain the high moral, ethical standard of behavior that requires that I keep the information confidential, that I refrain from passing judgment, and that I use their sensitive information only for the benefit of the sick. Our contemporary concepts of confidentiality and ethical standards of physician behavior can be traced directly from the Hippocratic Oath. I agree with L. Davey when he says that,

> This Oath, more than any single discovery or any one person, time, or place, unites even the most diverse specialties in the unique effort to serve mankind (7).

As physicians, we owe a great debt to this Hippocratic tradition, although there are several portions of the Oath that are either controversial or anachronistic. Scholars have cited that the Oath's bans on abortion and use of the knife for persons "laboring under the stone" were inserted centuries later. These insertions were intended to depict Christian ethics, since other texts in the Hippocratic Corpus are at odds with a ban on such practices. Other researchers have stated that the Oath's ban on abortion was not an attempt to insert Christian ethics but a practical statement (8). Moreover, the Oath does not ban all abortions, only those using a pessary (9). A Hippocratic physician had only his reputation to maintain a fee-for-service practice. There were no certification exams, licensing boards, or legal restraints on physicians in ancient Greece. Ludwig Edelstein, an eminent classicist, put forth the idea that the development of the high standard for Greek medical ethics was simply a practical one—to set Hippocratic physicians apart from others pretending to practice the healing trade (10). Charlatans pretending to practice healing, often in collusion with nefarious acquaintances of a patient, would actually poison individuals. The Oath may have been a functional method to ensure that Hippocratic physicians could be trusted to only help patients. Even with a skeptical nod to the notion that the Hippocratic moral standards of physician behavior were merely pragmatic, this moral bar was set high. Physicians (and patients) have been reaping its benefits since the time of ancient Greece.

THE HIPPOCRATIC CORPUS: EPIDEMICS

To understand the insights of the Hippocratic physicians into what we know as infectious diseases, their theories of disease must be taken into consideration. These physicians knew nothing of microorganisms. Even the concept of infection is in opposition to the humoral theory. An alteration of the balance of humors was considered to be a process inside the body. Despite this antagonism, Hippocrates and his colleagues described cases of malaria, typhus, and meningitis but did not blame these diseases on infectious organisms invading the human host. The concept of contagion, at least as we know it, was foreign to Hippocrates. Even though the title of seven of the Hippocratic texts is *Epidemics*, in ancient Greece the term epidemic did not have the same meaning as it does today. In the Hippocratic texts, the adjective epidemios (literally, on the people) came to mean diseases that circulate or propagate in a country (11). In *Epidemics*, books I and III are simply a series of case vignettes. Since Hippocratic medicine focused on the prognosis, syndromes with similar symptoms became diseases. To the Hippocratic physician, fever was a disease. There were multiple febrile diseases, each with characteristics that forecast the prognosis. Acute, protracted, and intermittent, e.g., quartian (every fourth day) or tertian (every third day), were all descriptions of febrile diseases. For example, "In acute fevers, spasms and strong pains about the bowels are bad symptoms" or "If in a fever not of the intermittent type, a rigor seizes a person already much debilitated, it is mortal." These pithy statements demonstrate so much of the Hippocratic mode of medicine. Fever was a diagnosis that was carefully correlated with other symptoms to project the all-important prognosis. A number of cases of what we now would call an infectious disease can be recognized from study of the Hippocratic Corpus, including meningitis, tonsillitis, pneumonia, and malaria (12). Malaria deserves special mention, since Hippocratic physicians were the first to describe the disease manifestations, including separating the fevers as tertian from quartian. We know today that several species of parasites, known collectively as *Plasmodium*, can cause malaria after a bite from a mosquito. One particular species, *Plasmodium malariae*, has a life cycle 1 day longer (quartian) than the other *Plasmodium* species, *Plasmodium vivax* and *Plasmodium ovale*, that regularly infect humans and have an established fever cycle (tertian). One of the most serious causes of malaria, *Plasmodium falciparum*, usually has no established fever cycle. The Hippocratic physicians were astute enough merely from clinical observation to see differences in the various forms of this disease. Only *Plasmodium falciparum* poses a significant threat of death. The careful observations of the ancient Greeks allowed them to forecast from the periodic fevers whether (and often when) the patient would recover. The Hippocratic physicians believed that the cause of malaria was a disruption in the four humors. In the text *On Airs, Waters, and Places*, the author relates malaria to the time of year and ingestion of or proximity to stagnant waters (13). The humoral theory often implicated the time of year and ingestion of foodstuffs, in this case, stagnant water, as the reasons for the appearance of a disease. The stagnant-water association was strong enough to cause the Romans to drain swamps centuries later. As a result, the Romans decreased the frequency of malaria, unknowingly by decreasing the breeding ground of the mosquitoes. Contagion was not considered, perhaps because the Greeks were attempting to elucidate the universe in microcosm, i.e., the human, and move away from invisible, unknowable explanations for disease.

THE LASTING INFLUENCE OF THE HUMORAL THEORY ON MEDICINE

For all the contributions that the Hippocratic physicians provided, little of therapeutics is contained in the Corpus. With limited knowledge of the anatomy, physiology, and pathology of the human body, perhaps this inadequacy is understandable. Theirs was a gentle art of therapy, by and large. Diet, exercise, and rest were foremost for treatments. Baths were used for many fevers. Care of wounds, often with wine, was carefully described. Purging of humors was another means by which the doctor could help the ill, such as the use of emetics to purge phlegm. Interventions beyond purges were confined to bleeding, cautery, and draining of fluids and pus. The notion of a physician actually bleeding a patient is one that has become a symbol for all that is appalling in noncontemporary medical practice. But bleeding had a rationale. In the case of the Hippocratic physician, bleeding was recommended to alleviate difficulty in breathing and to relieve pain. In a patient with fever, removing some of the humor responsible, namely, blood, could alleviate the excess of innate heat. Oddly, this recommendation was made when the fever was connected with some other signs of inflammation, a swollen, red leg, for example, and not a primary fever. Bleeding a patient was especially indicated in pleurisy, inflammation of the lining of the cavity around the lungs, the pleura. Bleeding was also recommended in quinsy, pus formation in the area around the tonsils following tonsillitis. The peritonsillitis, or quinsy (a term no longer in use), could lead to abscess formation with dangerous extension of the infection to other parts of the body, including the neck, brain, and chest. America's first president, George Washington, was believed to have had quinsy just before he died in 1799. The influence of the humoral theory, developed 2,200 years earlier, was so powerful that its impact could still be seen in Washington's demise. George Washington Custis, the adopted son of George and Martha Washington, lived at Mount Vernon and was eyewitness to the episode. He was 19 years old when Washington died and described the events (13).

On the morning of the thirteenth [of December 1799], the general was engaged in making some improvements in the front of Mount Vernon. The day became rainy with sleet, and the improver remained so long exposed to the inclemency of the weather as to be considerably wetted before his return to the house. About one o'clock, he sat down to his indoor work. . . . At night on joining his family circle, the general complained of a slight indisposition, and after a single cup of tea, repaired to his library, where he remained writing until between eleven and twelve o'clock.

Having first covered the fire with care, the man of mighty labors sought repose . . . The night passed in feverish restlessness and pain. The manly sufferer uttered no complaint, would permit no one to be disturbed in their rest on his account, and it was only at daybreak he could consent that the overseer might be called in, and bleeding resorted to. A vein was opened, but no relief afforded.

George Washington died on 14 December 1799 at age 67, slightly over 2 years after completion of his second term as the first U.S. president (Fig. 2.2). Historians have argued over the exact diagnosis and how much blood was removed from the febrile man. Some sources say the blood volume removed was as much as 3.75 liters, which would have been more than half of his blood volume (7). Whatever the cause of Washington's acute fever/infection, so long lasting was the belief in the

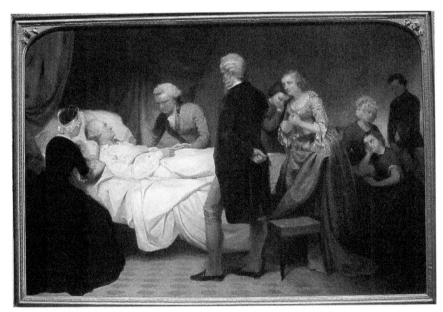

FIGURE 2.2 *Washington on His Deathbed, 1851. Junius Brutus Stearns (1810–1885), American painter. Oil on canvas, 37 1/4 by 54 1/8 in. The Dayton Art Institute; gift of Robert Badenhop, accession number 1954.16.*

Hippocratic Corpus that in 1799, the removal of a large amount of blood was performed for precisely the same reasons that Hippocrates might have performed the procedure, although one wonders if Hippocrates would have been more cautious.

EVOLUTION OF THE HUMORAL THEORY THROUGH THE CENTURIES AFTER HIPPOCRATES

While the theory of four humors was the basis of medicine, including the approach to the febrile patient, for over 2,000 years, the philosophy did not live unchanged for two millennia. In the centuries following the death of Hippocrates, little was added to the practice of medicine, even with notables such as Plato and Aristotle writing extensively on the workings of the human body. These writings were largely philosophical and theoretical. Numerous sects developed, such as the Empirics, the Dogmatists, the Methodists, and the Pneumatists, splitting the direction of the teachings of the Corpus in several philosophical paths. While the underlying principles of the four humors remained, the theory was twisted and distorted. In the centuries after Alexander the Great's death, the Alexandrian medical school developed in the Egyptian city at the mouth of the Nile. There, the humoral theory remained the base of medicine, but the importance of pneuma was expansively considered. The attitude towards the dissection of the human body was advanced in the Alexandrian school, with some noteworthy progress in the description of human anatomy, but dissection was still not widely performed. By the first century, it was generally believed that veins alone carried blood and arteries carried pneuma. Perhaps this belief was bolstered by the observation that, at death, the great artery of the body, the aorta, extending directly from the left ventricle of the heart, had no blood at dissection, since it pools in veins and dependent areas of the body. With the decline of the Alexandrian school, the period of the first century CE was one of some practical improvements but mostly was filled with comments on the Hippocratic Corpus (10).

GALEN: HIS LEGACY

By the second century, the last of the great ancient Greek physicians, Galen, combined the medical theories and practices of the various sects in ancient Greek medicine into a more unified doctrine. Galen was born in the city of Pergamon (in present-day Turkey) but spent most of his career as a physician in Rome. Galen's role in the history of medicine is one of paradox. Galen made a great number of advances. He built on the Hippocratic humoral theory, adding importance to anatomic systems in disease. Galen believed that health depended upon the balance of humors in these organs or organ systems, not simply the body as a whole.

To Galen, the human body consisted of three interconnected organ systems: (i) the brain and nervous system, responsible for thought and sensation; (ii) the heart and arterial system, responsible for transporting the vital spirit from pneuma to the rest of the body; and (iii) the liver and venous system, responsible for nutrition. According to Galen, blood formation occurred in the liver from food that was absorbed from the stomach and intestines and passed to the liver via the portal vein. This liver-formed blood was carried to the body via the venous system in a tidal, or "ebb and flow," action, much the way waves lap against the beach shore. The body, then, consumed the blood, converting it to the flesh. The heart's function as a pump was unknown to Galen. Rather, Galen believed that this organ's task was to force pneuma, the vital spirit, which came into the body via the lungs, to the rest of the body. Attributing an importance of anatomy in disease came from the many careful dissections that Galen performed. For example, he showed that arteries, in addition to veins, carry blood, not exclusively pneuma. To Galen's thinking, the pneuma combined with blood in the arterial system and the two systems never really intermingled. How did the venous blood get to the arteries to combine with pneuma? Galen postulated, or more accurately, he described, tiny pores in the heart that allowed liver-formed blood to migrate to the arterial side—pores that do not actually exist, but that Galen claimed he saw, an error that was perpetuated for centuries. Not all of his anatomy was error prone, however. Galen also showed that cutting nerves would lead to loss of function, e.g., cutting the recurrent laryngeal nerve would cause loss of phonation or vocalization. Galen's advances were the result of experimentation; he was one of the first physicians to advocate and use the experimental approach in medicine. Galen combined two methods to describe disease that seemingly had nothing to do with each other: the humoral theory from Hippocrates and the study of anatomy that was acquired from the Alexandrian school. Notably, Galen did not dissect human cadavers but inferred human anatomy from animal dissections. Unfortunately, Galen introduced numerous anatomic errors into his medical texts from these inferences that remained unchallenged for centuries.

Unlike Hippocrates, we know a great deal about Galen, largely because he was a prolific writer and often wrote about himself, usually in a self-promoting style. Despite living a large portion of his life in Rome, Galen wrote in ancient Greek. He had a philosophy that went along with his medical teachings. Galen believed that the human body has a predetermined, master plan. The physician/scientist's role was to describe this plan. He also regarded the body as a vehicle of the soul. This predeterministic philosophy met with the approval of the great religions, Christianity and Islam, in subsequent centuries, adding to his influence. In his writings, Galen had an answer for every question, a solution for every problem. He was a formidable adversary and proclaimed that all that one needed to know about medicine was contained in his texts. His reputation, philosophy, and innate wisdom became so great that one did not question this authority. In the centuries that followed his death, if someone did question Galen's teachings, he was in danger of being labeled a heretic. For all the advances attributed to Galen, his medical legacy was ultimately disastrous for progress in the medical profession. His writings were blindly accepted as the ultimate authority, stagnating medical progress for nearly 1,500 years.

REFERENCES

1. **Wilcox RA, Whitham EM.** 2003. The symbol of modern medicine: why one snake is more than two. *Ann Intern Med* **138:**673–677. http://dx.doi.org/10.7326/0003-4819-138-8-200304150-00016.
2. **Adams F.** 1985. *1849. The Genuine Works of Hippocrates. Birmingham: Classics of Medicine Library.* New Syndenham Society, London, United Kingdom.
3. **Lyons AS, Petrucelli RJ.** 1987. *Medicine: an Illustrated History.* Harry N Abrams, Inc, New York, NY.
4. **Phillips ED.** 1973. *Greek Medicine.* Camelot Press Ltd, London, United Kingdom.
5. **Libby W.** 1922. p 23–45. *In The History of Medicine in Its Salient Features.* Houghton Mifflin Company, Boston, MA.
6. **Davey LM.** 2001. The oath of Hippocrates: an historical review. *Neurosurgery* **49:**554–566.
7. **Lewis FO.** 1932. Washington's last illness. *Ann Med Hist* **4:**245–248.
8. **Nuland S.** 1988. *Doctors: the Biography of Medicine.* Knopf, New York, NY.
9. **Riddle JM.** 1991. Oral contraceptives and early-term abortifacients during classical antiquity and the Middle Ages. *Past Present* **132:**3–32. http://dx.doi.org/10.1093/past/132.1.3.
10. **Edelstein L.** 1967. *Ancient Medicine.* Johns Hopkins University Press, Baltimore, MD.
11. **Martin PMV, Martin-Granel E.** 2006. 2,500-year evolution of the term epidemic. *Emerg Infect Dis* **12:**976–980. http://dx.doi.org/10.3201/eid1206.051263.
12. **Pappas G, Kiriaze IJ, Falagas ME.** 2008. Insights into infectious disease in the era of Hippocrates. *Int J Infect Dis* **12:**347–350. http://dx.doi.org/10.1016/j.ijid.2007.11.003.
13. **Custis GW.** 1860. Recollections and private memoirs of Washington. Derby & Jackson, New York, NY. As quoted in The Death of George Washington, 1799. EyeWitness to History (website). Available from http://www.eyewitnesstohistory.com/washington.htm.

3 Avicenna, a Thousand Years Ahead of His Time

Historians regard the fall of Rome in 476 CE and the fall of Constantinople in 1453 as the years that serve as bookends to the Middle Ages in Europe. For medicine in Western Europe, these Dark Ages were, indeed, years of little intellectual progress. Despite recurrent disastrous illnesses, including the so-called Black Death, no real advancements occurred in diagnosis or treatment in Western Europe during the Middle Ages. The Christian Church's anti-Hellenism policy led to a loss of many of the important Greek and Roman writings that served as the basis of medicine. In many Germanic areas of Western Europe during the Middle Ages, a belief in supernatural forces dominated thought on the diagnosis and treatment of disease, resulting in a kind of folk medicine. With the increasing domination of the Christian Church, some documents, particularly Galen's works, were translated into Latin and studied in monasteries. But these translations did not always influence the medicine practiced within them. The church dominated physicians. Much emphasis was placed on religious medicine, which was more concerned with the soul than the body. Many unscientific practices, such as exorcisms and other rituals with supernatural elements, were performed during this time.

BUBONIC PLAGUE IN THE MIDDLE AGES

One outbreak became synonymous with disease in the Middle Ages in Western Europe: bubonic plague. While we understand the connection between the microbe associated with bubonic plague, *Yersinia pestis*, and its transmission via

Germ Theory: Medical Pioneers in Infectious Diseases, Second Edition. Robert P. Gaynes.
© 2023 American Society for Microbiology.

fleas from rodents to humans, a thorough grasp of the complex relationships that were responsible for a disease that wiped out millions of people during the Middle Ages eludes us. Since no effective treatment or prevention existed, we are uncertain about the explanations for the periodic emergence, disappearance, and reemergence of plague in cities throughout Western Europe, especially in the 14th century. Biological, climatic, cultural, economic, and social factors affect epidemics. Changes in these factors, not medical/public health advances, are the likely explanation for the ultimate disappearance of the disease. Complicating our understanding, historical accounts of plague are sometimes difficult to reconcile with modern knowledge of the disease, suggesting to some historians that the Black Death of the Middle Ages may have had other causes. However, most medieval historians view the catastrophic illness as caused at least in large part by *Y. pestis* (1). The devastation from this disease reminds us of how disastrous an infection can be to our planet. But we must also recognize the complexities involved in explaining the appearance and disappearance of an epidemic, since medical recommendations of the period could not have been terribly helpful.

The contradictory advice that physicians offered during outbreaks of the plague was indicative of the chaotic medical thinking of the age. Fleeing affected areas was often advised to avoid exposure to tainted air, or miasma (from the Greek, meaning pollution or bad air). Decaying matter, or miasmata, was thought to be responsible for this volatile miasma. If escape was not possible, doctors suggested moving about slowly to avoid breathing in the tainted air or inhaling through scented sponges. Bathing was regarded as more dangerous than helpful because it was thought that bathing opened pores. The corrupted air would better penetrate the body. For those who could seek medical attention, a variety of regimens were devised to remove impurities and bad humors. Physicians prescribed bleeding, cauterization, and plasters to treat painful swellings, known as buboes; these treatments did not help the afflicted and likely made a person's health worse. Medieval medicine in Western Europe was, if anything, a step backwards from the advances of Greco-Roman medicine.

MEDICINE IN THE MEDIEVAL ISLAMIC WORLD

In another part of the globe, it was the Arab world that preserved the Greco-Roman tradition of medicine for hundreds of years. The intellectuals, collectively known as Arabists, included Islamic scholars, Christians, Jews, Persians, and others who resided in Muslim countries. Some historians view the Arabist contribution to medicine as merely one of placeholder status where nothing original was added and only the Greek tradition of medicine, according to mostly Galen, was continued. But the medieval Arabic influence on medicine deserves much more than placeholder status. A number of innovative ideas and practices had their roots in medieval Islamic medicine and were the prototype for Western medical practice.

With the rise of Islam in 600 CE, the lands over which the religion dominated grew until the Muslim world extended from Spain and North Africa in the west to Central Asia and India to the east. For a period of about nine centuries, from about the 7th until the end of the 15th century, intellectual activity flourished in this region. While there were multiple approaches to medicine over such a large area and over such a long period, three important developments can be discerned. First, the critical importance of translation of medical texts from a variety of cultures, but notably ancient Greek, led to the emergence of medieval Islamic medicine. Second, the system for delivering medical care and training doctors took on an innovative quality. Third, the first comprehensive medical reference, the *Canon of Medicine*, was written, which was the culmination of medical thought to date. Out of many high intellects of this period, this book's author would rise above others. In the Muslim world he was known as Abu Ali al-Husayn Ibn Sina, although he was known in the West as Avicenna (Fig. 3.1).

ROLE OF TRANSLATION IN THE ESTABLISHMENT OF MEDIEVAL ISLAMIC MEDICINE

Before discussing Avicenna, we should examine the emergence of medieval Islamic medicine. In the early medieval Islamic period, Hellenistic influences from the time of Alexander the Great existed alongside Bedouin and other cultural influences in this region of the world. Medieval Islamic medicine had influences of many cultures with different languages, but Arabic was the predominant language. Like Latin in Western Europe, in the Islamic world, Arabic became the language of learning or scholarly pursuits. The role of translation from other languages to Arabic was fundamental to the establishment of the medieval Islamic medical practice. Ancient Greek medical theory, which dominated medical thinking at the time, was transmitted to the Arab world in a circuitous route. When the Muslim world overran Asia Minor, a Christian sect in Syria, the Nestorians, had already translated important medical works of the ancient Greeks and Galen into Syriac. From this language, these texts were translated into Arabic and spread through the Middle East. The systematic Arabic translation appears to have begun around 754, when the center of power of the Abbasid dynasty moved from Damascus to Baghdad (2). The translated Greek medical texts began to transform Islamic culture. New words had to be created in Arabic to correspond to ancient Greek medical terminology. The Hippocratic Oath was curiously transformed to be suitable to Islamic culture. Of the ancient Greek gods and goddesses mentioned in

FIGURE 3.1 *Avicenna. G. P. Busch sculpture. Courtesy of the National Library of Medicine (NLM image ID B01242).*

Hippocrates' time, only Asclepius's name made it through the translation process, "I swear by God (Allah), lord of life and death, giver of health, creator of healing and all cures; I swear by Asclepius; and I swear by all male and female 'friends' of God . . ." (1).

Galen's writings dominated the translations, so the humoral theory remained the basis of Islamic medicine. While Islamic tradition was overwhelmingly an extension of ancient Greek medical tradition, other elements such as Syriac, Persian, Indian, and, later, Chinese medical traditions found their way into Islamic medicine via vigorous, systematic translations of texts from these regions. The humoral theory was supplemented by pharmacology, a science in which the Arabists excelled. They developed many chemical techniques such as distillation, crystallization, and sublimation, which were to become essential to the developing science of pharmacology. The translated Greek works of Dioscorides and Galen supplied the initial pharmacological armamentarium. The wide scope of the Islamic world and their continual translation of texts brought them into contact with new drugs. By 1248, the *Comprehensive Book on Simple Drugs and Foodstuffs* by Ibn al-Baytar had over 1,400 medicines arranged alphabetically using over 260 sources, many of which were translations from other languages (3).

CHANGES TO MEDICAL EDUCATION IN MEDIEVAL ISLAMIC MEDICINE

During the period of medieval Islamic medicine, the process by which one could become a physician changed. Early in medieval Islamic times, one could become a physician by scholarly, independent study and by apprenticing with another physician. These methods were borrowed from the ancient Greek system. An alternative system of medical education developed in this period—one that seems startlingly familiar to modern medical schools—training in hospitals. Hospitals became highly developed by the 10th century in the Middle East. The first hospital in Western Europe, Hôtel-Dieu, was founded in Paris in 651 but did not offer medical treatment, such as wound care, until the 13th century. The exact origin of the hospital is uncertain. Hospitals may have emerged from the Nestorian Christian traditions in the late eighth or early ninth century in present-day Iraq (3). Caliphs supported the development of hospitals in Baghdad. These institutions were run by the elite physicians of the day. One hospital of this era in Baghdad, the Adudi hospital, boasted a staff of 25 doctors. Architectural plans from this period survive and support the notion that these facilities were complex in design, with lecture halls, pharmacies, separate wards for male and female patients, mental health wards, and outpatient clinics. The Islamic hospital carried on a wide variety of functions that were much closer to their modern counterparts than to the practices in Christian monasteries and hospices of the same period. During this period, medical education in hospitals enhanced the apprentice system with formal lectures, demonstrations with patients, and group discussions. In 931 CE, the first formal system for examining and licensing physicians was established in the Middle East (1). The ethical standards introduced by the ancient Greeks, including a form of the Hippocratic Oath, were continued. The medical profession attracted the most distinguished of men. There were a few limits to advancement of physician education. The social anathema in Islam of a man touching a strange female's pelvic area led to midwife practice. Islamic law prohibited human dissection during this period, so Galen's writing remained the primary source of anatomic knowledge.

AVICENNA, THE PRINCE OF PHYSICIANS

Persians of this era had a prescription for a great scholar:

> Be born in a great time, when learning is honored and generously sustained, when genius abounds and talent is common, when serious subjects are held in high repute. Also, take care to be born of parents who hold learning in reverence as a supreme value, and have a teacher who will be more than a father to you. Furthermore, travel. (4)

The Persians must have had Avicenna in mind when they wrote such a passage. Avicenna was born into this developing Islamic intellectual world in 980 CE in the village of Afshana near Bukhara, which is in Uzbekistan today. Avicenna dictated an autobiography of the first 21 years of his life to his friend and pupil, Juzjani. Thus, we know a good deal more about his early life than those of most historic figures of the time. His father held an important government position and moved the family to Bukhara when Avicenna was still a small child. In that city, which was under Samanid rule, the boy had a good education, primarily from his father, and quickly began to demonstrate a remarkable intellect. For example, Avicenna said that by age 10, he had memorized the Koran and could recite Arabic and Persian literature by heart. He quickly learned mathematics, law, astronomy, and philosophy. When he reached the age of 13, he began to study medicine. In the course of about 3 years, he had learned enough to begin to see patients. After Avicenna cured the Samanid ruler of an illness, he was rewarded with the use of the Royal Library, a development that Avicenna viewed as significant in his academic pursuits on a large range of subjects. Avicenna did not suffer from modesty. In his autobiography, he stated that he thought that medicine was not difficult and that after a short period of time, he excelled at it. Unfortunately for Avicenna, two momentous events coincided and changed his life forever—political instability resulting in the defeat of the Samanids and the death of his father when he was 21 years old, the year in which his autobiography ends. These two events forced Avicenna into wandering around from town to town in the region. His medical knowledge helped him make a living during the day, leaving Avicenna free to gather students for philosophical and scientific discussions in the evening. Despite his nomadic existence, he was able to continue his academic activity, including writing the first part of the *Canon of Medicine*, his most famous work. After a period of wandering, Avicenna finally settled in Hamadan in western Iran. Having gained something of a reputation as a physician, he became the court physician to the ruling prince, who appointed Avicenna as vizier, a reward, in part, for curing the prince of an ailment. Unlike medicine, politics did not come easily to him. Avicenna ran afoul of the army, probably because of his philosophy. Troops ransacked his home. Avicenna was forced into hiding at a friend's home for a time, where he continued to work on the *Canon*. When the ruling prince was stricken with colic, Avicenna was summoned, effected successful treatment, and was reinstated. Avicenna was a man of extraordinary energy and varied interests. He had a love of music and wine and shared both with anyone who was willing to engage in intellectual discussions with him. This conduct must have seemed appalling to others in Islamic society, especially for someone in government service. But Avicenna had a faculty for completely disregarding public thought in pursuit of his own interests.

When his employer eventually died and his son took over rule, Avicenna began secret correspondence with the ruler in Isfahan, a city with periodic conflicts with Hamadan. Avicenna quietly worked, but when his correspondence with the ruler of Isfahan was discovered, he was thrown in prison for several months. Eventually, the ruler of Isfahan captured Hamadan. Avicenna was freed and departed for Isfahan (3). He was offered a position in the court of the local prince and entered a period of comparative calm. However, after some time, the political position of the ruler of Isfahan became grave and he fled the city. Avicenna accompanied him. The sequence of events following the departure from Isfahan is unclear. Avicenna became ill; whether it was a natural illness or the result of poisoning is not certain. The expedition marched on Hamadan. Avicenna had a lingering illness and died soon after his return to Hamadan at age 57. He was buried outside the town in 1037. While both Isfahan and Hamadan have claimed Avicenna as their citizen, a mausoleum to Avicenna was built in Hamadan, where it stands today.

Avicenna led a chaotic, disjointed life. He was often lonely, as he never married. He was reported to have had a violent temper. Like Galen, Avicenna could be a formidable opponent in any argument. He relentlessly pursued truth and was scornful of those who settled for mediocrity. As a physician, he could be charming and devoted to his patients. The physician in Avicenna paid particular attention to a patient's appearance and pulse and to examination of the urine. His passion appeared more towards writing about medicine, among other topics, than in attending to patients, however.

THE *CANON OF MEDICINE*

In Isfahan, Avicenna completed two major works, the *Book of Healing* and the *Canon of Medicine*. The former is a philosophical work covering logic, sciences, psychology, astronomy, mathematics, and even music. The great physician William

Osler described the *Canon of Medicine* as the most famous single book in the history of medicine (5). Osler wrote,

> Canon was a medical bible for longer than any other work. It stands for the epitome of all precedent development, the final codification of all Greco-Arabic medicine. . . . Avicenna imparted to contemporary medical science the appearance of almost mathematical accuracy, whilst the art of therapeutics, although empiricism did not wholly lack recognition, was deduced as a logical sequence from theoretical premises.

Avicenna wrote many medical texts, but the *Canon of Medicine* stands alone (6). For this book, he is often referred to as the Prince of Physicians or the Doctor of Doctors. The word canon (from the Persian, Qanun, meaning law) is originally derived from the Greek word canon, which essentially means a standard. The *Canon of Medicine* contains over one million words and is divided into five books: (i) *General Principles of Medical Practice*, (ii) *Simple Drugs*, (iii) *Local Disease*, (iv) *General Diseases*, and (v) *Compound Medicines*. This book was the compendium of medical knowledge to date. Much of the first four books consists of descriptions of Hippocratic and Galenic principles, with additional teachings from other Arabic physicians and Avicenna's own observations. The fifth book is a formulary of medicines that was a true Arabic innovation. The *Canon* is elaborately divided and subdivided, perhaps to allow for better memorization. Even though the *Canon* is largely an extension of Greco-Roman medical thought, the book contains Avicenna's remarkably insightful observations, such as his recommendations for dealing with malignancy. The only hope, he wrote, was to surgically excise the disease in its early stages with an incision that is wide and bold, including all the veins running to the tumor. Even then, he cautioned that cure was uncertain. This was written around 1020 CE, and no one has improved upon this advice for 1,000 years.

BEGINNINGS OF EVIDENCE-BASED CLINICAL TRIALS

Avicenna gave medicine its first codified lessons on evidence-based clinical trials when he listed principles for reliable experimental investigations of drugs in humans. These principles include the following:

> The drug must be free from any extraneous accidental quality. It must be used on a simple, not a composite, disease. The drug must be tested with two contrary types of diseases, because sometimes a drug cures one disease by its essential qualities and another by its accidental ones. The effect of the drug must be seen to occur constantly or in many cases, for if this did not happen, it was an accidental effect (7).

These principles are every bit as valid today as in 1022 CE, when they were written.

Avicenna's emphasis on evidence-based, pharmacological principles is no coincidence. First, physicians of his day were using hundreds of new drugs, as the science of pharmacology was separating from the art of medicine under Islamic culture. It should not be surprising that a man of Avicenna's intellect should formulate some orderly manner for using and evaluating medicines. He could see in his practice that touted effects of so-called medicines might have a psychological explanation. Avicenna had a great understanding of the effect of the mind upon a person. One account of Avicenna's treatment of a young woman who was bent over at the waist and would not straighten up shows his insight. No one seemed able to help. Avicenna was challenged to cure her within the hour. After assessing the situation, he accepted the challenge. He commanded that the woman's veil be removed. While this caused considerable embarrassment to those attending her, the woman remained bent. Avicenna then shouted, "Remove the skirt." She immediately straightened up and left the room. The cure was effected within the hour. The challenge was met. Avicenna wrote extensively about the interplay between the mind and body in the *Canon of Medicine*. This interplay suggested to Avicenna that the world around a person affects their health, a concept that may have led to one of the most surprising insights in the *Canon*.

THE CONCEPT OF CONTAGION

The *Canon of Medicine* contains the most extraordinary supposition about the contagious nature of phthisis, an archaic term for tuberculosis. The word phthisis is from the Greek, meaning to waste away. Wasting away, or consumption, was evident in patients with pulmonary tuberculosis as described by Hippocrates. It may seem curious that an intellect who spent hours considering philosophy and psychology would develop an entirely new way of considering the cause of disease. Under the humoral theory, an imbalance of the four humors was responsible for illness. Among many factors, pneuma, or vital air, could affect the four humors, but the imbalance of humors was thought to be largely an internal process in the body. The concept of contagion involves a cause of disease that was entirely external to the body. Avicenna wrote,

> Very often the air itself is the seat of the beginning of the decomposition changes—either because it is contaminated by adjoining impure air, or by some "celestial" agent of a quality at present unknown to man (8).

By this passage, he came astonishingly close to the development of the germ theory, six centuries before the development of the microscope. The great historian of science George Sarton offered an explanation on how Avicenna came to this concept.

A philosophy of self-centeredness, under whatever disguise, would be both incomprehensible and reprehensible to the Muslim mind. That mind was incapable of viewing man, whether in health or sickness as isolated from God, from fellow men, and from the world around him. It was probably inevitable that the Muslims should have discovered that disease need not be born within the patient himself but may reach from outside, in other words, that they should have been the first to clearly establish the existence of contagion (9).

The *Canon* provides additional food for thought for infectious disease practice. It distinguishes mediastinitis from pleurisy, describes the contagious nature of sexual diseases, and proposes the use of quarantine to limit the spread of phthisis (tuberculosis). The *Canon* also mentions worms, including tapeworms and nematodes (roundworms), as the cause of several clinical syndromes. Avicenna promoted the use of wine in tending to wounds, inferring its antiseptic nature.

THE *CANON OF MEDICINE*'S INFLUENCE IN WESTERN MEDICINE

The medical influence of Avicenna's *Canon* was long lasting. Gerard of Cremona was the first to translate the *Canon of Medicine* into Latin in Toledo, Spain, in the 12th century, making it available to European scholars. Questions have been raised about accuracy of the translation, but its influence remained great. The *Canon* was translated and improved by Andrea Alpago, a physician and scholar. In 1527, the superior version was published in Venice. It was reprinted more than 30 times in the 16th and 17th centuries. The *Canon* remained an authoritative text at medical schools in Western Europe, such as the ones in Montpellier, France, and Leuven, Belgium, through the 17th century. However, the concept of contagion became a contentious one in Renaissance Italy. Despite the advances in medicine that the Renaissance brought, much of Western Europe fought against Avicenna's idea of contagion until the 19th century.

Avicenna's *Canon of Medicine* offered Western Europe an additional and valuable lesson. While his compendium of medical knowledge to date was written by a man of great energy and intellectual prowess, the *Canon* was the product of a society with a tolerance and openness to outside cultures. Such openness fostered an atmosphere of remarkable intellectual growth and creativity, including the formation of hospitals, changes in training physicians, and the foundations of pharmacology and evidence-based medicine. The political instability that the crusades brought with them aided in the decline of this noteworthy period. But crusaders brought to Western Europe some of the Islamic medical innovations, such as the foundations for the modern hospital. Others would have to wait until translations from Arabic to Latin during the renaissance of the medicine in Europe in the 16th century. Avicenna's concept of contagion of phthisis, a true innovation, received a tepid reception in Western medicine, which persistently held to the Galenic view of medicine and its balance of humors. An imbalance of humors, which remained the basis for disease, was a process that was viewed as internal to the human body. An external agent was somewhat counter to the Galenic view. But the innovation may have suffered more because it was not a European innovation. As we shall see, medieval Islamic thought was not always viewed with favor by some authorities in the West. As a result, instead of building on a variety of advances from medieval Islamic medicine, Western medicine chose to ignore them. A threat from the East in the 13th century, the Mongol invasion, helped spell the end of this distinguished period. Never again would ideas flow from the East to Europe in the same way.

REFERENCES

1. **Magner LN.** 2005. p 135–195. *In The History of Medicine.* Taylor and Francis, Boca Raton, FL.
2. **Pormann PE, Savage-Smith E.** 2007. p 6–41. *In Medieval Islamic Medicine.* Georgetown University Press, Washington, DC. http://dx.doi.org/10.1515/9780748629244-005.
3. **Afnan S.** 1958. p 57–82. *In Avicenna: His Life and Works.* George Allen & Unwin Ltd, London, United Kingdom.
4. **Pope AU.** 1955. Avicenna and his cultural background. *Bull N Y Acad Med* **31:**318–333.
5. **Osler W.** 1921. p 84–126. *In The Evolution of Modern Medicine.* Yale University Press, New Haven, CT.
6. **Smith RD.** 1980. Avicenna and the Canon of Medicine: a millennial tribute. *West J Med* **133:**367–370.
7. **Wickens GM (ed).** 1952. *Avicenna: Scientist and Philosopher.* Luzac & Company, Ltd, London, United Kingdom.
8. **Gruner OC.** 1930. *A Treatise on the Canon of Medicine of Avicenna Incorporating a Translation of the First Book.* Luzac and Co, London, United Kingdom.
9. **Sarton G.** 1927. *Introduction to the History of Science,* vol I. Williams and Wilkins, Baltimore, MD.

4 Girolamo Fracastoro and Contagion in Renaissance Medicine

The rise of the universities in Western Europe is closely coupled with formalization of medical education and general progress in medicine. But medieval progress on this road was painfully slow. One of the oldest Western European-style medical schools developed from a Salerno, Italy, monastery in the ninth century. Its coastal location near Naples offered the opportunity to receive works from all over the Mediterranean. Many Greek medical texts accumulated in the library, where they were translated into Latin. The Salerno school was one of the chief portals by which Arabic texts reached the West. In the 12th century, the medical school at Montpellier began to equal the Salerno school in stature. By the 15th century, a number of universities with medical schools had sprung up throughout Western Europe and began to change medicine. The invention of printing and the revival of interest in Greek learning changed the structure of academic medical instruction and the medical profession (1). These schools were given a shot in the arm when the capital of the Byzantine Empire, Constantinople, fell to the Ottomans in 1453. Byzantine medicine had developed between 400 and 1453. Some ancient Greek and Roman documents unavailable to scholars in the Western European regions were accessible in Byzantine cities and translated into Latin. Some historians claim that Byzantine medicine formed the first hospitals in the eighth and ninth centuries, preceding those in the medieval Islamic world. But these Byzantine facilities were not as complex in design as their Islamic counterparts. With the fall of Constantinople in 1453, many Byzantine scholars fled to the West, particularly to Italy, taking their texts with them. A new climate of intellectual creativity began

Germ Theory: Medical Pioneers in Infectious Diseases, Second Edition. Robert P. Gaynes.
© 2023 American Society for Microbiology.

to bloom—humanism. Humanism in the Renaissance attached importance to human dignity, interests, and potential, particularly rationality. Humanism affixed great weight to rediscovering ancient Greek texts, whose ideals were in step with the prevailing Renaissance view of the human condition. This look back was not done out of nostalgia but with great reverence to a golden age. Humanism had to overcome hundreds of years of intellectual stagnation, which was as evident in medicine as any other academic discipline.

For centuries, Galen had been the ultimate authoritative source of medical knowledge in Western Europe, and questioning this authority was a dangerous and daunting undertaking. But gradually, the intellectual climate in Europe began to change. Avicenna's *Canon of Medicine* began to have influence in the West in the latter part of the 13th century in schools such as Padua and Bologna, Italy, and Montpellier (2). The *Canon's* all-encompassing scope and its reliance on Galenic fundamentals made it an acceptable medical reference for most in the West. But by the 1520s, Western criticism of Arabic authors was evident based upon commentaries from scholars at these schools. These commentaries appeared to be shared among medical schools as far away as Paris. Much of the dislike of Arabic medical advice may have been based upon religious and moral grounds rather than scientific considerations. But there were some practical and specific denunciations, primarily in the pharmacological parts of the text (1). These were the parts where the Latin translation was the least intelligible. By the mid-1500s, accessibility to Latin-translated, ancient Greek medical texts and the commentaries condemning Avicenna led to a decreased reliance on his work in Western Europe, although by no means was the *Canon of Medicine* disregarded. The decreased dependence on the *Canon* led to disregard of many of its innovative concepts, including contagion, in Western Europe.

PLAGUE IN THE RENAISSANCE

To the modern reader, a disregard of the concept of contagion seems hard to reconcile with the presence of two diseases of the period: plague and syphilis. Black Death continued its periodic appearances during the Renaissance. Some consideration of its transmissibility was considered as early as in 549 during the first pandemic of plague, which occurred in and around Constantinople. The Emperor Justinian called for the isolation of travelers approaching from regions where the disease was evident. Historians have suggested that the Justinian plague, as it became known, and several subsequent recurrences of plague through the seventh century so disrupted trade in the Mediterranean that this disease was in large part responsible for a shift to a bartering economy, sinking Southern Europe into an economic abyss and precluding its importance as a center for trade for centuries (3). The first pandemic ended abruptly and mysteriously in 755. The second pandemic of plague was ferociously introduced to Europe from the city of Kaffa in 1347. During a siege of the city, the Mongol army was reportedly suffering plague. The Mongols catapulted the infected corpses over the city walls to infect the inhabitants. The Genoese who were trading in Kaffa fled and took the plague by ship back to Southern Europe (4). While the contagious nature of plague may have been considered, transmission of bubonic plague is not person to person, although the rarer, pneumonic form can be passed from human to human. More often, plague involves not one but two spreading agents, rats and fleas. Quarantine was attempted during this period, again suggesting some consideration of contagion. The word quarantine comes from a Venetian dialect form of the Italian *quaranta giorni*, meaning 40 days, the time sailors were kept on ship or away from a city's inhabitants to see if they developed plague. The first well-documented use of quarantine was in Dubrovnik, Croatia, in the 14th century. While passengers and sailors were quarantined, the infected rats passed onto land and happily passed the infection to the crowded, filthy city dwellers. Proponents of contagion must have had difficulty explaining the complex transmission of plague and the repeated futility of quarantine. However, one of the most important contributions of the second pandemic of plague in Europe was a series of measures that were adopted to prevent plague. Using the concept of quarantine, northern Italian cities such as Venice and Genoa took extraordinary action by establishing maritime quarantine to insulate the city's population from invasion by the sea. In Venice, lazarettos, which are buildings (or sometimes the ships themselves) used to house individuals in quarantine, were established in the 15th century on outlying islands in the lagoon where arriving ships were directed. Vessels from suspect areas were impounded, cargo scrubbed, and crew and passengers were taken ashore under guard and sequestered for forty days. The time frame was taken not from medical theory but from biblical scripture (e.g., 40 days/nights of the flood in Genesis, 40 days Moses spent on Mount Sinai). Still, the time frame was well beyond the incubation period of plague and provided sufficient time to guarantee that any person in good health could enter the city free from plague. This quarantine requirement also spurred the development of an institutional framework of officials that were given nearly unlimited authority to defend the community in an emergency: boards of health. Officials in France, Spain, and Northern Europe imitated the Italian initiative. Becoming increasingly sophisticated and comprehensive in the 17th and 18th centuries, these anti-plague actions marked humanity's first major conquest of a disease: exclusion of plague from Western Europe by the end of the 18th century (5). As discussed in chapter 3, the medical explanations for the cause of plague and its spread during this period were fit into Galenic medical theory of miasma affecting the balance of humors in the body. But an apparently new disease affected Europe during the Renaissance that seemed to defy explanation: syphilis.

SYPHILIS IN THE RENAISSANCE

Whether some form of syphilis existed in Europe before 1492 has been the subject of considerable controversy. However, substantial evidence exists to suggest that something new arrived in Renaissance Europe soon after Columbus's voyages to the New World (6). In 1495, as French soldiers under Charles VIII laid siege to the Italian city of Naples, an epidemic of disease broke out among the troops. This disease took on remarkable virulence. Soldiers were described with pustules covering their bodies. So distressing was the disease in this outbreak that the army, comprised largely of mercenaries, dispersed due to the disease, not military defeat. The mercenaries left for many of the Western European countries, bringing this new ailment with them. The disease during this period became known by many names, including the Great Pox. The cause was unknown, but the Great Pox was associated with unclean foreigners. Blame on foreigners, always popular scapegoats, ascended to new heights. The Italians called it the French disease; the French called it the disease of Naples; the English called it the Spanish disease; and the Russians called it the Polish disease. Even the Asians entered the name game, or, more accurately, the blame game. The Japanese referred to it as the Chinese disease. In the early years of the disease which we have come to know as syphilis, the severity was far greater than just 25 years into the epidemic. For example, in 1591 Ulrich von Hutten, himself a sufferer, wrote,

> When it first began, it was so horrible to behold that one would scarce think the Disease that now reigneth to be of the same kind. They [sufferers] had Boils that stood out like Acorns, from whence issue such filth stinking matter, that whosoever came within the scent, believed himself infected. The colour of these was of a dark green and the very aspect as shocking as the pain itself, which yet was as if the sick had laid upon a fire (7).

By the 1600s the symptoms of the disease were dramatically less, changing from a severe, debilitating, and obvious disease to a milder illness more similar to modern syphilis.

STAGES OF SYPHILIS

Syphilis, as we have come to understand it, has three stages. In the first, or primary, stage, wherever the causative organism, *Treponema pallidum*, comes in contact with mucous membranes or any microscopic breaks in the skin, the infected person experiences a change known as a chancre, usually about 3 weeks (range,

3 to 90 days) after exposure. A chancre, a raised, reddish change in the skin, is normally painless and lasts 4 to 8 weeks, if untreated. The chancre usually goes away without treatment, but the infection does not. The treponemal bacteria disseminate within hours of infection to lymph nodes and, via blood vessels, to more distant sites, but the distant sites of infection are usually not evident until the secondary stage, usually 6 weeks (range, 2 to 12 weeks) after the chancre has appeared. This stage is often associated with fever, a sense of fatigue, and skin rashes of all varieties. The skin rashes themselves can be teeming with treponemes. The disease in the secondary stage does not produce the severity described by von Hutten but can affect internal organs, including the brain. Surprisingly, the symptoms in the secondary phase usually disappear fairly quickly without treatment. Syphilis can enter a latent phase of infection where the individual has no symptoms. For 70% of people infected with syphilis, the disease never progresses to the third stage of clinical illness. For the remainder, in the third, or tertiary, stage, syphilis can produce damage to a variety of organs in the body, but this process can take years or even decades. Prior to the availability of effective treatment in the 20th century, tertiary syphilis produced serious damage to the central nervous system, resulting in psychiatric symptoms that could lead to madness. It is estimated that by the end of the 19th century, nearly one-third of the patients in European mental health hospitals were there because of damage from syphilis.

Physicians of the Renaissance were at a loss to explain the appearance of an apparently new disease, its rapid spread across Europe, and its seemingly inexplicable evolution from severe to more mild illness. Modern historians and physicians have struggled for biological explanations. The nationalistic and social connotations that were associated with study of this particular disorder have complicated our understanding. How could all of this have happened? With the initial outbreak, a review of commentaries by one of the medical scholars at Padua, Giambattista da Monte, in the late 15th century suggested that contagion was considered and even endorsed as the means whereby *morbus gallicus*, or French disease, was transmitted. However, da Monte tried to show that the concept of contagion could be fit into the Galenic humoral theory (2). The concept of contagion would need a stronger, more independent-thinking advocate to overcome the hold of Galen. Contagion would find its greatest proponent during the Renaissance from the very same university only a few years later.

GIROLAMO FRACASTORO: EARLY INFLUENCES

Girolamo Fracastoro, also known by his Latin name, Hieronymus Fracastorius, was born in 1478 in Verona, which was, at that time, still part of the Republic of Venice (Fig. 4.1). He was the sixth of seven brothers in a noble family. Lightning

FIGURE 4.1 *Girolamo Fracastoro. Portrait by Titian from the National Gallery (inventory number NG3949).*

killed his mother when he was very young. Fracastoro was in his mother's arms when she was struck. Remarkably, he escaped unhurt, lending credibility to an ancient Greek legend: Apollo saves from lightning the heads of poets—a legend that portended Fracastoro's future. Fracastoro suffered from poor health as a child and spent considerable time with his father at the family's villa at Incaffi, near Verona. His father saw to it that he was well educated, introducing him to literature and philosophy. Since Verona had no university, Fracastoro was sent to nearby Padua, a city also under Venetian control. Fracastoro had a liberal arts education that included astronomy, mathematics, literature, and philosophy. He also made many friends with individuals from all over Europe. With the guidance of a family friend, he turned his education towards medicine. Fracastoro had a strictly ortho-dox upbringing, a factor that contributed to his cautious temperament. But the relative freedom of thought in Padua influenced Fracastoro's nature. He was an extremely quiet man who rarely spoke unless directly addressed. He was dignified and generally cheerful. However, his taciturn manner led many individuals to con-sider him austere and unfriendly. Around 1500, Fracastoro married Elena de Clavis, with whom he had a daughter and four sons. Fracastoro excelled academi-cally. He graduated in 1502 with a degree in medicine. Following his graduation, he initially taught anatomy at Padua but was quickly appointed Professor and Chair of Logic and Philosophy, a post he held for about 6 years. He had several friends and learned colleagues at Padua, including Pietro Bembo and Gaspare Contarini, both of whom were later named Cardinals of the Catholic Church. His Padua associations were to influence his path later in life. Fracastoro had the good fortune of working with a doctor-philosopher, Pietro Pomponazzi. Medicine was his main passion, although he also was interested in astronomy, mathematics, physics, botany, geology, geography, and composition of verses. During his time at Padua, he befriended Nicolaus Copernicus, the Polish astronomer, who came to Padua in 1501 to study medicine. Fracastoro himself made contributions to astronomy. For example, with Pietro Apiano, he described the direction of a com-et's tail in relation to the sun.

FRACASTORO'S EPIC POEM—*SYPHILIS SIVE MORBUS GALLICUS*

Political instability resulting from a war between the Roman Emperor Maximilian I and Venice forced Fracastoro to leave the University at Padua, which had to tem-porarily close its doors. Expatriates from Padua, including Fracastoro, formed an informal academy called Academia Forojuliensis at Udine. Fracastoro wrote verses while at the academy for which he became well known in Italy. The academy had a brief existence. In 1509 he moved to Verona. The death of his father left Fracastoro to manage the villa at nearby Incaffi, which had been damaged and neglected. Verona lived under German domination for the next 8 years. Fracastoro scratched out a living with a medical practice. The unstable political situation left Verona an impoverished, filthy city close to anarchy. When disease broke out in 1510 in Verona, Fracastoro retreated to his family villa at Incaffi some distance away. He continued writing, including the beginnings of one of his most famous works, *Syphilis sive morbus Gallicus*. In 1516, two sons died from unspecified causes, but Fracastoro expressed grief in his later writings over his inability to treat them. In 1518 Verona went back to Venetian control and a relative calm. Fracastoro moved back to the city with his wife, daughter, and new son, now 1 year old. He worked as a physician and wrote in Verona until 1530.

Fracastoro was witness to the beginning of the outbreak of the French disease in Italy and saw the terrifying effect it had on the population. In 1510, combining his interests in literature, medicine, and philosophy, Fracastoro began composing an epic poem that would make him famous throughout Europe. In 1525 Fracastoro completed an earlier version of the poem, *Syphilis sive morbus Gallicus*. This version contained two books that may have been printed or given as a manuscript to a famous humanist of the time. Based on some comments he received, Fracastoro enlarged and rearranged the piece over the next 5 years until it reached the 1,300-verse epic poem, *Syphilis sive morbus Gallicus*, that was published in 1530 in three books. It was a peculiar mixture of fact and fiction. Befitting his cautious, pondering nature, Fracastoro tried to advance intellectual thought about a frightening disease without seeming too controversial. The poem succeeded admirably in using mythical characters to convey medically accurate but potentially alarming symptoms in a nonthreatening way. In Book I, Fracastoro used fictional characters in the poem to describe the causes, stages, and symptoms of illness of the French disease. Early elements of his theory of contagion are present in the poem. Fracastoro suggests that *semina*, or seeds, of certain diseases, such as distemper, can attack goats. For the French sickness, these *semina* seem only to attack humans, resulting in symptoms of the malady. He even weighs in on whether Columbus brought the malady back from the New World. Fracastoro actually rejects this idea for several reasons. First, the disease seemed to break out in many European countries simultaneously. Second, he cited cases early in the outbreak that appeared to have contracted the disease spontaneously, without any apparent contacts. Finally, Fracastoro cited an alignment of Mars, Jupiter, and Saturn, a phenomenon that he was convinced regularly foreshadowed maladies. He cited just such a planetary alignment occurring two centuries earlier in the 1300s, immediately prior to the plague's launch in Europe. An astrological reference to explain disease states had been cited since the time of Hippocrates, but this idea may also have arisen from Fracastoro's interest in astronomy. Book II describes remedies for the disease, including exercise, exclusion of fresh air, and appropriate diet. Fracastoro suggests that mercury, in the form of quicksilver, is the best remedy. He tells the myth of a Syrian hunter who was punished with the disease for shooting a sacred stag. He is told to visit a nymph who is the guardian of all metals. She dips the hunter in mercury salts beneath Mount Etna and his symptoms disappear. Perhaps this story is the origin of the saying, "A night in the arms of Venus can lead to a lifetime on Mercury." Book III is devoted to treatment with guaiac, a resin derived from the wood of the Hyacus tree. In the poem, natives of Haiti are nearly all afflicted with the great malady and use guaiac as treatment. The third book contains two mythical stories. One describes the Spaniards killing a bird sacred to Apollo and their punishment with the terrible disease. (Note the ironic and repeated appearance of the Greek god Apollo in Fracastoro's poem, considering the legend of Apollo saving the heads of poets from lightning, mentioned above in connection with the death of his mother.) Treatment with guaiac saves the day, although the treatment was eventually determined to be ineffective. The more famous of the two stories of Book III involves a shepherd to the King of Haiti named Syphilis, who blasphemes Apollo when drought kills off the herd. Apollo sends a new disease, of which Syphilis is the first victim. Others are affected and the natives call the disease after its first victim, Syphilis. Fracastoro never divulged the derivation of the name Syphilis. Historians have attempted to determine the source, or root, of the name. The most likely is the account by Boll, a 16th century historian, suggesting that Fracastoro borrowed the name from Ovid's *Metamorphoses V*, where a lad named Sipylus was slain by Apollo because his mother had insulted Apollo's mother, Leto (8). Whatever its derivation, Syphilis's disease was associated with the new Renaissance illness. Eventually, just the word syphilis became the accepted name for the disease due to the success of the poem and the lack of association with any nation or socioeconomic class. Fracastoro dedicated his poem to his old Padua friend, Pietro Bembo, now Cardinal of the Catholic Church, who claimed it was the "most precious gift he had ever received." Fracastoro received lavish praise and considerable fame for his epic poem, with comments comparing him to Lucretius and Virgil. Since every educated person could read Latin, the poem reached a wide academic audience and secured Fracastoro's reputation both as a poet and, to a lesser extent, as a physician. The praise was primarily from scholars, not physicians, however. The success of the poem gave Fracastoro the luxury of a patron, freeing him to write.

FRACASTORO: ON CONTAGION AND CONTAGIOUS DISEASE

For the next 16 years after the publication of *Syphilis sive morbus Gallicus,* Fracastoro worked on a book that he asserted was "not as a poet but as a doctor," *De Contagione et Contagiosis Morbis et Eorum Curatione,* or *On Contagion, Contagious Disease and Their Cure,* published in 1546. Precisely where his ideas emerged for *De Contagione* is in doubt. Fracastoro was influenced by the philosophies of Hippocrates, Aristotle, and Avicenna, referring to the last at least five times in his book. The notion that disease could be initiated from outside the body was not Fracastoro's concept. Fracastoro viewed Galen as the authority of the age and reviewed his writings carefully. Whether he actually took the notion of seeds of contagion from Galen has been proposed (9). But Galen's dialogue about seeds of contagion was disconnected. Additionally, Fracastoro dared to take on Galen's clout in his book, particularly in a chapter on pestilent fevers, suggesting Galen's thinking to be imperfect. Fitting his nature, he was cautious in his approach. Fracastoro dedicated the book to Cardinal Alexander Farnese. This dedication may seem curious since the Church was often at odds with science. But the dedication may have been an attempt at appeasing a possible critic. Fracastoro was also careful in his justification for challenging conventional thinking: "Galen had written much on the subject but ... omitted much that greatly needed investigation" (8). Indeed, there may be a grain of truth in what he wrote. Galen did comment on contagion in his writings, but the comments were brief and did not present a coherent theory of contagious infectious diseases (9). It seems likely he was also influenced by Avicenna's conception of contagion. But Fracastoro advanced a theory beyond whatever Galen or Avicenna considered.

In his book *De Contagione*, Fracastoro described seminaria, or seeds of contagion, in three ways: (i) those that infect by direct contact; (ii) those that infect by contact and by fomites, which he defined as clothes, linens, etc., which were not themselves corrupt but could foster the essential seeds of contagion and therefore cause contagion; and (iii) those that not only act by direct contact or fomites but also could be transmitted to a distance. In his own (translated) words,

> The term contagion is more correctly used when infection originates in very small imperceptible particles.... There are, it seems three fundamentally different types of contagion: The first infects by direct contact only; the second does the same but, in addition, leaves fomes and this contagion may spread by means of the fomes ... by fomes, I mean clothes, wooden objects, and things of that sort, which though not themselves corrupted can, nevertheless, preserve the original germs of the contagion and infect by means of these. Thirdly, there is a kind of contagion which is transmitted not only by direct contact or fomes as intermediary, but also infects at a distance (8).

Fracastoro's word choice is noteworthy. The term *seminaria* has been translated in English as meaning germs, but that is a generous translation (8). Translated after the development of the germ theory, Fracastoro's concepts seem to accurately depict modes of spread of microorganisms. However, it is erroneous to think that this is what Fracastoro had in mind. Microorganisms were unknown during his day. The Latin derivation of *seminaria* is actually closer to "seed bed" or "seedlets." One historian describes the derivation of the Latin, *seminarium*, as meaning plantation or nursery (10). Exactly what Fracastoro understood and believed about these "small imperceptible particles" is uncertain, but the use of *seminaria* as agents of contagion is clearly Fracastoro's innovation (11). He never implied that they were living organisms, referring to them as "hard," "viscous," or "a combination of diverse elements." Fracastoro compared them to and differentiated them from poisons, suggesting that *seminaria* were chemical in nature. His perception of contagion, accurate even by our standards, is astonishing when one considers that it was almost entirely philosophical and not based upon any experiment. But the former Professor of Logic at Padua presented reasoning that was plain and logical. The term fomes is from the Latin, meaning tinder or wood chip. Fracastoro did not intend the literal use of fome. Rather, he was using fome as a metaphor for the term used in theology for the small portion of original sin left behind after baptism, which might burst into fire of sin upon temptation (9). The modern medical use of the term is fomites, the plural form of fome; it depicts inanimate objects serving as an intermediary for microorganisms to pass from an infected person to a healthy host.

REACTION TO *DE CONTAGIONE*

Fracastoro's fame from his epic poem helped ensure the initial publication in 1546 and at least two additional printings of De Contagione by 1555. The reaction to Fracastoro's De Contagione in his own time ranged from hostile to favorable, but even the positive responses were tepid. In his day, most medical theorists preferred the Galenic theory of miasma to the seeds of contagion. Giambattista da Monte at Padua initiated a vigorous debate on the contagion theory in Padua, even before Fracastoro published his book. Fracastoro even delayed publication to address some of da Monte's objections. da Monte was especially troubled by the idea that disease could occur without direct contact. Upon publication, da Monte's published commentary in 1548 attacked Fracastoro's theories. da Monte's views were conditioned on his Galenic beliefs, although certain individual criticisms of Fracastoro's position were accurate and carefully constructed (11). Fracastoro had dared to attack Galen and had, at least in da Monte's eyes, faulty logic. The "imaginary seedlets" were no match for bad air and individual receptivity to disease in da Monte's Galenic view of the world. The debate became more personal: da Monte called Fracastoro, "our foolish and stupid friend from Verona" (11). The comments had a devastating effect for Fracastoro. da Monte, speaking from a position of authority in Padua, influenced students from across the continent of Europe. His discourse became less personal over time but was always critical. Although Fracastoro had some supporters, including Bassiano Landi, a professor of medicine in Padua, and Francois Valleriola of Arles, over the next two decades, the Galenic view of contagion held, emphasizing bad air as the reason for the occurrence of diseases such as plague. With the appearance of plague throughout Western Europe, particularly the Venetian outbreak from 1555 to 1557, the debate continued. Certain writings make it clear that in the 1570s, Fracastoro's notions on contagion made their way across Northern Europe to such countries as Germany (11). da Monte's personal attack aside, debate over contagion was academic and courteous and failed to produce major controversy. Fracastoro's cautious approach to questioning Galen may have been responsible for his theory's failure to ignite change. He was careful not to challenge Galen openly or blatantly. Some of Fracastoro's contemporaries began to use his terminology interchangeably with what they had known from Galen. For example, in 1579, Alessandro Marssaria chose to describe an agent of contagion as putrid particles, vapors, spirits, or whatever one chose to call them when it was clear that a healthy body had become infected (11). References to Fracastoro's concepts of contagion are difficult to find by the 1650s until the mid-1800s, with the advent of the germ theory.

FRACASTORO AND THE COUNCIL OF TRENT

Following publication of *De Contagione*, Fracastoro continued to write, but he never achieved further success, at least not to the level of his poem and this book. Because of his prominence, Fracastoro became involved in an incident of history—the Council of Trent—that would test his fame, his knowledge, and his medical and personal skills. The Council of Trent rose out of intrigue between Emperor Charles V and Pope Paul III. Both men called for the Council to counter the demands from Protestants to reform the Catholic Church. Charles V, the most powerful monarch of the day, ruled over extensive regions in Central, Western, and Southern Europe and was embroiled in multiple assaults on his power, including those from the Protestants. Pope Paul III, however, was fearful that a Council would weaken the Papacy, especially with the dominant presence of the Holy Roman Emperor, Charles V, at the table. The Pope schemed to hold the summit in a city far away from Charles V's influence, hopefully in Rome or at least in a nearby city like Bologna. Eventually, he had to yield to the mighty ruler. On 13 December 1545, the Council was held in the city of Trent, inside the Holy Roman Empire in northern Italy. Two individuals were appointed medical advisors to the Council—Balduino Balduini, a private physician to a cardinal who happened to be president of the Council, and Fracastoro. Charles V was cognizant of Fracastoro due to the success of his Syphilis poem. For example, in 1541, Charles V was passing through the town of Peschiera, near Milan, Italy. Fracastoro stood among the crowd on the hot day. The Emperor, who did not respond to the crowd at all, was told of Fracastoro's presence. Charles V halted, looked intently at Fracastoro, gestured to show recognition, and moved on. Fracastoro openly detested Charles V, who he believed was an oppressor of Italy and an enemy of the Pope but had to recognize the respect that the monarch paid to him from the brief incident. However, Fracastoro's fame was not the main reason he was present at one of the most important Councils in the Church's history. The Cardinal Bishop of Trent was his patient and friend. Additionally, the Pope's grandson, Cardinal Farnese, was the individual to whom *De Contagione* was dedicated. Both men ensured Fracastoro's appointment to the Council. Illness played a central role in the Council's first 2 years. Farnese fell ill on the way to the Council, and Fracastoro had to rush to attend to him. The health status of the city of Trent itself became a pivotal feature of the drama. In March 1547, an outbreak of "spotted fever" (likely typhus) broke out in the city. On 6 March, a bishop/delegate died of the disease. Three days later, the matter was brought up before the Council. Twelve delegates had already departed, and others declared their intentions to leave due to the danger of the disease. A proposal to move the meeting to Bologna was met with some opposition. Fracastoro, who seemed to carry more clout than Balduini, his medical colleague at the meeting, made a stunning statement. He told the Council that he came to treat members of the meeting, not to deal with an outbreak of plague, lenticular (typhoid) fever, or typhus. He demanded permission to leave Trent. Fracastoro's comments created a panic among the delegates. Whether that was his intent has been a matter of some debate. Some historians pegged Fracastoro as "the Pope's man," where others suggest that Fracastoro saw the danger of contagious epidemic as very real (11). The Council was moved to Bologna but failed to reach an effective conclusion. The Council was suspended on 17 September 1549. The incident had the effect of ending Fracastoro's connections to the Council, which reconvened in Bologna 2 years later, in 1551, under Pope Julius III.

The years after the Council of Trent for Fracastoro were spent writing, but the tone of his books changed. The tone for everyone changed. The Protestant Reformation caused the Catholic Church to become more determined to suppress ideological thoughts apart from its thinking. It became more and more difficult to publish works that the Church might consider heresy, a risk the wary Fracastoro was aware of and unwilling to undertake. His writings descended into convention, avoiding controversy. In 1553, Fracastoro died of a stroke at his villa in Incaffi at the age of 78.

THE TUMULT OVER FRACASTORO'S BURIAL PLACE

For a man who tried to avoid controversy, his burial place has provided a storm of debate. While Fracastoro was universally mourned with a magnificent funeral, no one is certain where he is buried. One version has that his body was carried to Verona and buried in a church near where he had lived in the city. The church was destroyed around 1740; his body was exhumed, and his bones were likely scattered. Some reports suggested that he was buried near Incaffi, and others say that he is buried south of the Porta Vittoria in Verona where monuments to the most famous Veronese citizens exist, including one with "Fracastoro" carved over it. The city of Verona voted 2 years after he died to erect a statue that is now in the Piazza dei Signori. A bronze statue of the man was erected in Padua.

THE REDISCOVERY OF *DE CONTAGIONE*

A renewed appreciation of Fracastoro's view of contagion was gained after the development of the germ theory in the 19th century. Fielding Garrison wrote the clearest modern statement on the contribution of *De Contagione* in *Science* in 1910:

> His work contains the first scientific statement of the true nature of contagion, of infection, of disease germs and the modes of transmission of infectious diseases. . . . Fracastoro shows himself to be a highly original thinker, far in advance of the pathological knowledge of his time . . . But it is his remarkable account of the true nature of disease germs . . . that we find him towering above his contemporaries. He seems, by some remarkable power of divination or clairvoyance, to have seen morbid processes in terms of bacteriology more than a hundred years before Leeuwenhook and the other men who worked with magnifying glass or microscope (12).

Garrison's laudatory praise slightly overstates Fracastoro's contribution since it was seen through a modern lens. Fracastoro's intellect, while great, did not quite achieve the standing that Garrison claims. Fracastoro held onto many beliefs that were no different than the contemporary Galenic thought and had some beliefs, such as the astrologic notion of planetary alignment as a cause of the Great Pox, that are nearly impossible to reconcile with his own theory of seeds of contagion. His guarded condemnation of Galen may have been responsible for the lukewarm reception of his contagion theory and kept him from receiving the highly regarded recognition in his own time that Garrison gives him. The stranglehold that Galen had on Renaissance medical thought stifled full understanding of Fracastoro's theory, which was subsumed into the existing medical thought of the period. Fracastoro's theory of contagion did not rock the Renaissance scientific world as a contemporary, Andreas Vesalius, did. One historian has argued that his concepts of seeds of contagion were not even original (9). But Fracastoro did go far beyond previous writers in systematically applying the notions of contagion and seeds to a variety of diseases. Fracastoro also had one other great disadvantage. He could not produce any observational or experimental evidence for the seeds of contagion so central to his theory. That evidence would have to wait nearly 100 years. For all its limitations, it is extraordinary how well Fracastoro described modes of transmission for microorganisms, having no idea of their existence.

REFERENCES

1. **Siraisi NG.** 1982. Some recent work on Western European medical learning, ca. 1200–ca. 1500. *Hist Univ* **2:**225–238.
2. **Siraisi NG.** 1987. *Avicenna in Renaissance Italy.* Princeton University Press, Princeton, NJ. http://dx.doi.org/10.1515/9781400858651.

3. **Sherman IW.** 2007. p 68–82. *In Twelve Diseases That Changed Our World.* ASM Press, Washington, DC. http://dx.doi.org/10.1128/9781555816346.ch5

4. **Naphy W, Spicer A.** 2004. *Plague: Black Death and Pestilence in Europe.* Tempus Publishing Limited, Stroud, United Kingdom.

5. **Snowden F.** 2020. p 69–78. *In Epidemics and Society.* Yale University Press, New Haven, CT.

6. **Harper KN, Ocampo PS, Steiner BM, George RW, Silverman MS, Bolotin S, Pillay A, Saunders NJ, Armelagos GJ.** 2008. On the origin of the treponematoses: a phylogenetic approach. *PLoS Negl Trop Dis* **2:**e148. http://dx.doi.org/10.1371/journal.pntd.0000148.

7. **Knell RJ.** 2004. Syphilis in renaissance Europe: rapid evolution of an introduced sexually transmitted disease? *Proc Biol Sci* **271**(Suppl 4)**:**S174–S176.

8. **Fracastoro G.** 1930. *Hieronymi Fracastorii—De Contagione et Contagiosis Morbis et Eorum Curatione, Libri III. Translation and notes by Wilmer C. Wright.* G P Putnam's Sons, New York, NY.

9. **Nutton V.** 1983. The seeds of disease: an explanation of contagion and infection from the Greeks to the Renaissance. *Med Hist* **27:**1–34. http://dx.doi.org/10.1017/S0025727300042241.

10. **Howard-Jones N.** 1977. Fracastoro and Henle: a re-appraisal of their contribution to the concept of communicable diseases. *Med Hist* **21:**61–68. http://dx.doi.org/10.1017/S0025727300037170.

11. **Nutton V.** 1990. The reception of Fracastoro's theory of contagion: the seed that fell among thorns? *Osiris* **6:**196–34. http://dx.doi.org/10.1086/368701.

12. **Garrison FH.** 1910. Fracastorius, Athanasius Kircher and the germ theory of disease. *Science* **31:**500–502. http://dx.doi.org/10.1126/science.31.796.500.b.

5 Antony van Leeuwenhoek and the Birth of Microscopy

A prevailing view of Antony van Leeuwenhoek is that he was an amateur scientist who was lacking in all scientific skills except his ability to grind lenses. Using lenses that he required hours to perfect, he invented the microscope. He dabbled in his newly discovered world of microscopy with no direction or refinement, scribbling images in his notebooks without ever presenting them to the outside world. This narrative of the man was widely believed for years. None of it is true. Recent investigations into van Leeuwenhoek's life have shown him to be "one of the most imperfectly understood figures" in the history of science (1). This examination of his life's efforts revealed methods to the way he fashioned lenses and prepared and viewed specimens that van Leeuwenhoek managed to keep secret for centuries.

THE DISCOVERY OF THE MICROSCOPIC WORLD

Many aspects of Antony van Leeuwenhoek's life are exceptions for this book. He was not a doctor. He never published a scientific paper. He did not begin making contributions to science until he was 40. He did not invent the microscope. However, he did markedly refine it, devising a high-powered microscope that led to a momentous breakthrough. His discoveries opened human consciousness to a world that may have few parallels in recorded history—the microscopic world. Only the invention of the telescope at roughly the same time is an adequate comparison. Extending our consciousness beyond what could be seen with the naked eye to directly observe an entirely new universe, either in the heavens or in the

Germ Theory: Medical Pioneers in Infectious Diseases, Second Edition. Robert P. Gaynes.
© 2023 American Society for Microbiology.

water of a pond, is a breathtaking advance. Others may have seen deeper into this microscopic world, but van Leeuwenhoek was the first to witness it. Consider that prior to van Leeuwenhoek, no one had any idea about the cellular makeup of human organs or single-celled organisms. Without such knowledge, the theory of contagious disease was a theoretical exercise. A study into the way this advance unfolded and the scientific culture that received it in the 1600s explains why the march to move medicine towards the germ theory stalled for well over a century in spite of one of the most significant technical developments in the history of science.

ANTONY VAN LEEUWENHOEK: EARLY INFLUENCES

Antony van Leeuwenhoek was born in 1632, a noteworthy year. This was also the year that several other great men of the 17th century entered the world—John Locke, Baruch de Spinoza, the architect Christopher Wren, and the Dutch painter Jan Vermeer. Vermeer and van Leeuwenhoek were both born in a small village in Holland named Delft. Their lives would take different paths but intermingle. Little is known of van Leeuwenhoek's parents. His father was a basket maker. His mother was the daughter of a Delft brewer. Brewers' families had important social standing in Holland. His mother and her family's standing in town were a great help to van Leeuwenhoek as he grew up. Antony was the first son but the fifth child of the family. When he was only 5, his father died. Three years later, his mother remarried. Antony's stepfather was a painter and lived only 8 years with the family before he died. During his mother's second marriage, Antony was sent to school at Warmond, a town near Leiden. He stayed with an uncle who was an attorney, but Antony never progressed far in his education. He learned no language other than Dutch and had no university training. At age 16 he was sent to Amsterdam, one of the great commercial cities of the age, just as it is today. Antony learned the linen-draper business from a shopkeeper. How he landed this apprenticeship is not clear, but it was a business that required careful and close study of the quality of cloth, particularly the number of fibers in a stretch of cloth. In his authoritative biography of van Leeuwenhoek, Clifford Dobell speculates that in Amsterdam, van Leeuwenhoek became acquainted with Jan Swammerdam, a man known in later years to have fashioned early microscopes (2). After about 6 years in Amsterdam, van Leeuwenhoek returned to the town of Delft, where he remained for the remainder of his life. Delft was hardly a center of academic learning. Antony had little contact with educated people in his native town despite Holland's place as one of the more learned countries in Europe at the time.

Shortly after his return to Delft, van Leeuwenhoek married. He had five children, but all died in infancy except his daughter Maria. Maria never married and stayed with her father until his death in 1723, tending to his house and to him.

After a marriage of only 12 years, van Leeuwenhoek's first wife died. Five years later, he remarried. His second marriage lasted 23 years until 1694, when his second wife died.

In 1654, Antony bought a house and shop in Delft and began in the drapery business. He had this business for many years and appeared to be reasonably successful. In 1660, he was made Chamberlain to the Sheriffs of Delft, essentially a local government official; he held this post for 39 years, including some responsibilities in the local treasury. This post apparently led van Leeuwenhoek to be appointed as curator or official receiver of the estate of the painter Jan Vermeer, including some extraordinary paintings. Vermeer died when he was only 43 years old, leaving his widow and eight children in desperate financial straits. Although van Leeuwenhoek did not seem to profit by this appointment, it suggests the high regard with which the townspeople held him. Some historians have speculated that Vermeer was also a personal acquaintance. Indeed, art historians have hypothesized that van Leeuwenhoek may have been the subject in two of Vermeer's paintings—*The Geographer* and *The Astronomer*. Antony van Leeuwenhoek also learned to be a surveyor in the town, an activity that required calculations illuminating his interest in and aptitude for a more academic and scientific approach to the world than might be expected of a simple businessman. On the surface, life seemed to be good but mundane for van Leeuwenhoek from the time he was appointed to his local government post in 1660 until 1673. The post of Chamberlain in Delft afforded him a consistent salary. Along with an inheritance following his mother's death in 1664, this salary allowed van Leeuwenhoek to pursue interests outside his drapery business. Through his letter to the Royal Society in London, a surprising and unexpected capacity became evident in 1673 (2).

VAN LEEUWENHOEK AND LENS MAKING

During this 13-year period, van Leeuwenhoek began making lenses. Presumably, this initial interest came from the need to examine the fiber count on fabrics for his business. In 1668, van Leeuwenhoek made his only journey to London, England. During his visit he likely obtained a copy of *Micrographia*, written by Robert Hooke. The book contained mostly observations about insects, but while he could not read the English text, the illustrations must have impressed the Dutchman. In the preface, Hooke suggested a design of a simple microscope. van Leeuwenhoek's designs of microscopes matched Hooke's suggestion, placing lenses between two metal plates. With this hand-held device, van Leeuwenhoek began to examine all sorts of objects. His skill at producing high-quality lenses for these microscopes caught the attention of Reinier de Graaf, a physician in Delft. de Graaf's acquaintance with the secretary of the British Royal Society, Henry

Oldenburg, was to change van Leeuwenhoek's place in history. de Graaf viewed some of the objects that van Leeuwenhoek had examined with his microscope. When de Graaf read *Philosophical Transactions* in 1668, the proceedings of the Royal Society, he saw an account using a primitive microscope from an Italian, Eustachio Divini, describing an "animal lesser than any of those seen hitherto" (2). So-called microscopes up to this point were essentially magnifying glasses, capable of enlarging 20 times. Robert Hooke in England and Jan Swammerdam used microscopes of this magnifying capacity and are sometimes given credit for inventing the microscope. The first recorded use of using two lenses to create what is known as a compound microscope was in 1590 by two Swiss spectacle makers, Zaccharias Janssen and his son, Han. Compound microscopes of the time had several technical limitations. The most important was chromatic aberration, a distortion of the image clarity and color, which severely limited the use for high-power magnification. But van Leeuwenhoek created single-lens, or simple, microscopes that could magnify 270 times and some as much as 500 times, a magnification far greater than that of any previous microscope and required for seeing bacteria and other single-celled organisms. de Graaf wrote Henry Oldenburg that van Leeuwenhoek's observations far surpassed those described by Divini. The Royal Society, founded a few years earlier for the purpose of promoting natural knowledge, owed a great deal of its communication outside of Great Britain to its energetic Secretary, Henry Oldenburg. Oldenburg was a German by birth. He communicated with hundreds of foreigners, translating their correspondences, while editing *Philosophical Transactions*. When de Graaf wrote Oldenburg in April 1673, he enclosed some specimens, parts of a bee, mold, and a louse, with crude observations that van Leeuwenhoek made. The Fellows of the Royal Society were impressed and instructed Oldenburg to communicate directly with van Leeuwenhoek.

VAN LEEUWENHOEK AND THE ROYAL SOCIETY IN LONDON

When van Leeuwenhoek received Oldenburg's letter in the summer of 1673, he wrote back to the Royal Society resolving to "express my thoughts properly" (2). While he stated that he could not draw, he would write and have some of his observations drawn and communicate them to the Society. For the next 50 years until his death, van Leeuwenhoek sent hundreds of letters to the Royal Society. These letters conveyed vast areas of interest—in zoology, botany, and chemical and physiological matters. The letters were often translated into English, sometimes into Latin, and published. Some were issued in Dutch. van Leeuwenhoek never wrote a book or a scientific paper. Many letters were somewhat incoherent, but the letters were van Leeuwenhoek's sole work. He worked alone, though he received some help from contemporaries. He distrusted others and disliked, even resented, interference in his efforts. In 1675, in a letter to Oldenburg, he wrote,

> Your Excellency recommends me to make use of the services of other people, who are in a position to form a proper judgment of such things. Sir, I must say that there are few persons in this Town from whom I can get any help; and among those who can come to visit me from abroad, I have just lately had one who was much rather inclined to deck himself out with my feathers, than to offer me a helping hand.

Up until van Leeuwenhoek's observations, the use of a microscope by his contemporaries such as Hooke and Swammerdam was directed towards creatures that could be seen by the naked eye but could be better detailed with magnification, such as bees or fleas. Not so for van Leeuwenhoek. He was more interested in life that was not visible to the naked eye. van Leeuwenhoek's observations on single-celled organisms began in 1674 when he began to look through his microscope at pond water. In a letter to Oldenburg, he describes what he saw:

> . . . there were many small green globules. Among these there were, besides, very many little animalcules, some were roundish, while others, a bit bigger, consisted of an oval. On these last I saw two little legs near the head and two little fins at the hindmost end of the body (3).

There is no doubt that van Leeuwenhoek saw single-cell protozoa, making him the first man to ever see creatures this small. He repeated the observation in rainwater and promised Oldenburg further notes at a later date. In October 1676, he kept his promise. In his 18th letter to Oldenburg, in what has become known as the "Letter on the Protozoa," van Leeuwenhoek detailed to the Royal Society a remarkably clear, organized, and thoughtful account of these little "animalcules," their appearance, means of locomotion, and relative size—suggesting that they were 10,000 times smaller [in bulk] than the animalcule which Swammerdam had portrayed (3). He carefully repeated these observations on at least six occasions over different days to verify their consistency. He was beginning not just to observe but also to experiment. He changed the sources of the water he examined and began adding elements to the water such as pepper, various spices, and vinegar to see if the creatures changed. His experimental approach was perhaps most evident years later when he systematically examined algal material from a barrel of water used for a garden (1). He saw what appeared to be a papery sample in the water, but under microscopic examination, he saw dried algae. He successfully re-created the conditions under which the papery sample had formed. He then moved on to a different source of material and repeated the procedure.

Each time, he confirmed microscopically the algal growth. van Leeuwenhoek prepared sections of living tissue for the microscope, including the capsule of the spleen, striated muscle, and structures of the eye. These experimental methods are not the work of a dabbler but of a scientist, albeit self-taught. van Leeuwenhoek examined seeds and other prepared specimens, some of which were sent to the Royal Society in London. In 1981, several of the specimens were discovered, stored in the Society's archives (4). These 307-year-old specimens were found remarkably intact and allowed modern-day scientists to confirm van Leeuwenhoek's findings, even using some of van Leeuwenhoek's surviving microscopes. To the surprise of many, the images obtained through these microscopes, nine of which survive today, were exceedingly clear and remove any doubt about the veracity of the man's 17th-century observations. And there were plenty of doubters at the time.

REACTIONS TO VAN LEEUWENHOEK'S MICROSCOPIC DISCOVERIES

The prospect of descriptions of animalcules too small to be seen without considerable magnification caused van Leeuwenhoek's credibility to be questioned by the Royal Society in 1676. van Leeuwenhoek refused to back down on any of his claims. Eventually, in 1680, the Royal Society sent a team of individuals, including a vicar, doctors, and members of the Society. They fully corroborated van Leeuwenhoek's observations.

THE FIRST DESCRIPTION OF BACTERIA

van Leeuwenhoek continued to pursue the little animalcules. In September 1683, he reported to the Royal Society what he had seen in plaque from a healthy person's mouth:

> With great wonder, that, in the said matter there were many very little living animalcules very prettily a-moving. The biggest sort had the shape of [a rod]: these had a very strong and swift motion and shot through the water like a pike does through water. These were most always few in number. The second sort had the shape of [a small rod]. These often spun round like a top, and every now and then took a [spiral] course and were far more in number (5).

These words are the first known descriptions of bacteria. He repeated these observations with material from the mouths of several persons, including his own. In 1680, van Leeuwenhoek was proposed for Fellowship of the Royal Society, even though van Leeuwenhoek had never attended (and never did attend) a meeting. His work was recognized as singular and acknowledged by his acceptance by the

Society. van Leeuwenhoek was humbled and honored by the admittance. He continued his correspondence with the Royal Society that published his letters, making him well known in Europe.

VAN LEEUWENHOEK'S MICROSCOPES

How did van Leeuwenhoek manage to see these tiny creatures? His simple microscopes produced wondrous images but were exceedingly difficult to use. The instrument was about 3 in. (7.5 cm) long, consisting of two metal plates (usually brass or silver), a lens placed in a hole made in the plates, and a small metal pointer on which the specimen was held. The pointer could be moved into position via two screws set up at right angles for focusing. The hole in one of the plates was slightly smaller than the lens that aided in reducing chromatic aberration. Using this hand-held device was no easy feat. The focal length of the lens was so short that the individual needed to place an eye so close to the lens that the eyeball was nearly touching it. One needed excellent eyesight and infinite patience. van Leeuwenhoek exhibited both qualities. However, the fatigue that set in using these simple microscopes probably contributed to their sparse use by many scientists at the time. But there were other reasons that few scientists worked with these simple microscopes. van Leeuwenhoek had a businessman's mind and wanted to keep trade secrets on his microscope's formation and use. During his initial correspondence to the Royal Society, he wrote,

> My method for seeing the very smallest animalcules and minute eels, I do not impart to others; nor how to see very many animalcules at one time. That I keep for myself alone (3).

Even when visited by members of the Royal Society to verify his work, van Leeuwenhoek remained cagey about showing his best microscopes, but one aspect of his lens making was obvious: the clarity. Thomas Molyneux wrote to the Society in 1685,

> As for the microscopes I looked through, they do not magnify much, if any thing, more than several glasses I have seen, both in England, and Ireland: but in one particular, I must needs say, they far surpass them all, that is in their extreme clearness, and their representing all objects so extraordinary distinctly. For I remember we were in a dark room with only one window, and the sun too was then off of that, yet the objects appeared more fair and clear, than any I have seen through microscopes, though the sun shone full upon them, or though they received more than ordinary light by help of reflective specula or otherwise: So that I imagine 'tis chiefly, if not alone in this particular, that his glasses exceeds all others, which

generally the more they magnify, the more obscure they represent the object; and his only secret, I believe, is making clearer glasses, and giving them a better polish than others can do (6).

MODERN INVESTIGATIONS INTO VAN LEEUWENHOEK'S MICROSCOPES

Modern investigators have unlocked many of his secrets. van Leeuwenhoek's secretive efforts fell into three categories: lens formation, illumination, and specimen preparation. Lens grinding was a technique well known at the time. But one surviving van Leeuwenhoek microscope, now in Utrecht, had a lens that was made by a different technique, one that his contemporaries seemed not to appreciate at the time—glass blowing. van Leeuwenhoek made some of his smallest lenses, with highest magnification, by taking a glass tube and heating the center over a concentrated flame until the glass could be pulled apart like taffy until it broke. One section with a strand of glass at the end could be placed back into the flame, where the hair-like glass would seize up to form a sphere. The sphere could be carefully blown to a desired size for use. There is substantial evidence that van Leeuwenhoek learned glass blowing techniques and used them to make numerous lenses (7). These lenses were necessary for the magnification required to see bacteria. Unfortunately, van Leeuwenhoek worked alone and never taught students, so this knowledge does not appear to have been directly passed on to anyone from van Leeuwenhoek himself.

Using these unwieldy microscopes required careful and precise illumination. Dobell suggests that his method of illumination was the best guarded of van Leeuwenhoek's secrets.

> I am convinced that Leeuwenhoek had, in the course of his experiments, hit upon some simple method of *dark-ground illumination*. Such a discovery—possibly inspired by observing the motes in a sunbeam—would at once explain all his otherwise inexplicable observations, without supposing him to have possessed any apparatus other than that which we now know he had (8).

van Leeuwenhoek's third closely guarded secret was how he prepared his specimens. He was one of the first people to prepare sections for the microscope and did so on a wide range of specimens. He performed microdissections of insects (1). In his investigations of the original specimens in the Royal Society, Brian Ford found that van Leeuwenhoek had used a new technique, serial sectioning, to determine the nature of a "paper" sample that proved to have an entirely different origin—algae (7).

van Leeuwenhoek's secrecy meant that his achievements in microscopy remained unmatched until the 19th century. The usual explosion of interest after a technological advance failed to happen. The widespread use of a high-powered microscope was aborted by the secrets kept by someone with the mind of a businessman more than that of a scientist.

SPONTANEOUS GENERATION AND VAN LEEUWENHOEK

The discovery of microorganisms raised a puzzling question: where did these organisms come from? The generation of these animalcules became a hotly debated subject and was considered alongside of the process known as putrefaction, how substances go bad, or decompose. Why did insects suddenly appear in decaying meats or fruits? The theory of spontaneous generation, first proposed by Aristotle, stated that living matter could spring forth from inorganic matter. Both philosophically and experimentally, spontaneous generation was being put to the test in the 17th and 18th centuries. Francesco Redi, an Italian physician, attacked the theory in 1668. Conventional wisdom of the time held that maggots would arise spontaneously in rotting meat. Redi hypothesized that flies laid eggs in the rotting meat, and it was the eggs that were the source of the maggots, not spontaneously generated life from the decaying meat. He devised what he considered to be the definitive experiment to test his hypothesis. Using a variety of flasks, he set out meat in flasks that were open to air, completely sealed, and, for some flasks, open to air through gauze. Maggots in the meat appeared in only the flasks that were open to air. This well-designed, controlled experiment did not silence all critics, who said that the sealed jars were deprived of air so life in those jars could not emerge. It might seem that the discovery of the existence of microorganisms would be a further attack on spontaneous generation. The reverse tended to be true. To generate these new animalcules, all one needed to do was to place inanimate material such as hay in water and wait a few days, when these creatures could be seen with microscopy. The theory of spontaneous generation had its powerful advocates of the time, including William Harvey and Robert Hooke (9). van Leeuwenhoek became a staunch opponent of the theory of spontaneous generation. His letters to the Royal Society explicitly pointed out his objections to the idea that creatures could arise from inorganic matter. The invisibility of processes in decaying matter often provided spontaneous-generation supporters with a durable defense. Now the microscope could help to answer questions. The use of a microscope did not lead everyone to the same conclusion. Athanasius Kircher, a German Jesuit scholar, reported using a microscope to describe the spontaneous generation of swarms of nematodes previously unseen (10). van Leeuwenhoek countered that had Kircher used a good microscope, he would not have made such claims. As van Leeuwenhoek's observations progressed into the microscopic intricacies of life, he became a stronger opponent to spontaneous generation even for the smallest of organisms. To van Leeuwenhoek, each of these little animalcules had to have a parent (1). The theory hung on for a century after van Leeuwenhoek's demise and would die a slow death until it was put to rest in the middle of the 19th century.

MICROORGANISMS AND DISEASE IN THE ENLIGHTENMENT

While a new world of microorganisms had been revealed, van Leeuwenhoek did not make any connection between these microorganisms and disease. While van Leeuwenhoek was the first to witness the living organisms, bacteria, in association with humans, these persons were not ill. Some historians have taken van Leeuwenhoek to task for not making an association between microorganisms and the nature of infectious illnesses. In hindsight, it seems clear that van Leeuwenhoek had observed the vital but missing experimental evidence that could have provided the key to Fracastoro's theory of contagion! The speculation was no longer theoretical. One could at last observe these "seeds." van Leeuwenhoek, in fact, did observe them but did not make the connection. How could he? Thomas Molyneux wrote this about van Leeuwenhoek in 1685:

> A very civil compleasant man, and doubtless of great natural abilities. But contrary to my expectations, quite a stranger to letters which is a great hindrance to him in his reasonings upon his observations, for being ignorant of all other men's thoughts, he is wholly trusting to his own (10).

van Leeuwenhoek, who spoke and read only Dutch, was unaware of Fracastoro's book on contagion. Other contemporary scientists may have been aware of the writings, although little was written about Fracastoro's theory between 1650 and the 1800s. It is not van Leeuwenhoek who should bear the criticisms for the failure to connect his observations to infectious diseases but the scientific and medical communities of the time. No one made the connection. The microscopic world that van Leeuwenhoek had uncovered seemed to be merely a world of curiosity. Despite his penchant for working alone and disdaining the social niceties, he found himself to be a bit of a 1600s celebrity, receiving visitors from all over Europe, including Queen Mary of England and Peter the Great, Czar of Russia. van Leeuwenhoek gave both of them a microscope that he had made as a souvenir. The curious microscopic findings were not of interest to physicians of the time. Medicine continued to hold on to the humoral theory through the

17th and much of the 18th centuries. Physicians had no reason to look for microorganisms as causes of illness when miasma and alterations in the balance of humors would provide perfectly satisfactory explanations for disease. On what grounds was there to think that these little animalcules were responsible for causing disease and bringing down humans millions of times their size? Microorganisms viewed by van Leeuwenhoek were from perfectly healthy people. Physicians of the time had no reason to suspect that they had anything to do with diseases. Medicine was about to undergo change from a multitude of forces—the philosophy of the Age of Enlightenment, experimentation in science spilling over into medicine, and the new field of anatomic pathology. These forces produced slow but relentless progress in medicine and permitted the field to finally reconsider its 2,000-year belief of the humoral theory of disease. The transition to modern medicine, which I consider in the next chapter, was an essential antecedent to the development of the germ theory and the pathological nature of bacteria and other microorganisms in human disease. Until the seat of disease was reassessed from a change in the balance of humors to our modern medical pathophysiological approach, the existence of bacteria was only a novelty of nature.

The new world that the microscope had revealed challenged the basic understanding of life and of old teachings. A door had been forever opened, yet scientists of the 17th and 18th centuries were slow to walk through it. Despite the technical advance, cumbersome instruments were difficult to use. van Leeuwenhoek's secrecy and unwillingness to teach were, in part, to blame. He viewed himself as exceptional, not without reason. He had his critics, too. Perhaps out of jealousy, Nicolaas Hartsoeker, a Dutch scientist of the time, stated that van Leeuwenhoek saw a thousand things through the microscope that were never really there at all (10). But there were other critics, too, who tarnished van Leeuwenhoek's reputation in academic circles. van Leeuwenhoek continued to make observations and communicate with the Royal Society until his death in 1723 (Fig. 5.1). But for all his discoveries, the academic climate in Europe in the 18th century lacked the spark to include microscopy in research. Some universities, including Leiden in Holland, had included microscopes among their teaching instruments, but medical instruction in microscopy appeared to actually decline during the 18th century (10). Simple microscopes would eventually yield to compound microscopes when technical problems such as chromatic aberration were solved in the 19th century. The waning interest in microscopy in the 18th century cannot diminish van Leeuwenhoek's 50 years of pioneering contributions that changed the way we think of the living world.

FIGURE 5.1 *Antony van Leeuwenhoek. Portrait by Jan Verkolje from the Rijksmuseum (accession number SK-A-957).*

REFERENCES

1. **Ford BJ.** 1992. From dilettante to diligent experimenter, a reappraisal of Leeuwenhoek as microscopist and investigator. *Biol Hist* **5:**1–12.
2. **Dobell C.** 1932. p 19–55. *In Antony van Leeuwenhoek and His Little Animals.* Harcourt, Brace and Company, New York, NY.
3. **Dobell C.** 1932. p 110–166. *In Antony van Leeuwenhoek and His Little Animals.* Harcourt, Brace and Company, New York, NY.
4. **Ford BJ.** 1981. Specimens from the dawn of microscopy. *Biologist* **28:**180–181.
5. **Dobell C.** 1932. p 239–240. *In Antony van Leeuwenhoek and His Little Animals.* Harcourt, Brace and Company, New York, NY.

6. **Dobell C.** 1932. p 56–60. *In Antony van Leeuwenhoek and His Little Animals.* Harcourt, Brace and Company, New York, NY.
7. **Ford BJ.** 1991. p 127–140. *In The Leeuwenhoek Legacy.* Biopress and Farrand Press, London, United Kingdom.
8. **Dobell C.** 1932. p 313–331. *In Antony van Leeuwenhoek and His Little Animals.* Harcourt, Brace and Company, New York, NY.
9. **Ruestow EG.** 1996. p 202–222. *In The Microscope in the Dutch Republic.* Cambridge University Press, Cambridge, United Kingdom.
10. **Dobell C.** 1932. p 2–18. *In Antony van Leeuwenhoek and His Little Animals.* Harcourt, Brace and Company, New York, NY.

6 The Demise of the Humoral Theory of Medicine

In the 16th, 17th, and 18th centuries, the progress that occurred in medicine could have led to the germ theory of human disease. Avicenna's concept of disease contagion, Fracastoro's seeds of contagion, and van Leeuwenhoek's animalcules seen via microscopy could have put physicians centuries ahead of a theory that waited till the 19th century. But they didn't. Why didn't anyone put all these elements together? A paradigm had to be torn down first, which was not an easy task. For centuries, physicians were trained in the basic humoral theory of disease according to Galen. They were not open to considering how the animalcules could enter the body and cause disease. Indeed, medical innovators were often treated unfairly or badly during their lifetimes. How, then, did physicians relinquish the entrenched ideas of Hippocrates, Galen, and Avicenna? There was no single incident but a series of events that challenged long-held views on anatomy, physiology, and eventually the foundation of human disease, pathology. The epidemics of plague, smallpox, and cholera in the 14th through 19th centuries challenged the medical establishment's reliance on humoral medicine to provide an explanation for the sudden and mass fatalities. If a disease was due to an imbalance of humors in an individual, how could the theory account for the abrupt and simultaneous imbalances in an entire community? Doubts about humoralism from these epidemics prepared the way for alternative philosophies. Advancement, however, was dreadfully sluggish, and modernizers of these fields often suffered for their efforts.

Germ Theory: Medical Pioneers in Infectious Diseases, Second Edition. Robert P. Gaynes.
© 2023 American Society for Microbiology.

ANDREAS VESALIUS AND HIS CHALLENGE TO GALENIC ANATOMY

The 16th century brought some needed change to medicine after the intellectually stultifying medieval age. However, Renaissance thinking was more preoccupied with the resurrection and admiration of ancient Greek medicine than the development of medical advances. For example, Renaissance discoveries of ancient Greek texts fostered renewed interest in human anatomy. The effect of this renewed interest thrust the first dagger into the humoral theory—an assault on Galenic anatomy that came in the form of a book, *De Humani Corporis Fabrica* (*The Structure of the Human Body*), in 1543, by Andreas Vesalius. To understand its impact, consider that anatomic knowledge had been passed down from ancient times but largely came from the writings of Galen. Galen never actually dissected a human cadaver. He made anatomic observations on humans who had been injured, but his anatomy was inferred from animal dissection, mainly of apes, sheep, pigs, and goats (1). He introduced in his writings egregious errors in human anatomy that remained unchallenged for centuries. For example, dissections of the base of the calf brain show a delicate network of blood vessels and nerves termed the *rete mirabile*, Latin for "wonderful net." Galen attached great significance to the network as the site where vital air was turned into animal spirits. He simply assumed the same network must exist in humans, though it does not. Galenic texts included other errors such as assuming the human liver to have five lobes when it only has three. As Galen became accepted as the ultimate medical authority, no one dared question the master. Correcting these errors required not only assaulting Galen's authority but also careful comparison of Galen's text with actual human dissection.

In medieval and Renaissance universities human dissection was a difficult matter. There were few bodies for dissection, which could only be performed in the colder months of the year. Professors would leave the few human dissections (often executed criminals) to assistants while they would sit on high lecterns, reading from Galenic texts. Enter Andreas Vesalius. He was among the first professors to perform human dissection himself. Born in Brussels, Belgium, in 1514, he showed an unusual interest in animal anatomy as a child, dissecting everything from insects to mammals. He studied medicine in Paris, where he acquired skills in human dissection. A war between Charles V and Francis I of France in 1536 sent Vesalius scurrying from Paris, eventually to the University of Padua, where he completed his doctorate in 1537. While dissection had been introduced at the University in Padua, it was primarily for surgical training and optional for physicians. Immediately upon graduating from the university in 1537, Vesalius was appointed professor of anatomy at Padua. Vesalius, an excellent teacher, became an extremely popular professor. Doing his own dissections, Vesalius began to have the disquieting sense that anatomy as written in Galenic texts was not always the human anatomy that he observed. He continually failed to find the *rete mirabile* in humans, contrary to Galen's writings. He found other errors, too, such as the point of insertion of certain muscles such as the rectus abdominis muscle. After acquiring a sufficient supply of cadavers, he established a close working relationship with Jan van Calcar, a protégé of the artist Titian, to prepare the first truly accurate book on human anatomy, complete with a number of magnificent artist plates. Unlike the cautious Fracastoro, Vesalius openly challenged Galenic anatomy. He wrote,

> How much has been attributed to Galen, easily the leader of the professors of dissection, by those physicians and anatomists who have followed him, and often against reason! Indeed, I myself cannot wonder enough at my own stupidity and too great trust in the writings of Galen and other anatomists (2).

For his attack on Galen, Vesalius was assailed by critics of the day. The most vicious criticism came from his former Paris professor. Vesalius was so angry over the attacks that he resigned his professor post in Padua and burned all his papers. But a dent in the humoral theory had been made. The book was so well produced, so stunning in its art and detail, that within his lifetime, Vesalius would have the satisfaction in knowing that his efforts changed the approach to human dissection—all anatomic assertions from Galen's texts were subjected to observational tests in human cadavers. Still, medicine could advance only so far in this period. Vesalius had wounded Galenic anatomy, but he had not opposed Galenic physiology, i.e., the way the body was understood to work. Vesalius's book did not fundamentally change humoral theory. However, the broader significance of *De Humani Corporis Fabrica* was to produce a profound change in thought: believe in what you can observe. More importantly, the academic medical community allowed itself to consider the idea that the ultimate authority, Galen, could be wrong. But convincing change to Galenic physiology would have to wait another 100 years.

THE CHALLENGE TO GALENIC PHYSIOLOGY: WILLIAM HARVEY, THE DISCOVERY OF THE CIRCULATION OF BLOOD, AND THE SCIENTIFIC METHOD IN MEDICINE

A new medical paradigm came in England in the 1600s. William Harvey, born in 1578, revolutionized medical thinking as much as or more than anyone in the history of the field. While Vesalius had planted the seeds of doubt in Galenic theory of medicine, Harvey can be viewed as a seminal figure in medicine. So great was the change in medical investigation that it is worth detailing his work even though Harvey's contribution did not deal with the field of infectious diseases. First, recall where the medical world stood in the late 1500s on the topic of the circulation of blood. The Galenic view was that there were two types of blood: venous and arterial. The pathways for passage through the body were separate. They had distinct functions as well. The venous blood, made in the liver, was for nutrition and growth. The arterial blood, originating from the heart, combined with pneuma (from the lungs), was for vitality. Venous blood ebbed and flowed through tidal action and was consumed by the tissues. The arterial blood was also expended in the body rather than returning to the heart. In fact, in Galenic physiology, the heart did not even act like a pump. The movement of blood was from the pulsating action of the arteries themselves. As previously mentioned (chapter 2), the means by which blood appeared in the left ventricle and was changed into arterial blood, combined with pneuma, was not a strong point of the theory. Hidden pores in the heart's interventricular septum, or the wall separating the right and left ventricles, were described by Galen, but anatomists (until Vesalius) had to thoroughly convince themselves that they could see them—because they were not there! Galen's explanation for how air got to the left ventricle, via the pulmonary vein, was also a bit problematic. Once in the left ventricle, the venous blood, oozing through the ventricular septum, would combine with pneuma, producing a by-product— "sooty vapours" (3). These vapors traveled back to the lungs via the same pulmonary vein, where they were exhaled. This two-way street of the pulmonary vein was, at best, a clumsy explanation, but the theory held until Harvey.

Harvey did not set out to discover the nature of human circulation but came at the discovery through a series of events and his inquisitive nature. Harvey grew up with an interest in anatomy and science. He attended Cambridge University. In 1600, he studied at Padua, working under Hieronymus Fabricius ab Aquapendente (or Fabricius), one of the greatest comparative anatomists of the time. Fabricius became Anatomy Chair at Padua after succeeding Gabriele Falloppio, who had learned anatomy directly from Vesalius. Fabricius was also interested in function of the body, rather than just Vesalian structure or architecture, a concept that Harvey incorporated from his time in Padua. Fabricius dissected various animals but carefully related his findings to observations of human dissections. Fabricius discovered venous valves, i.e., small structures in veins that prevent backflow of blood. This discovery would prove crucial for Harvey. It suggested that blood flowed only one way, not as a tidal ebb and flow as Galen had suggested. Fabricius showed Harvey that by temporarily occluding a surface vein of blood, one could determine the direction from which the segment filled. Harvey learned to observe the function of this new anatomic finding.

With his newly found anatomic training from Padua, Harvey headed back to England convinced that new ideas were needed. He, like Vesalius and Fabricius, believed that there were no holes in the septum of the heart to allow blood flow from the right side to the left. Additionally, the discovery of the venous valves left Harvey to consider the anatomic finding in physiological terms—blood flowed only one way. How could Galen be right? In Padua, Harvey had been introduced to a new kind of medical investigation—experimentation. Observation was not enough. Harvey combined keen observation with computation and experimentation. Harvey was inspired by Galileo's motto,

> Measure all that is measurable and make things measurable which have hitherto not been measured. (4)

But Harvey would forge a path with little guidance from others. Shortly after his return to England, Harvey was granted membership in the College of Physicians in 1604. In the same year, he married Elizabeth Browne, daughter of Lancelot Browne, the physician to King James I. His wife's connection to royalty proved essential in Harvey's work and influence. In 1607, Harvey became Assistant Physician to St. Bartholomew's Hospital. He had a flourishing practice, which included members of the aristocracy, eventually to include King Charles I. Such a practice provided a sufficient source of income that Harvey could maintain an independent interest in investigation in anatomy and physiology. He vivisected countless animals to determine the nature of the heartbeat and its relationship to the pulse, eventually theorizing that it was the heart's contraction that was the source of the arteries' pulsation. But in 1616, Harvey landed at the crux of his argument on circulation of blood. First, he measured the total amount of blood that could be drained from an animal such as a sheep or pig. Then, Harvey measured the amount of blood that could fill the left ventricle of each animal. Using this measured quantity of blood in the heart, about 2 oz, Harvey calculated the amount of blood, beating at 72 times per minute, that the body would need to consume if the Galenic ebb-and-flow model of circulation was accurate. This amount, 8,640 oz, or 540 lb, was far in excess of the animal's blood volume. It was even in excess of the entire body weight of each animal! The use of this quantitative evidence was compelling and new. The scientific method and introduction of quantitative evidence into physiological problems were Harvey's greatest contribution to medicine. Such investigations eventually led to the 1628 publication of one of the great books in Western medicine, *Exercitatio Anatomica de Motu Cordis et Sanguinis in Animalibus*, or *Anatomical Studies on the Motion of the Heart and Blood in Animals*—often referred to simply as *De Motu Cordis*. The first half of this relatively short book details the relationship between the heart and the pulse. For the second half of the book, Harvey took on the more complex explanation of the circulation of blood. Harvey had observed that the atria, the upper chambers of the heart, contract just prior to the ventricles, the thick-walled lower chambers.

He demonstrated that blood passes through the atria into the ventricles. The next step was to show that the blood in the right ventricle passes out to the lungs and the blood in the left ventricle passes to the rest of the body. Using his quantitative measurements, Harvey effectively showed that Galen's tidal, ebb-and-flow theory of circulation was wrong, but he realized he needed an alternative explanation for circulation. Harvey observed that blood came to the heart by the largest of veins, the vena cava. Blood passed through the right atria into the right ventricle. Blood then passed through the lungs, as he clearly observed and described in *De Motu Cordis*. Upon returning to the left side of the heart, blood passed through the left atrium, then the left ventricle, and out to the tissues via the arteries. So how did blood complete a circuit to get back to the heart? If tissues did not consume blood, how did it go through tissues and return to the veins, eventually to the great vena cava? To answer this question, Harvey returned to the finding and the physiological approach to anatomy from his former teacher, Fabricius, the man who discovered the venous valves. The unidirectional flow of blood suggested by the venous valves, a finding that can easily be confirmed by any observer, said to Harvey that there was a sort of motion as in a circle. Harvey's words seem simple, but their power cannot be denied:

> It has been shown by reason and experiment that blood by the beat of the ventricles flows through the lungs and heart and is pumped to the whole body. There it passes through pores in the flesh into the veins through which it returns from the periphery everywhere to the center, from the smaller veins into larger ones, finally coming to the vena cavae and right atrium. . . . It must therefore be concluded that the blood in the animal body moves around in a circle continuously, and that the action of the heart is to accomplish this by pumping. (5)

There is a hint of irony in the last element to complete the theory. What was the pathway by which the blood passed through tissues from the smallest of arteries to the smallest of veins? Harvey postulated "pores" in the tissues. We know them as capillaries. He fully anticipated that someone would find these pores. Still, such postulation must have bothered Harvey. These pores were all too reminiscent of the invisible interventricular holes that Galen proclaimed existed, but Harvey and others did not believe did exist. Harvey even stated in the introduction to *De Motu Cordis*, "Damn it, no such pores exist, nor can they be demonstrated" (5). But within 40 years, Marcello Malpighi of Bologna proved the existence of capillaries using a microscope. Harvey's circuit was complete.

As compelling as the reasoning was, *De Motu Cordis* was greeted with some angry opposition for disagreeing with Galen. The book initially generated little effect on medical practice. This lack of effect was probably because Harvey's ideas could not be translated into cures or treatments for patients' symptoms any better than the long-accepted Galenic ebb-and-flow theory. The bleeding, cupping, emetics, or other treatments dictated by the long-held humoral theory were still used on patients, even by Harvey himself. Harvey had helped explain how the body worked but not how disease occurred. Better understanding of disease processes was needed before physicians' daily practice could change. That would take the better part of a century. But medicine had taken a large step forward into experimental medicine, a step that would eventually cause the downfall of humoral medicine, although it would take nearly two more centuries. Despite the initial dissent, Harvey lived to see his doctrine on the circulation of blood accepted by the scientific and medical community. Later in his life, he was honored and applauded. In the next hundred years, others would discover the medical utilities associated with the knowledge of the circulation of blood, but Harvey's words stand out: reason and experiment. With these words, Harvey forever changed medicine, introducing the scientific method into the field.

For doctors to completely discard the humoral theory, with its attention to humors, innate heat, or pneuma, some viable alternative was needed to not just reject Galenic anatomy and physiology but to explain disease processes or, more specifically, the seat or starting place for disease. The Galenic approach to medicine had serious flaws. Surely others besides Vesalius and Harvey had been capable of pointing them out. So, what took so long for medical science to discard such a problematic theory? For centuries, Galen's status as a demigod held back any doubters. By the 17th century, Vesalius, Harvey, and others had chipped away at Galen's authoritative status. But that was not sufficient to take down the humoral theory. According to Thomas Kuhn,

> The decision to reject one paradigm is always simultaneously the decision to accept another, and the judgment leading to that decision involves the comparison of both paradigms with nature and with each other. (6)

The 18th-century experimental approach to anatomy and physiology did not offer many clinical successes to take to the bedside. A gulf seemed to develop between research and clinical practice. The bedside physician in the 18th century had little new knowledge to which to turn until a new paradigm emerged in Italy that allowed medical science to fully reject the humoral theory. Rejection was not the result of a single breakthrough but materialized from the meticulous, lifelong efforts of one man. His efforts resulted in a new concept that disease began not as a disruption of humors but because of anatomic changes in the very organs that Vesalius and other anatomists had so carefully described.

MORGAGNI AND THE ANATOMIC BASIS OF DISEASE

For hundreds of years, doctors knew that organs at death would vary in their appearance from one person to another. But no one had correlated the symptoms of a person before death with the findings at autopsy in any systematic way. The first attempt to do so was by Theophil Bonet, a Swiss-born physician who wrote a book in 1679 relating findings at autopsy to discrete diseases. However, the book was too inaccurate and disorganized to have much influence on the practice of medicine. It did influence the man who was to develop the so-called anatomic concept of disease: Giovanni Battista Morgagni. Developing the concept would require most of his professional life to produce a publication that would finally dispel the humoral theory once and for all.

Morgagni was born in 1682. As he completed his studies, he became Chair of Anatomy at the University of Padua, but Morgagni also treated patients at the famous university. There, he carefully catalogued patients' symptoms prior to death and correlated them with morbid anatomy at autopsy. His discoveries described correlations that to us seem so obvious and commonplace that physicians and the general public are shocked that the connections had not always been known and were made by Morgagni. These findings included chest pain known as angina pectoris and the findings of heart muscle degeneration and clots in coronary arteries after death; symptoms of stroke due to alterations in cerebral blood vessels with atherosclerosis, or so-called hardening of arteries; breathing difficulties and lung findings of emphysema; and symptoms of abdominal upsets and the pathological findings in gastric ulcers and appendicitis. Morgagni showed that disease, producing symptoms in a patient, could be located in certain organs. More succinctly stated by Morgagni himself, "Symptoms are the cries of suffering organs" (7). Morgagni patiently collected over 700 case histories and autopsy findings until he believed he had amassed enough evidence to convince fellow physicians. He did. In 1761, Morgagni published *De Sedibus et Causis Morborum per Anatomen Indagatis*, or *The Seats and Causes of Disease Investigated by Anatomy*. Unlike Harvey, Morgagni did not have to wait or suffer through criticism for his book to find favor. His book was an instant success. The anatomic concept of disease sent the humoral theory packing. With this new concept that disease could be located in specific organs at death, doctors began the hunt for clues to pathological anatomy before a patient's death. At first, doctors performed this search by listening to a patient's symptoms. Symptoms were of little use to the humoral theory of medicine. The nature or character of a patient's abdominal pain did not help a physician understand a disruption in the balance of humors. But the localization of symptoms to a diseased organ required careful correlation. The more specific the lead, i.e., the nature of the symptom, the better the correlation.

THE PARIS SCHOOL OF MEDICINE

Morgagni's work stimulated the correlation of symptoms to pathologic anatomy at death but also revived interest in the physical examination. Doctors began to examine more than a patient's pulse. They examined the organs, as well as they could, before the patient died. The late 18th and early 19th centuries saw the beginning of the modern physical examination—touching, percussing, and listening—all done to predict the anatomic findings at autopsy. Such predictions were often inexact. Even today, with all the technology such as computerized axial tomography scans, magnetic resonance imaging, and various laboratory-based tests, autopsies regularly yield surprises. But after Morgagni's theory, medicine's framework evolved further. Nowhere were these efforts better epitomized than in Paris. Paris had some of the largest hospitals in the world at the end of the 18th century, but the confluence of intellectual and philosophical underpinnings and the upheaval of the French Revolution provided an opportunity to reconstruct the medical profession. Paris became the undisputed center of medicine for the first half of the 19th century. The French Revolution demolished the notion of privilege and birthright in favor of ability and merit. Hospitals came under state control with the Paris Hospital Council, which regulated all aspects of care. The new world of Parisian medicine saw the locus of education shift from the study of books, i.e., "library medicine," to "hospital medicine." Trainees spent nearly all their time on the wards of a huge, interlocked, and prestigious educational network that was different from anything that had previously existed.

The intellectual climate fostered numerous advances in medicine. For example, Marie Francois Xavier Bichat (1771 to 1802) refined Morgagni's work. Bichat's advance was the concept of tissues, such as connective tissue and muscle and nerve tissues, being the building blocks of organs. Diseases affected specific tissues, not simply organs. Bichat's focus on tissues and their properties began to move medicine toward the cellular basis of disease, although elucidation of cellular pathology would have to wait until the mid-19th century. Bichat's view of the vital properties of tissues formed the basis of a new physiology in medicine. He believed that the tissues forming structures exhibiting animal life, such as voluntary muscle and sensory organs, were distinguished from those of organic life, such as the tissues comprising lungs and the circulatory system. This distinction led to a new understanding of the workings of the body in health (physiology) and disease (pathophysiology), separate from what could be discovered at autopsy. Doctors had always known that there were deaths that could not be readily explained by changes in organs at autopsy. Could some physiological or pathophysiological change be responsible for illness, change that could be discovered before death? This notion produced further interest in examining the living. The physical examination was perfected, especially by the introduction of the stethoscope by René Laënnec in the early 19th century in Paris. Doctors began to accurately predict pathological and anatomic causes of a patient's demise and prove themselves correct by postmortem examination.

These efforts gave rise to the concept of the specificity of disease, that disease was not a holistic phenomenon as Galen and Hippocrates had stated, but specific to pathologic changes in an organ or tissue. This view led to extensive classification of diseases, known as nosology. As a central feature of the Paris school, disease specificity and its classification suggested that diseases did not morph into one another. For example, cholera did not arise as a heightened form of summertime diarrhea, as had previously been thought, but was its own disease. Disease specificity was a critical concept for the emergence of the germ theory of disease (8). The Paris school became the model for international medical reform. However, the school had a serious weakness—therapeutics. While diagnoses could be made premortem, patients did not see significant benefits, since diagnosis is of limited utility without therapy. The medical profession continued to rely on traditional therapeutics and practices such as bloodletting that were remnants of humoral medicine.

THE RISE OF MODERN HOSPITALS IN WESTERN MEDICINE

With the demise of the humoral theory, medicine moved toward a more modern age in the late 18th century. Careful documentation of a patient's symptoms, a physical examination, and correlation with changes in organs and tissues at autopsy to define the seat of disease are the foundation of modern medicine. No longer were doctors searching for disruptions of humors, pneuma, or miasma. However, therapeutics had little advancement through the 18th and early 19th centuries. In fact, certain innovations proved enormously detrimental. The explosive development of the urban hospital occurred during the 18th century. Many of the European hospitals that are still operating today were founded in the late 1700s. These institutions provided ample diseased patients and, eventually, cadavers for study. But the clustering of patients, often with infectious illnesses, gave disease transmission a place to happen. Increased mortality was most evident in the new lying-in hospitals for childbirth. Mortality rates of 10 to 30%, usually from puerperal sepsis or childbed fever, were evident in delivering women. Puerperal sepsis was rarely seen before the 18th-century hospitals. The disease and its transmission were poorly understood until the mid-19th century, when that changed largely through the work of Ignaz Semmelweis, as we shall see in chapter 8.

CONTAGION AND 18TH-CENTURY MEDICINE

As the 18th century drew to a close, the concept of contagion was still not widely accepted. Even more alien to medical theory was a role for microorganisms in disease. The demise of the humoral theory was not enough to bring about change in physicians' notions that these little animalcules, discovered in the mid-17th century, played any role at all in human disease. In addition, since autopsies were performed with the naked eye in the 18th and early 19th centuries, physicians had no reason to consider the microbiological causation of disease when examining diseased organs. Indeed, the role of microorganisms was hotly debated in processes unrelated to human disease such as the causes of putrefaction, or how substances such as meat go bad. Were microorganisms to blame? If so, how did they get there? Many scientists held to the theory of spontaneous generation. If microorganisms could spontaneously appear, say, inside the human, how could a cogent theory of contagion even be considered? Dispelling the concept of spontaneous generation would be crucial for the development and acceptance of the germ theory of disease. Since spontaneous generation breached the church doctrine that God alone could create life, the debate even entered the metaphysical. Experiments to prove or disprove spontaneous generation during the 17th and 18th centuries remained controversial, both experimentally and philosophically. Until this debate was settled, the idea that a microorganism could be the cause of human disease, let alone take down a human, was not seriously considered.

THE SANITARY MOVEMENT, CONTAGION, AND 19TH-CENTURY MEDICINE

As Europe moved into the 19th century, urbanization and industrialization caused a population shift and, more importantly, a dramatic increase in the size of the populace; e.g., the population of England and Wales doubled in the first 50 years of the century. Diseases, particularly typhoid and cholera, caused massive mortality in the impoverished areas of industrialized European cities. Earlier measures of isolation and quarantine during specific disease outbreaks were clearly inadequate in an urban society. Other measures were needed. The early 19th century witnessed a change in perception that there was a public responsibility for the health of the population (9). Historians postulate that it was not medical science that was responsible for the uplift in human existence but advances in nutrition, wages, and sanitation (10). The improvement in sanitation came before the germ theory of disease was an accepted notion; reformers exploited the last remnant of the humoral theory of medicine to justify their sanitation methodology. The idea for improved hygiene began in Paris when reformers in the Paris school of medicine correlated certain categories of disease with the social and geographic backgrounds of the patients. The growing awareness that the intolerable stench of the French capital could be associated with disease was an affront to the sensualists in Paris and led to a hypothesis that miasma emanating from the filth caused disease. The Paris Board of Health worked to clean the city but had a limited role in the rest of the country.

The greatest early-19th-century advocate of sanitation was Edwin Chadwick (pictured on p. 71) in Great Britain. Chadwick had gained notoriety in reforming "Poor Laws" in England, aiding the welfare of indigent citizens. By 1834, he had gathered extensive data examining the devastation caused by disease in the impoverished cities of the time. He leaned heavily on the medical theory of his associate, a physician named Thomas Southwood Smith. His remedy also assumed that diseases were caused by miasma coming from the decomposition of waste. To remove disease, therefore, Chadwick proposed building a nationwide drainage network to remove sewage and waste. Further, Chadwick proposed that a national board of health, local boards in each district, and district medical officers be appointed to accomplish this goal. Many of his suggestions were adopted in the Public Health Act of 1848. The report documented the extent of disease and suffering in the population, promoted cleanliness (sanitation) and engineering as means of controlling disease, but laid the responsibility on the national government as the only organization with the financial means and authority to ensure that measures were realized. This report laid the foundation for public infrastructure for combatting and preventing disease, influencing later developments in public health in many countries including the United States. The sanitary movement in Britain transformed the country over the next 50 years, although it is impossible to calculate the numbers of deaths or illness prevented as a result (11). Still, the sanitation movement narrowed the focus of public health to removal of filth, not the improvement of living/working conditions or the diet of the people. But the success of the sanitary movement and its concentration on the removal of filth set up a furious debate in the mid-19th century on contagion.

CONTAGIONISM VERSUS ANTICONTAGIONISM IN THE 19TH CENTURY

The debate centered on the issue of whether diseases are contagious or not. But it didn't involve all disease. No one doubted, in the 19th century, that syphilis, for example, was contagious. The debate was about a subset of diseases: leprosy, typhus, plague, yellow fever, and cholera. European outbreaks of cholera in the 1840s and 1850s spurred the most fervent disputes since there was widespread panic, and depending upon which side of the debate one was on, the practical public health strategies to combat cholera differed. Prior to the discovery of the germ theory, the contagionist view was that some particle or poison, perhaps even a living "animalcule," was transmitted in some manner, by contact with a sick person or objects that had been in contact with that person. At the very time Semmelweis was postulating a cause of childbed fever (chapter 7), John Snow was theorizing a cause of a London cholera outbreak. Using epidemiology, Snow deduced that residents within 250 yards of a public pump on Broad Street in Soho were acquiring cholera. He showed that more than 500 deaths from cholera had occurred in a 10-day period and the victims had consumed water from that pump. He persuaded authorities to remove the pump's handle, and the outbreak ceased. Snow wrote:

> I am of opinion that the contamination of the water of the pump-wells of large towns is a matter of vital importance. Most of the pumps in this neighbourhood yield water that is very impure: and I believe it is merely to the accident of the cholera evacuations not having passed along the sewers nearest to the wells that many localities in London near a favourite pump have escaped a catastrophe similar to that which has just occurred in this parish. (12)

The great contagionist public health measure was quarantine and isolation, which required state power and involvement. But there were others who were skeptical of this approach. The most passionate anticontagionist was Max von Pettenkofer, who became involved with a cholera outbreak in Bavaria, where he worked. To von Pettenkofer's mind, the idea that the cholera poison was some sort of chemical substance made it entirely improbable that the disease could be spread in the manner postulated by Snow. Looking at London and the River Thames, how could a poison contaminating the Thames, if it was a chemical, not be enormously diluted? Further, to von Pettenkofer's thinking, the water-based transmission of cholera was a poor explanation for the seasonal aspect of the disease. Doctors who cared for cholera cases rarely became ill, another reason von Pettenkofer dismissed contagion. For Max von Pettenkofer, the explanation lay in a hovering miasma, springing from filth. Even after the discovery of the causative agent, *Vibrio cholerae*, by Robert Koch (chapter 10), he clung to this anticontagionist view. Cholera occurred, he postulated, when the *Vibrio* got into the soil beneath a major city. Pettenkofer believed in germination or fermentation after the *Vibrio* gained access to the soil. Under certain conditions it could germinate like a plant and taint the air above, air that the population inhaled, and those who were susceptible fell ill in large numbers. This theory was sometimes called the groundwater theory. He viewed groundwater as important because he considered variations in the water table beneath a city to be a measure of the capacity of a soil to support fermentation and give rise to cholera epidemics. This theory was the basis for the rebuilding of Naples following a devastating cholera epidemic in the mid-19th century. Von Pettenkofer's views provided a major impetus to undertake needed sanitary reform throughout Europe and can probably be credited with saving hundreds of thousands of lives. Even when the germ theory of disease became accepted, von Pettenkofer took this anticontagionist belief to his grave when he died in 1901. As Margret Pelling wrote,

> Anticontagionism therefore, presents . . . a paradox, that of an eminently progressive movement based on a wrong [and therefore retrogressive] scientific theory. (13)

Despite the invention of the microscope, the discovery of microorganisms, and the establishment of the modern medical approach to the patient, the germ theory of disease would have to wait. But before the debates over spontaneous generation and contagionism/anticontagionism were settled and a theory encompassing the role of microorganisms in human infection was developed, the early 19th century saw two of the most momentous advances in the history of medicine: vaccination and the successful prevention of childbed fever. But once again, innovators paid a powerful price.

REFERENCES

1. **Porter R.** 1997. *The Greatest Benefit to Mankind: a Medical History of Humanity from Antiquity to the Present*, p 44–82. HarperCollins, London, United Kingdom.
2. **Porter R.** 1997. *The Greatest Benefit to Mankind: a Medical History of Humanity from Antiquity to the Present*, p 163–200. HarperCollins, London, United Kingdom.
3. **Porter R.** 1997. *The Greatest Benefit to Mankind: a Medical History of Humanity from Antiquity to the Present*, p 201–244. HarperCollins, London, United Kingdom.
4. **Schultz SG.** 2002. William Harvey and the circulation of the blood: the birth of a scientific revolution and modern physiology. *News Physiol Sci* **17:**175–180. http://dx.doi.org/10.1152/nips.01391.2002.
5. **Harvey W.** 1931. *Anatomical Studies on the Motion of the Heart and Blood in Animals.* Translated by Chauncey D. Leake. Charles C Thomas, Springfield, IL.
6. **Kuhn TS.** 1962. *The Structure of Scientific Revolutions.* University of Chicago Press, Chicago, IL.
7. **Nuland S.** 1995. *Doctors: the Biography of Medicine*, p 145–170. Vintage Books, New York, NY.
8. **Snowden F.** 2020. *Epidemics and Society: from the Black Death to the Present*, p 168–183. Yale University Press, New Haven, CT.
9. **Goudsblom J.** 1986. Public health and the civilizing process. *Milbank Q* **64:**161–188. http://dx.doi.org/10.2307/3349969.
10. **McKeown T.** 1976. *The Role of Medicine: Dream, Mirage or Nemesis?* Nuffield Provincial Hospital Trust, London, United Kingdom.
11. **Snowden F.** 2020. *Epidemics and Society: from the Black Death to the Present*, p 184–203. Yale University Press, New Haven, CT.
12. **Thomas KB.** 1968. John Snow, 1813–1858. *J R Coll Gen Pract* **16:**85–94.
13. **Pelling M.** 1978. *Cholera, Fever and English Medicine, 1825–1865.* Oxford University Press, Oxford, United Kingdom.

7 Edward Jenner and the Discovery of Vaccination

Medicine has made a few extraordinary contributions that have dramatically improved human existence, but vaccination remains its greatest contribution. What other discovery has eradicated a disease from the face of the earth? Not just any disease but one of the most devastating illnesses in human history—smallpox. As magnificent an accomplishment as global eradication of smallpox was, the achievement that made it all possible began in the 18th century, even before the germ theory of disease was an accepted medical theory. One of the greatest achievements in medicine, smallpox vaccination, is worth scrutinizing. To understand the genesis and the impact of this discovery, one must first study the horrifying history of smallpox, especially in the 18th century, when the disease was at its peak of devastation in Europe. The 18th-century theories that attempted to explain the disease are emblematic of the hodgepodge of medical thought that occurred with the closing stages of the humoral theory of medicine but preceded the germ theory of disease, which was developed in the 19th century. Understanding these theories about smallpox in the 18th century helps to explain what Edward Jenner and others were up against in order to hypothesize, test, and prove the theory of vaccination to a skeptical and critical medical world.

THE DISEASE OF SMALLPOX

Smallpox was a disease (I use the past tense for the malady since the last natural, non-laboratory-acquired case occurred in 1977 and in spite of the danger of smallpox as a bioterrorism threat) caused by the variola virus, which affected only

Germ Theory: Medical Pioneers in Infectious Diseases, Second Edition. Robert P. Gaynes.
© 2023 American Society for Microbiology.

human hosts. The virus entered the human by the respiratory tract in most cases. After an incubation period of about 12 days, the virus produced an acute febrile illness. While there are two variants, variola major and variola minor, the former produced the more severe disease and was more common. The initial symptoms were fever, malaise, muscle aches, and headache and extreme prostration. By the third day or so, spots began to appear in the mouth, nasal passages, tongue, and throat. Large amounts of virus were present in saliva during this time. One or two days later, a skin rash appeared, first as spots but then as vesicles, or fluid-filled, raised skin lesions. These lesions tended to occur over the face, the distal parts of the extremities, and less on the trunk. These vesicles were filled with cellular debris and by 7 to 10 days after onset of the illness appeared to be pustules, although they were not actually filled with pus. Characteristically, these firm pustules appeared simultaneously and were of equal size, usually about ½ cm, around the body. Death was due to circulatory collapse most commonly and occurred between the 10th and 16th days of illness. By 2 weeks into the illness, pustules began to regress and scab. By day 21, the lesions scabbed over, flaked off, and left permanent scars, assuming that the person survived. Since the face was affected, the scars permanently disfigured many, if not most, people in areas where smallpox was endemic. About 2 to 5% of victims were left blind due to scarring over the cornea.

The overall smallpox mortality rate was 30%, but there were variations that affected that number. Children less than 1 year of age had a higher mortality rate, about 45%. For those persons whose pustules became confluent or joined together, the mortality rate could be as high as 50 to 75%. One form of smallpox where the pustules never formed as raised skin lesions but remained flat, so-called malignant smallpox, carried 90% mortality rates. And if there was severe bleeding from the skin or internal lesions, known as hemorrhagic smallpox, death was a certainty.

A SHORT HISTORY OF SMALLPOX

Smallpox has been traced back with some certainty to about 1200 BCE. It appeared to result in the death of the Egyptian Pharaoh Ramses V. The disease may have been the cause of the Athenian plague described by Thucydides. Smallpox was likely present during Roman times and may have contributed to the decline of the empire around 180 CE (1). The Chinese described smallpox in 300 CE. The first clear clinical description was by a predecessor of Avicenna, Rhazes, who first described and differentiated smallpox from measles in the 10th century in the Middle East. The expansion of Arabs and the Crusades contributed to the spread of smallpox in medieval Europe. But it was unknown in the New World until the Spanish and Portuguese introduced smallpox in the 1500s. To say that this disease had a devastating effect on the Aztecs is an absurd understatement. In 1518, when the Spanish arrived in Mexico, an estimated 25 million Aztecs lived there. By 1620, only 1.6 million survived (2). Similar wreckage of the Native North American population occurred in the 1600s, although not quite as rapidly as in Mexico.

The rich and poor were equally affected by smallpox. All ages and social classes were susceptible. Smallpox killed an impressive array of monarchs and political leaders, including Marcus Aurelius in 180 CE, the Aztec emperor Cuitláhuac in 1520, Emperor Ferdinand IV of Austria in 1654, Queen Mary II of England in 1694, Tsar Peter II of Russia in 1730, and King Louis XV of France in 1774, to name but a few. In the 18th century, four reigning European monarchs died of smallpox. During the same period, the line of succession to the Habsburg Empire changed four times because of the disease. The high-profile deaths among nobility led to a perception that the upper social class was more at risk. Many believed lifestyles rich in luxury led to smallpox. It wasn't true. In Europe alone, smallpox claimed over 400,000 deaths per year in the 18th century, an estimated 10% of all deaths. That's 40 million deaths during the 1700s in Europe alone. Those individuals who survived were left disfigured and sometimes blind. Because so many people were affected by smallpox in the 18th century, it was the leading cause of blindness in Europe.

SMALLPOX IN THE 18TH CENTURY

The way the populace viewed smallpox depended, in large part, on where one lived in the 18th century. For those in large cities such as London, smallpox was endemic, generally affecting the young and those who migrated to the city. The disease was viewed as an unfortunate fact of life. Many authorities believed that everyone would eventually develop smallpox. Many Londoners considered themselves fortunate to have had a mild case in childhood since it was recognized that afterwards, one no longer had to fear the devastating nature of the disease. Smallpox was so common among the young in London's 18th century that parents did not even count their children until they had survived the illness (3). In the countryside, however, the view was quite different. Smallpox was not always present in rural villages. Periodic flares of the malady in large cities like London, producing staggering death tolls there, led to epidemics in the countryside in that or the following year. Once introduced, the disease could lead to explosive outbreaks of illness and death in a village, resulting in terror among the inhabitants and desperate attempts to put a stop to it. Epidemics of smallpox across England were recorded in 1710, 1714, 1716, and 1719, producing a growing alarm over their increasing frequency.

With our modern eye, it is difficult to imagine that 18th-century scientists could view smallpox as anything but a contagious disease as they observed the various epidemics every few years. But at the beginning of the 18th century, the humoral theory still held. Despite the efforts of Fracastoro, the microscopic discoveries of van Leeuwenhoek, and the downfall of Galenic anatomy and physiology, the concept of contagion was still not widely accepted. However, during the 18th century, theories about smallpox evolved from a disease whose origin was inside the human body to one whose origin was outside the body. Rhazes described classical, humoral explanations for smallpox in the 10th century. These remained the views of most until the 18th century. Rhazes' view was that the seat of the disease was in the blood. In everyone something came from the mother that was responsible for smallpox. The mother's blood seeded the fetus with some factor that required blood to go through a process similar to the fermentation of grape juice into wine. The blood had to be purified of some innate seed via the pores of the skin. The innate seed from the mother offered a satisfactory explanation for the nearly universal affliction of smallpox. The theory held that sooner or later something in the air or in the diet of the individual caused this fermentation to occur, disengaging the innate factor from the blood. Since initial symptoms of smallpox often involved nausea and vomiting, these symptoms explained the body's attempt to begin the expulsion of the innate factor. But the onset of fever served as the explanation that all the remaining factor or humor was disengaged from blood. The sweating that occurred was the body's way of trying to rid itself of the factor. When the rash began, the explanation was that the excretory ducts were blocked. Eventually, if the person survived, the humors that were caught up in the pustules fermented and dried up, and the poisonous factor was discharged from the body. Adults, as the theory held, had built up more of the humor than children, so their disease tended to be more severe. While this theory explained much about smallpox, 18th-century physicians observed two facts that conflicted with the theory. First, records showed that smallpox first appeared in Europe in the Middle Ages. How could something innate from the mother be the explanation? Second, during the 18th century, there was a shift from smallpox's appearance as an early, often minor, childhood illness to one that was a serious threat to the entire populace. What explained the increasing virulence? Some authorities suggested that the change in diet or perhaps other habits of people explained both the appearance of smallpox during European medieval times and the change in virulence.

The concept changed where smallpox was thought to occur because of an external agent or influence that activated an innate propensity to the disease. Fracastoro discussed smallpox in his book on contagion, but he considered smallpox a minor disease of childhood, suggesting a contagious substance in the atmosphere but being quite general as to its nature. The discoveries of the properties of air in the 17th century by Robert Boyle and others tended to popularize air or changes in the atmosphere in the causative theory. Others believed the cause was something in the diet and/or some facet of immoderate living.

CONTAGION AND SMALLPOX IN THE 18TH CENTURY

By the early years of the 18th century, the repeated epidemics of smallpox could not be ignored, and physicians began to seriously consider a contagious agent for smallpox. At first, some poison or "noxious atom" was considered (4). To be certain that the noxious atom was not animate, attempts were made to observe these particles with a microscope by members of the Royal Society in 1723, but to no avail. There were other suggestions offered by animalculists, but the inability to find anything microscopically limited the appeal of such explanations. More appealing to the 18th-century theorists was the idea of venomous corpuscles. The origin of such corpuscles was never well explained, but these particles could cause similar disease in anyone who was unfortunate enough to encounter them. If poisonous substances were responsible, wouldn't the body react to the poison each time it came in contact with it? What was the explanation for the fact that people seemed to only suffer from smallpox once? In 1730, Thomas Fuller combined the idea of an innate seed, which explained why nearly everyone got smallpox, with the contagious particle. His hypothesis was that smallpox, like any contagious disease, was caused by the fertilization of specific ovula in the blood by an "afflatus genitalis" which was introduced from outside the body (3). These ovula lay dormant until the right poisonous particle came along. When it did, the fertilized ovula caused the disease and eventually was expelled from the body, never to cause trouble again, even if the person encountered the poison. All these theories of smallpox were a combination of the old and the new medical explanations that attempted to describe observed phenomena. But these theories would be put to the test when a new method to control this dreaded disease was introduced to Europe: inoculation, also called variolation.

VARIOLATION AND THE "CONTROL" OF SMALLPOX

In the 18th century, Europeans began to hear of practices in Asia and the Near East of preventing smallpox by means of variolation, an outmoded term that I use to avoid further confusion with inoculation, a term still used today but one that was also used interchangeably with variolation. Variolation is rooted in the term variola, a word also used for smallpox. It is derived from the Latin, *varus*, meaning "mark on the skin." Smallpox came to common parlance from smallpox (meaning sacs), first used in the 15th century to distinguish the disease from syphilis, then known as the Great Pox. The process of variolation was to take a lancet with fresh material from a pustule on a person with a mild form of smallpox. The probe would be introduced into a healthy person, often above the thumb, in the forearm. The practice did indeed produce smallpox in the recipient, but the case was often mild, perhaps because of the portal of entry (the subcutaneous skin versus the respiratory tract), the inoculum (i.e., the numbers of virus introduced), or the lower virulence of the chosen virus introduced via variolation. However, variolation was not without its problems. Deaths still occurred following variolation, albeit at a lower rate, estimated to be between a 5 and 12% case fatality rate, compared with an average of 30% for naturally acquired smallpox. Other problems were evident, too. Some recipients got severe cases of smallpox, which did not enhance the reputation of the process. Lancets were not cleaned and carried other organisms, including syphilis bacteria and bacteria that caused wound infections. Even though the variolation recipients may have had mild cases of smallpox, they could still transmit virus to others; the practice could have been responsible for smallpox outbreaks.

The origin of variolation is a bit murky. The Chinese tried to induce immunity by drying material from ripe smallpox pustules and putting it up the noses of susceptible persons in the 10th century (5). This practice was not always consistent in its effect. A process was described in India in the 17th century in which blankets from a smallpox sufferer with mild illness were wrapped around children with the intention of transmitting mild disease. The actual practice of variolation described above appeared nearly simultaneously in Africa, India, and China prior to the 17th century. In the 17th century the Ottoman Turks frequently used variolation. Circassian traders introduced the process to the Ottoman Empire around 1670. They learned the practice from women in the Caucasus, who used it not just to lower mortality rates from the disease but as a means to circumvent the disfiguring scars left by smallpox. The women of the Caucasus were legendary for their beauty and in demand in the Turkish sultan's harem, in large part because of the success of this practice. Consider that most European women portrayed in portraits from the 18th century had facial scars from smallpox which were not depicted in paintings. Portrait painters were the airbrush artists of their day. Any practice that allowed a person to get through the illness without disfiguring scars would be eagerly considered. Variolation came to Europe in the 18th century from travelers returning from Istanbul. The Royal Society in London received several communications about the process in 1714 and 1716, both published in the Society's *Philosophical Transactions*. There is little doubt that the English medical community was aware of the procedure in the early part of the 18th century, but the conservative society did not respond. Physicians were careful not to risk their reputations on a new technique. There were fundamental theoretical concerns that prevented variolation from being introduced in England. There was a belief that the practice may work in the Middle East but that differences in diet, notably more wine consumption among the British, would affect the British view of the utility of variolation in England. Climate was also a particularly crucial consideration. The climatic conditions in England differed so significantly from those in Turkey that English physicians considered the environmental conditions in the cool, wet English climate to be very treacherous in trying variolation. Recall that the evolving theory about the cause of smallpox at the time owed much to the environmental influences external to the body.

LADY MARY WORTLEY MONTAGU

Variolation was eventually popularized in England, not by a physician but by the wife of a British diplomat, Lady Mary Wortley Montagu. Lady Mary has been incorrectly credited with introducing the practice to England. She did not, as the cursory deliberations of the Royal Society in 1714 and 1716 demonstrate. But she did increase the appeal of variolation to the British and helped to remove the hesitation that the medical society had for it. How she came to have this influence becomes evident by reviewing her background. Lady Mary had married at age 23. Her husband was elected to Parliament 2 years after the marriage. Upon their move to London, Lady Montagu won great favor in British society circles with her charm and beauty. But at age 26 she contracted smallpox. She survived, but the disease scarred her great beauty. She lost a brother to the disease, too. About a year later, her husband was appointed British Ambassador to Turkey. The family moved to Istanbul, where she became acquainted with the technique of variolation. She became determined to keep her children from suffering from smallpox as her brother or even as she had. In letters back to England, she describes the technique:

> The small-pox so fatal and so general amongst us is here entirely harmless by the invention of ingrafting [which is the term they give it]. There is a set of old women who make it their business to perform the operation. Every autumn in the month of September, when the great heat is abated, people send to one another to know if any of their family has a mind to have the smallpox. They make parties for this purpose, and when they are met [commonly fifteen or sixteen together] the old woman comes with a nutshell full of the matter of the best sort of smallpox and asks what veins you please to have opened. She immediately rips open that you offer to her with a large needle [which gives you no more pain than a common scratch] and puts into the vein as much venom as can lye upon the head of her needle, and after binds up the little wound with a hollow bit of shell, and in this manner opens four or five veins... The children or young patients play together all the rest of the day and are in perfect health till the eighth. Then the fever begins to seize them, and they keep their beds two days, very seldom three. They have very rarely above twenty or thirty in

their faces, which never mark, and in eight days time they are as well as before the illness. . . . There is no example of any one that has died in it, and you may believe I am very well satisfied of the safety of the experiment since I intend to try it on my dear little son. I am patriot enough to take pains to bring this useful invention into fashion in England, and I should not fail to write to some of our doctors very particularly about it if I knew any one of them that I thought had virtue enough to destroy such a considerable branch of their revenue for the good of mankind, but that distemper is too beneficial to them not to expose to all their resentment the hardy weight that should undertake to put an end to it. Perhaps if I live to return I may, however, have courage to war with them (6).

The procedure was performed on her son, aged 5, in Turkey. He recovered without incident. In 1718, Lady Montagu and her family returned to England. Three years later, in 1721, a smallpox epidemic struck London. At that point, she allowed friends to witness her daughter, then age 4, having the procedure. She also recovered uneventfully. The King's physician, Sir Hans Sloan, witnessed the process. Suitably impressed, he suggested to the Royal Family that it be considered in the midst of an epidemic. Rather than proceed straight ahead, the decision was made to test the procedure on six prisoners who were scheduled to be hanged in Newgate prison. In exchange for their freedom, variolation was performed on the six individuals in full witness of several prominent physicians. All six prisoners survived. Would this procedure be enough to protect the recipients from smallpox? To answer the question, one of the prisoners, a 19-year-old woman, was sent to a village with a severe smallpox epidemic to care for victims of the disease. She was exposed to the illness for 6 weeks and did not come down with any further illness herself. Newspapers carried accounts of the experiment. The Princess of Wales became so interested in the procedure that she instructed that a list of all orphaned children in St. James Parish who had not had smallpox be made so that they could have the variolation procedure at her expense. In 1722, sufficient confidence existed in variolation that it was performed on two daughters of the Prince and Princess of Wales. They recovered, to the great relief of the Royal physicians and with much fanfare from the newspapers. Slowly, the practice spread throughout England. Nearly a year after variolation was tried successfully on the Royals, a high-profile death, that of the son of the Earl of Sunderland, precipitated years of controversy over the procedure. One of the chief antagonists was William Wagstaffe, who argued in 1722 that variolation would not always produce smallpox, since the procedure was unreliable. This belief was supported by the notion that the state of the recipient, i.e., his or her diet and the state of that person's blood at the time of variolation, would help determine the outcome. Accordingly, a preparation of a special low-calorie diet and bleeding became the rule before variolation to bring the body to the "optimal" condition. The controversy did not go away despite attempts to show that mortality rates appeared to be somewhat improved using the procedure. There was at first a general acceptance in the mid-1720s, and then the controversy caused a decline in variolation use in the 1730s, a period of relative quiet for smallpox. But as the debate continued, so did epidemics of smallpox in England. By 1743, variolation was made compulsory for all children in the Foundling Hospital in London. In 1746 an epidemic of smallpox returned in full force to London. Amid this crisis, an entire hospital, appropriately called the Smallpox and Inoculation Hospital, was established for the procedure. Other countries, particularly France, were not keen on variolation. But in England, variolation became an accepted medical practice by 1750. The advent of the Smallpox and Inoculation Hospital did little to help the lower-class population (7). Some resistance to the procedure remained, largely among the clergy. However, a raging epidemic of smallpox in all of England in 1752, the worst year for the disease in England during the 18th century, removed all resistance to the procedure and ensured that variolation would be the accepted medical practice. Increased attention to preparation for variolation came during this period in England, including strict regulation of diet for at least 2 weeks before the procedure, bleeding, emetics, and purging. It is hard to imagine that this approach was considered the foundation for an improved constitution prior to deliberate infection with smallpox. But it was. Gradually, the practice spread across England, with the supreme "stamp of approval" coming in 1755 when the Royal College of Physicians unanimously approved this supporting statement:

> The College . . . judge it [variolation] to be a Practice of the utmost benefit to mankind (7).

This statement effectively ended any controversy over the use of variolation in England. We might be horrified at the idea that after 2 weeks of ghastly preparation, physicians began to make a sizable income from deliberately giving smallpox to someone. But an important positive came from variolation. This was the first procedure in medical history to introduce the idea that a contagious substance that was specific for smallpox produced the disease.

EDWARD JENNER: EARLY INFLUENCES

Amidst all of England's struggles with smallpox, Edward Jenner was born in Berkeley, Gloucestershire, near Bristol, on 17 May 1749. Many aspects of his childhood had a profound effect on him and on his discovery of vaccination. His father, Stephan, was a clergyman; his mother, Sarah, was the daughter of a clergyman. Edward was one of nine children, two of whom did not survive to adulthood. He was the eighth child in the family. His mother died in childbirth on delivery of the ninth child. Two months later, Edward's father died, leaving him an orphan at the tender age of 5 years. His older siblings struggled to raise the boy. When Edward was 8 years old, he was sent to a free boarding school in England, where events occurred that would shape his life. The school suffered an epidemic of smallpox. All children, including Edward, who had not had the disease, were pulled from the school, and prepared for variolation. Jenner was bled until his blood was "thin," purged repeatedly till his body was wasted to a skeleton and kept on a low-vegetable diet in a stable (8). After preparations, each boy had the variolation. Amazingly, none of the 12 students died from the procedure. Edward, of course, survived but was badly damaged because of the incident, suffering anxiety and insomnia for years. As a result, his family moved him to a small private school, where Edward recovered. His poor academics in certain areas prevented him from following in his father's footsteps: going to Oxford and becoming a clergyman. With his aptitude in science and his keen interest in nature, medicine became his chosen field.

MILKMAIDS, COWPOX, AND SMALLPOX

Eventually, Jenner became an apprentice to a country surgeon, a position he retained for 6 years. One account of Jenner's life suggests that during this period, he overheard a dairymaid say,

> I shall never have smallpox for I have had cowpox. I shall never have an ugly pock-marked face (9).

Indeed, there was widespread belief in rural England that dairymaids were, in some manner, shielded from smallpox and its deadly and disfiguring effects. This notion led to the 18th-century impression that the image of the purest complexion was one of a dairymaid. But the prevailing theories of smallpox did not provide a clear explanation for any protection from cowpox. Still, Jenner made careful note of the dairymaid's comment. Upon completion of his apprenticeship, Jenner headed for London at the age of 21. He had the remarkable good fortune to become a student of John Hunter, one of the most famous surgeons in England, actually, in history. Hunter's stature came from a lifetime of sound discoveries, including the way tendons and wounds heal and the first serious study of the process of inflammation (10). Hunter was responsible nearly single-handedly for upgrading British attitudes towards the surgeon as a medical professional rather than a slipshod technician. The Royal College of Surgeons each year on Hunter's birthday, 14 February, invites a renowned speaker to deliver the Hunterian Oration. But Hunter was a doer, not a thinker. He experimented constantly on various animals and, famously, himself, referring to the experiment in which Hunter gave himself syphilis to study it. Even with his growing fame, Hunter was not an easy man with whom to get along. He was quick to anger, combative, and outspoken. Still, there is ample evidence in his correspondence with Jenner that he had a genuine affection for Edward. Hunter imparted his insatiable curiosity and experimental methodology to Jenner. Jenner spent only 2 years with Hunter, but the two remained friends and corresponded for years. Hunter's influence on Jenner was immense. Hunter helped Jenner's proposed election to the Royal Society in 1787, not for a medical discovery but for Jenner's discovery of the nesting habits of cuckoo birds.

Following the 2 years in London with John Hunter, Jenner headed back to the countryside of England. This decision profoundly affected Jenner's career and life. Hunter had asked him to take a position in London at a school of anatomy and natural history, but Jenner refused. Jenner returned to Gloucestershire and set up practice while continuing his naturalist investigations. In 1789 when he went to London to present his cuckoo paper to the Royal Society, he began to sound out his professional colleagues, including Hunter, on an idea he had been hearing in the countryside: cowpoxed milkers could not "take" smallpox. This was rather common talk in the countryside where cowpox occurred when Jenner began to practice. But the association was not evident even among idle gossipers until the middle to late 1700s (11). Jenner had heard of the association even before working with Hunter, and the continued discussion among the locals had caught his attention. But Jenner had not made any personal observations on the topic. So, during his trip to the Royal Society, he decided to ask around. He got little help on whether the two diseases were linked, since cowpox was essentially unknown in London. But when Jenner told Hunter of his idea that cowpox could protect someone from smallpox, his mentor did offer one solid piece of advice:

I think your solution is just; but why think? Why not try the experiment (12)?

Jenner returned to Gloucestershire to consider his next step. However, history shows that he was not the first in the area to test the idea. The best-documented test of this notion occurred in 1774, when a dairy farmer named Benjamin Jesty performed the same procedure as variolation but administering material from a cow's sore teats into the arms of his wife and two children. No one is exactly sure what happened next; a doctor (not Jenner) was called in to treat his family, but no one in Jesty's family ever acquired smallpox. Area doctors who were more familiar with cowpox than Jenner assured him that they could produce case after case where a cowpox sufferer came down with smallpox. To Jenner, this legend about cowpox protecting against smallpox had to be proven one way or another. He saw the possibility of repeating the success of his cuckoo bird paper before the Royal Society. But how to make the association between cowpox and smallpox scientifically plausible? Jenner pondered, apparently for a while. Jenner mentioned nothing more on the subject of experiment with cowpox until 1794 in some correspondence with other medical professionals. While recuperating from a bout of typhoid fever in 1795, Jenner considered the idea of deliberately giving a perfectly healthy person cowpox and then exposing that person to smallpox to see if cowpox was protective. But cowpox outbreaks were intermittent. He would have to wait until the right moment.

THE FIRST INOCULATION AGAINST SMALLPOX

In 1796, Jenner finally acted. Upon hearing of an outbreak of cowpox among local milkmaids at a farm near Berkeley, he took off some fluid from a milkmaid named Sarah Nelmes, who became infected with coxpox after milking a cow named Blossom (that really was the cow's name!). On 14 May 1796, he inoculated it on the arms of an 8-year-old boy named James Phipps. Eight days after being injected with cowpox, the boy suffered a fever for 2 days but no great effects. Jenner then exposed the boy not once but twice to smallpox. Nothing happened to him. It had worked!

Today, if one were writing a protocol to test whether some microorganism, say, simian immunodeficiency virus, produced immunity in a human to another microorganism, e.g., human immunodeficiency virus, any investigator who would propose taking material directly from an infected person without purifying the causative agent, directly injecting the material in an 8-year-old child as the first test subject, and then deliberately injecting the child with human immunodeficiency virus to see if the inoculation worked, would be fortunate to avoid jail, besides rejection from an institutional review board. Yet, this progression was precisely what Jenner did. How could he have done this? Why was there no ethical outcry? Ethics vary with the times and the circumstances. While we would view Jenner's approach as highly unethical or even criminal, in its time and under the circumstances, it seems much less of a cause for an ethics violation. First, recall that the 18th century was one when smallpox was a nearly constant threat. Anything that saved someone from the smallpox threat would be viewed with favor. But I do not believe that it is an accident that the first subject was an 8-year-old child. Jenner himself was subjected to variolation at nearly the same age and suffered greatly. He must have had a strong desire to save his neighbor's son from his fate. But how could Jenner have deliberately injected the boy with smallpox? Jenner merely subjected the boy to variolation, an accepted technique, not once but twice in the summer following his "vaccination." Phipps did not develop smallpox with either attempt. Phipps lived to a ripe age and to demonstrate his immunity to smallpox, he had the variolation performed some 20 times during his life. Phipps remained friendly with Jenner, even attending his funeral.

Jenner wrote about the experience with James Phipps and 12 others in a paper to the Royal Society. To his surprise, it was rejected. The rejection remained a sore point for Jenner for years. The precise reasons for the rejection may never be known, especially since Jenner was well acquainted with the reviewers and was a member of the society. The findings were described as "in variance with established knowledge and incredible" (1). One of the greatest advances in medical history went unrecognized by the Royal Society! True, he had only presented 13 cases; 3 cases had not even been exposed to cowpox, but a disease called grease of horses, which Jenner believed was the source of cowpox. Only one case in the paper, James Phipps, had been deliberately given cowpox by Jenner; the rest had casually acquired cowpox.

PUBLICATION OF *AN INQUIRY INTO THE CAUSES AND EFFECTS OF THE VARIOLAE VACCINAE*

Jenner became determined to show that his experiment was not a fluke. Because of the intermittent nature of cowpox, Jenner once again had to wait. In 1798, when another outbreak of cowpox occurred in the countryside, Jenner added more cases, including his son, Robert, to his "experiment" to show that cowpox protected against smallpox. Jenner was not inclined to try publishing in the Royal Society's publication again. Instead, encouraged by some friends, Jenner used his own money to publish *An Inquiry into the Causes and Effects of the Variolae Vaccinae* (13). The full title was *An Inquiry into the Causes and Effects of the Variolae Vaccinae: a Disease Discovered in Some of the Western Counties of England, Particularly Gloucestershire, and Known by the Name of the Cow Pox*. Jenner used this publication, only 64 pages, to coin two new terms. The first, *variolae vaccinae*, literally means "smallpox of the cow" in Latin. Jenner invented the phrase *variolae vaccinae* to support his notion that the diseases smallpox and cowpox were related. However, cowpox was essentially unknown in most areas of England, so he took his chance with the name. The term didn't stick. His other term, vaccination, became part of our medical lexicon. Jenner wanted to distinguish his procedure from variolation. He took the Latin root, *vacca*, meaning "cow," and invented the term for his process. The 1798 publication, *An Inquiry*, had three parts (2). The first part discussed what Jenner considered to be the origin of cowpox, a hypothesis that was quickly discredited. Jenner came to believe that the only reliable vaccines to be used should be derived from cowpox-infected cows. The second part discussed the hypothesis that cowpox protects against smallpox. The third part of the publication includes a discussion of the findings when the hypothesis was tested along with a variety of issues related to smallpox, describing 23 cases. The first 16 cases were descriptions of people who had somehow acquired cowpox on their own and then were exposed to smallpox, mostly via variolation by Jenner. The remaining seven cases, including number 17, James Phipps, were case descriptions where he used cowpox to protect a healthy person from smallpox and included his own son, Robert Jenner. Only four of these cases were subsequently challenged with variolation. Importantly, he used material from the deliberately induced cowpox infection in one person to vaccinate another case. He concluded "that the cow-pox protects the human constitution from the infection of the smallpox (13)." Jenner believed that vaccination was safer than variolation and that the protection imparted by cowpox vaccination would be lifelong.

REACTION TO *AN INQUIRY INTO THE CAUSES AND EFFECTS OF THE VARIOLAE VACCINAE*

Convinced of his own correctness, Jenner headed off to London to convince his more urbane colleagues. In 3 months, Jenner was unable to secure any volunteers for his vaccination in London. A skeptical medical profession surprised and tested Jenner. A man who was described as kind and affable towards patients and colleagues in the English countryside became increasingly bad-tempered as he faced criticism over his beliefs. Some of the most vehement criticism was from physicians who had financial self-interests in variolation. Jenner gave cowpox material to a surgeon in London, Henry Cline, who began to try vaccination, with success. Cline communicated the success to other colleagues in London, an event that greatly helped publicize vaccination. A few other physicians followed, including one from London's Smallpox and Inoculation Hospital. Unfortunately, some patients at this hospital developed a generalized rash that differed substantially from what Jenner described. It is likely that the lancets that had been used for vaccination were also used for variolation and had become contaminated with smallpox, serving to cloud the success of vaccination in 1799.

Controversy dogged Jenner and vaccination. Financial self-interest and contaminated material were only part of the impediments. Another part of the problem of acceptance was that physicians who wished to perform vaccination had difficult access to the material. They had to get the material directly from Jenner since cowpox did not occur widely in England. There were those who challenged Jenner as the "Discoverer of Vaccination." His critics tried to discredit him by inviting Benjamin Jesty to London. Unlike Jenner, Jesty had never fully documented or published his efforts. Still, this debate soured Jenner. There were also public concerns that people injected with cowpox would grow cow parts, depicted in editorial cartoons of the time (Fig. 7.1).

FIGURE 7.1 *The Cow Pock—or—the Wonderful Effects of the New Inoculation! Drawing by James Gillray, 1756 to 1815. Courtesy of the National Library of Medicine (NLM image ID A021551).*

Many of Jenner's opponents were simply jealous, but they were an impressive group. William Woodville, Director of the Smallpox and Inoculation Hospital, and George Pearson of St. George's Hospital used their own cow-derived vaccine, inoculating thousands of people, many more than Jenner could have. Ironically, they inadvertently confirmed Jenner's claim. Pearson published in order to take credit for the discovery in 1798 even though he did not make any original contributions, reporting on the results of others with a self-promoting style and never acknowledging Jenner as the innovator of vaccination. In contrast to the selfishness of Pearson, Jenner refused to profit from his discovery. Instead, Jenner became obsessed with convincing his countrymen of the value of vaccination, so much so that his country practice suffered. He conducted a nationwide survey of all vaccination recipients or persons who had had cowpox and were exposed to smallpox. The results confirmed Jenner's theory (4). By 1800, the technique of vaccination had reached other countries, including the United States. In July 1800, Benjamin Waterhouse, a professor at Harvard, obtained some material for vaccination from John Haygarth of Bath, England, who had gotten the vaccine material from Jenner himself. Waterhouse introduced vaccination to the United States and persuaded Thomas Jefferson to consider it.

THOMAS JEFFERSON'S LETTER TO JENNER

Thomas Jefferson was impressed by Jenner's work, appointing Waterhouse as Vaccine Agent in the National Vaccine Institute, designed to implement a national vaccination program in the United States. Jefferson wrote Jenner in 1806:

To **Dr. Edward Jenner**

SIR, —I have received a copy of the evidence at large respecting the discovery of the vaccine inoculation which you have been pleased to send me, and for which I return you my thanks. Having been among the early converts, in this part of the globe, to its efficiency, I took an early part in recommending it to my countrymen.

I avail myself of this occasion of rendering you a portion of the tribute of gratitude due to you from the whole human family. Medicine has never before produced any single improvement of such utility. Harvey's discovery of the circulation of the blood was a beautiful addition to our knowledge of the animal economy, but on a review of the practice of medicine before and since that epoch, I do not see any great amelioration which has been derived from that discovery.

You have erased from the calendar of human afflictions one of its greatest. Yours is the comfortable reflection that mankind can never forget that you have lived. Future nations will know by history only that the loathsome smallpox has existed and by you has been extirpated.

Accept my fervent wishes for your health and happiness and assurances of the greatest respect and consideration. *Thomas Jefferson*

The controversy over vaccination, despite large-scale successes, simply would not die, draining Jenner. His efforts in championing his vaccination cause cost him financially. He was neglecting his practice and his family, spending much of his time in London. He was in serious debt but always maintained that he should not be paid for his efforts to promote vaccination. His supporters tried to help by petitioning the British Parliament for aid. In 1802, England finally afforded Jenner some recognition for his efforts when the British Parliament awarded him 10,000 pounds; 5 years later, he would be given 20,000 more. Jenner received honors from Harvard, Cambridge, Oxford, and numerous societies. Despite the cool reception of variolation, France embraced vaccination. In 1805, Napoleon required his troops be vaccinated; a year later he ordered vaccination of French civilians. Jenner's growing fame helped him negotiate the release of a number of British prisoners in England's war with Napoleon, who was reported to have said, "Ah, Jenner, I can refuse him nothing."

In 1803, in the wake of increasing financial burdens, Jenner returned to his country home and family. Other countries, notably India, sent Jenner monies to help. He built a clinic adjacent to his home where he gave free vaccination to anyone. Yet, the opposition to vaccination, in favor of variolation, continued. More English people had variolation than vaccination in the years after Jenner returned to his home. He continued his efforts for vaccination, responding to reports of failed vaccinations until 1809, when he "retired." He continued his medical practice, but events in his family took their toll on Jenner. In 1810, his son, Edward Jr., died of tuberculosis. In 1815, Jenner's wife died from tuberculosis, too. In 1823, Jenner suffered a stroke and died at age 73. His home in Gloucestershire is now the Jenner Museum. A statue to Edward Jenner stands in Kensington Gardens, though it originally stood in Trafalgar Square. Jenner has become known as the Father of Immunology, although he had no knowledge of how the immune system worked.

VACCINATION AND THE ERADICATION OF SMALLPOX

Despite the clear success, opposition to vaccination over variolation continued until 1840, when the British Parliament outlawed further variolation. The technique that Jenner used, placing material from a cowpox pustule into cuts in the forearm, was discarded in 1858 in favor of the use of lancets, the technique still in use for smallpox vaccination. In 1801, Jenner predicted the eradication of smallpox using vaccination. He did so without knowledge of the microorganisms that caused cowpox or smallpox, how they were related, or how they spread. It required two centuries before Jenner's prediction came true. But it did. Smallpox eradication came using essentially the same technique of a man who first experimented on an 8-year-old child. Through selfless perseverance and at great personal sacrifice, Jenner managed to convince a skeptical world that he was right. No more fitting tribute can be made than that from the Jenner Museum website:

> All that is known about disease prevention by vaccination, our understanding of allergy, autoimmune diseases [such as rheumatoid arthritis], transplantation and AIDS follows from this fundamental work by Edward Jenner (14).

In the last few years of his practice, Jenner used a strain of vaccine named for a patient, Ann Bumpus, who had over 300 eruptions following her vaccination in 1799. The material taken from her eruptions lasted months and was distributed around the globe. Subsequent vaccine material was taken from arm-to-arm transfer until the middle of the 19th century. Occasionally, fresh material from cows was introduced. After the 1840s, propagation of the vaccine material occurred in cows. The addition of glycerol to the bovine material allowed for longer storage. In 1898, human arm-to-arm passage was outlawed in Britain.

GLOBAL APPLICATION OF VACCINATION

Within 10 years of its inception, vaccination had reached around the globe. Over the next century, smallpox prevalence declined steadily. By the early 20th century, smallpox, while not eliminated, was a less important cause of childhood mortality than measles. Outbreaks continued periodically. World War II hampered control efforts for smallpox, with increasing numbers of countries reporting cases in the year after the war ended (89 countries) and large outbreaks hitting Africa and Asia. India, as one example, reported over 1 million cases in 1944. In 1948, the World Health Organization at its first meeting took on smallpox prevention. Subsequent efforts led to worldwide eradication of the disease, which has been detailed elsewhere (15). The vaccine's virus used for eradication is called vaccinia virus. The vaccine remains available for use today. However, in its present form, the vaccine virus differs from the cowpox virus used by Jenner and from the smallpox virus. Cowpox, found only in Britain and isolated areas of Western Europe, remains a rare disease.

VACCINIA VIRUS IN THE CONTEMPORARY SMALLPOX INOCULATION

Vaccinia virus represents a hybrid virus whose precise origin is unclear, but which likely arose from inadvertent mixing of cowpox and smallpox viruses in those early days of vaccination. Genetically, vaccinia virus is more closely related to smallpox than cowpox virus (16). Despite any genetic differences in virus and any modifications of the technique of vaccination, worldwide application of the basic technique from Jenner eliminated smallpox. Eradication was a bold goal when it was proposed in the 1950s. The global eradication program even encountered areas in Asia and Africa where variolation was still practiced up through the early 1970s. But the global effort achieved its goal through vaccination. Since smallpox only affects human hosts, eliminating the reservoir through vaccination eradicated the disease.

SUCCESS OF VACCINATION

The global eradication of smallpox began with one physician, Edward Jenner, making note of folklore, boldly testing the hypothesis, and, through perseverance at great personal cost, proving to a skeptical world that vaccination "protects the human constitution from the infection of the smallpox." Jenner's only major error was the assumption that lifetime immunity occurred after vaccination, a notion proven to be incorrect in the late 1800s, requiring vaccination every 5 years to maintain immunity to smallpox. Jenner laid the groundwork for the most important contribution to medical science, even before the development of the germ theory of disease. Vaccination would be extended to dozens of additional diseases by others. Despite some adverse effects of vaccination and the clatter of opponents, we should remind ourselves of the misery that has been prevented over the last century by application of Jenner's innovation (Table 7.1). Of the 10 infectious diseases listed in Table 7.1, vaccination reduced by over 95% the disease mortality and morbidity (another word for illness that does not produce death associated with the disease) in a century.

The advances made during Jenner's lifetime are not limited to vaccination alone. Medicine had accepted the concept that a contagious substance that was specific for smallpox produced the disease and that the disease could be prevented with a substance from another disease. Moreover, the protective substance could be taken from one human and given to another and then another and produce the same protection. This was a far cry from the unbalanced humors and miasma dominating the medical theory of smallpox at the beginning of the 1700s. By the early 1800s, contagion was becoming an accepted notion within medicine but was by no means universally accepted or considered in all diseases that we now know as infectious diseases. The role of microorganisms in producing disease remained dubious in the minds of many physicians of the early 1800s, which was evident when a second, pre-germ theory contribution was made: the importance of hand washing.

TABLE 7.1 Impact of immunizations, 1900 to 1999[a]

Infectious disease or agent	No. of cases in 1900	No. of cases in 1999	% Annual morbidity/ mortality decrease
Smallpox	48,165	0	100
Diphtheria	175,885	1	100
Pertussis	147,271	6,279	95.7
Tetanus	1,314	34	97.4
Poliomyelitis (paralytic)	16,316	0	100
Measles	503,282	89	100
Mumps	152,209	606	99.6
Rubella	47,745	345	99.3
Congenital rubella syndrome	823	5	99.4
Haemophilus influenzae type b	20,000	54	99.7

[a] Adapted from reference (17).

REFERENCES

1. **Barquet N, Domingo P.** 1997. Smallpox: the triumph over the most terrible of the ministers of death. *Ann Intern Med* **127**:635–642. http://dx.doi.org/10.7326/0003-4819-127-8_Part_1-199710150-00010.

2. **McNeil WH.** 1976. *Plagues and People*. Anchor Publications, Garden City, NY.

3. **Miller G.** 1957. p 26–44. *In The Adoption of Inoculation for Smallpox in England and France*. University of Pennsylvania Press, Philadelphia. http://dx.doi.org/10.9783/9781512818086-004.

4. **Miller G.** 1957. p 241–276. *In The Adoption of Inoculation for Smallpox in England and France*. University of Pennsylvania Press, Philadelphia. http://dx.doi.org/10.9783/9781512818086-011.

5. **Gross CP, Sepkowitz KA.** 1998. The myth of the medical breakthrough: smallpox, vaccination, and Jenner reconsidered. *Int J Infect Dis* **3**:54–60. http://dx.doi.org/10.1016/S1201-9712(98)90096-0.

6. **Wharncliffe L, Thomas WM (ed).** 1861. *The Letters and Works of Lady Mary Wortley Montagu*, vol I. Henry G Bohn, London, United Kingdom.

7. **Miller G.** 1957. p 134–171. *In The Adoption of Inoculation for Smallpox in England and France*. University of Pennsylvania Press, Philadelphia. http://dx.doi.org/10.9783/9781512818086-008.

8. **Behbehani AM.** 1983. The smallpox story: life and death of an old disease. *Microbiol Rev* **47**: 455–509. http://dx.doi.org/10.1128/mr.47.4.455-509.1983.

9. **Riedel S.** 2005. Edward Jenner and the history of smallpox and vaccination. *BUMC Proc* **18**:21–25.

10. **Hunter J.** 1982. *1794. A Treatise on the Blood, Inflammation, and Gunshot Wounds. Classics of Medicine Library (Facsimile edition)*. Gryphon Editions, Ltd, Birmingham, AL.

11. **Creighton C.** 1889. p 19–48. *In Jenner and Vaccination: A Strange Chapter of Medical History*. Swan Sonnenschein and Co, London, United Kingdom.

12. **Nuland SB.** 1988. p 171–199. *In Doctors: A Biography of Medicine*. Vintage Books, New York, NY.

13. **Jenner E.** 1798. *An Inquiry into the Causes and Effects of the Variolae Vaccinae: a Disease Discovered in Some of the Western Counties of England, Particularly Gloucestershire, and Known by the Name of the Cow Pox*. http://www.gutenberg.org/etext/29414.

14. **Edward Jenner Museum.** *The Final Conquest of the Speckled Monster*. Edward Jenner Museum, Berkeley, Gloucestershire, United Kingdom. http://www.jennermuseum.com/Jenner/finalconquest.html.

15. **World Health Organization.** 1980. *The Global Eradication of Smallpox. Final Report of the Global Commission for Certification of Smallpox Eradication*. World Health Organization, Geneva, Switzerland.

16. **Chernos VI, Surgai VV.** 1980. Study of genome homology of some orthopoxviruses. *Acta Virol* **24**:81–88.

17. **Centers for Disease Control and Prevention (CDC).** 1999. Impact of vaccines universally recommended for children–United States, 1990-1998. *MMWR Morb Mortal Wkly Rep* **48**:243–248.

8 Ignaz Semmelweis and the Control of Puerperal Sepsis

While few people delight at the prospect of going into a hospital, we know hospitals as institutions for healing, surgery, and other procedures that cannot be offered in the home or a doctor's office. But the hospital has not always been viewed as this therapeutic establishment. People have viewed hospitals with sadness, despair, and, at times, sheer terror during their long history. The very institution that was designed, in part, to diagnose and treat infectious diseases became a source of them. Understanding how one particular infectious disease, childbed fever, was determined to be acquired in hospitals during the 19th century is vital to the evolution of the germ theory but takes us to a dark chapter in medical history.

THE DEVELOPMENT OF HOSPITALS IN WESTERN MEDICINE

Before considering the impact of childbed fever in hospitals, we must consider how the institution developed in Western medicine. Facilities like Byzantine and medieval Islamic hospitals, already discussed in chapter 3, did not appear in Western Europe until after the Crusades. Until then, monasteries in Europe played the role of hospitals but were little more than places to die. The emphasis was on saving the soul, not the body. With the first crusade, the mission of the hospital in Western civilization took an abrupt shift. A Papal bull from Pope Paschall II, issued in 1113, placed the new St. John Order of the Hospital in Jerusalem under papal protection. Like many monasteries, the institution gave aid to Christian pilgrims and crusaders. With the security provided by the papal protection, St. John Order,

Germ Theory: Medical Pioneers in Infectious Diseases, Second Edition. Robert P. Gaynes.
© 2023 American Society for Microbiology.

in the 1150s, shifted from a monastic hostel to subordinating its staff to the wishes of the sick (1). To make tending to the sick, i.e., mending the body rather than the soul, a mission of the hospital required a specific, written charter at St. John's Order in the 1150s. Despite the end to Christian rule in Jerusalem in 1187, St. John's Hospital was allowed to remain open to care for its sick. St. John's Hospital became a model in design and in attitude towards the sick in Western Europe. Hôtel-Dieu in Paris, the first hospital in Western Europe, was founded in 651, more as a hospice than as a hospital. Later, it was heavily influenced by the healing mission of St. John's. The Parisian institution began offering surgical treatments, mostly wound care, in 1221 and medical treatments in 1230.

Through the medieval and Renaissance periods, some hospitals played distinctive roles. The prevalence of leprosy in medieval times led institutions to confine and isolate sufferers. Prevailing humoral theory of leprosy contended that close contact with those already ill produced the disease through repeated breathing of corrupted air, or miasma, from the leper. Since plague was associated with similar miasma theories, hospitals, notably in Italy, were set up to keep the sick away from the healthy, usually locating them outside the city and downwind.

During the Renaissance, European hospitals began to fill a social role to care for the poor and sick, although spiritual salvation remained a primary objective. The Church generally financed hospitals of the time. During the Enlightenment, the Edinburgh Infirmary for the Sick Poor in Scotland opened with a private and public financial endowment, a notable deviation from primarily religious institutions. Another change for the hospital mission occurred during the Enlightenment. Hospitals became places of learning. Universities for medicine had affiliated with hospitals since the 16th century. But during the 1700s, the role of bedside teaching was expanded. Since the rich often received care in their homes, it was the poor that huddled in hospitals, providing the ideal training ground for budding physicians. As rural residents flocked to European cities in the 18th century, new hospitals were constructed across the continent to deal with the rising numbers of urban poor. In England, for example, 32 facilities opened their doors between 1736 and the end of the century; 5 were in London alone, and most of the others were in urban settings.

The European hospital ideal was envisioned in Vienna, Austria, when Joseph I of Austria-Hungary centralized in one institution all hospital care in the area. In 1781, Vienna had some 20 hospitals scattered throughout the city, with about 1,000 beds. When Vienna's Allgemeines Krankenhaus, or the University of Vienna Hospital, opened in 1784, it instantly became the largest hospital in the world, with 2,000 beds. But the immense size and complexity of the institution led to problems not previously encountered, including hospital-associated diseases, an embarrassment to its founder but terrifying to its patients. And no group of people became more fearful of the hospital than women in childbirth.

THE TRAGEDY OF PUERPERAL FEVER

The development of hospitals for women in childbirth was a phenomenon that began in the 17th century, when there was sufficient interest among physicians in becoming obstetricians and wresting the occupation away from midwives, who delivered babies in the home. This change was largely the result of the discovery of forceps to aid in the delivery process. This innovation made obstetrics a distinct specialty in the 1600s. As so-called lying-in hospitals made their 17th-century appearance, so did a disease that was previously rare, namely, childbed fever, or puerperal fever. The word puerperal is from the Latin, *puerperal*, meaning "woman in childbirth." The disease, described since the time of Hippocrates, remained a dreaded and often deadly complication of childbirth but was relatively uncommon when women delivered at home. The first well-described epidemic of puerperal sepsis occurred at Hôtel-Dieu in 1746, with 20 cases; all were fatal. The institutional clustering of women in childbirth, usually poor women, in Europe during the 18th and 19th centuries changed the frequency of puerperal sepsis. The misery was repeated in hospitals in all the great European cities. For example, between 1831 and 1843, in London's General Lying-In Hospital the frequency of mortality from puerperal sepsis was 600 mothers per 10,000. In contrast, the Royal Maternity Charity recorded a frequency of only 10 per 10,000 when delivery occurred in the home (2). This extraordinary difference in frequency, evident elsewhere on the continent of Europe, should have signaled the concept of contagion to anyone considering the numbers. But it didn't. Physicians had no notion that puerperal fever had an infectious cause, or even a cause external to the patient, at the end of the 18th century.

THEORIES ABOUT THE CAUSES OF PUERPERAL FEVER IN THE 18TH AND 19TH CENTURIES

Like with smallpox, the theories about the causes of puerperal fever that flourished during this period mirrored medicine's disjointed evolution away from the humoral theory and towards the concept of contagion. Theories progressed during the 18th century from the belief that the disease had its origin inside the human body to one in which the origin was outside the body. In the late 1600s, the view was that puerperal sepsis had its beginnings not at the time of delivery but early in pregnancy, with tight stays that a woman might wear. The consequence of such a wardrobe was that fecal material might be retained in the intestines and the putrid parts absorbed into the blood and retained by the uterus. The concept of retention of material by the woman's uterus was consistent with Hippocratic humoral theory and opposed any notion that an external agent might be involved.

Under normal circumstances of delivery, a blood-tinged fluid called lochia is discharged from a woman's uterus for 4 to 6 weeks after birth of the infant.

A popular early 18th-century theory of puerperal fever was that the lochia was retained, stagnating to a putrid mess in the womb. At first, physicians reasoned that this retention was due to the retained humors that built up in early pregnancy due to tight clothing, but the reasons for retention evolved. One theory suggested that retention occurred when a woman's blood became "too thick" or when the vessels of the uterus narrowed. A later concept was that cold air inadvertently exposed to the uterus near the time of delivery could cause uterine constriction and the retention of fluid. Even drinking cold water or exposing one's feet to the cold might begin the process. Nearly any shock to a woman's system late in pregnancy was deemed a possible trigger to childbed fever.

Another theory of puerperal fever of the time was the milk metastasis theory. This hypothesis required some imagination. The development of breast milk was thought to occur because of a transformation of menstrual fluid from the uterus to the breast following delivery. As strange as that sounds, physicians were convinced that an anatomic connection existed between the top of the uterus and the breast. Leonardo da Vinci even depicted the duct in a drawing, even though he was never able to visualize it in his dissections. When autopsies were performed on women who died of puerperal sepsis, the abdomens were filled with a fluid so similar in appearance to breast milk that it was assumed that the woman had pathologically retained the precursor to breast milk in the uterus. This conjecture explained why women suffering from childbed fever would stop lactating. According to the theory, eventually, the milky fluid would build up, obstructed from taking its normal path to the breast for unknown reasons, and spill over, or metastasize to other organs in the abdomen, pelvis, or elsewhere in the body via the bloodstream. This retained milky fluid seen throughout an afflicted woman's body at autopsy was, in reality, pus—inflamed fluid with white blood cells, bacteria, and decayed tissue. This pus-filled fluid was evident all over a woman's abdomen after a death from puerperal fever. Autopsies of women who died from puerperal sepsis were not for the faint of heart. The stench released from the affected uterus was sufficient to cause fainting among those who had not previously experienced it.

ALEXANDER GORDON AND PUERPERAL FEVER IN ENGLAND

Despite outbreaks of the disease in the 18th century, these prevailing theories about the cause of puerperal fever did not alert physicians to suspect agents outside the woman's body, except for a few scattered writings, until a Scottish physician, Alexander Gordon, questioned the retained-lochia theory in 1795 (3). Gordon was witness to an outbreak of the disease beginning in 1789. Surprisingly, the outbreak was not in the lying-in hospital of Aberdeen but involved a few midwives and the practice of a single obstetrician—Gordon himself. Gordon carefully detailed all the cases that he saw over a 3-year period. In his treatise, Gordon argued that puerperal sepsis was an inflammatory disease, not one of putrid retention of lochia. His tabulation of all the cases allowed Gordon to trace any links between the cases. He made a remarkable leap, suggesting that puerperal sepsis had a "specific contagion or infection," citing examples of how the disease could be traced from one patient to another via a midwife or himself.

> The midwife who delivered No. 1 in the table carried the infection to No. 2, the next woman she delivered. The physician, who attended Nos 1 and 2, carried the infection to Nos 5 and 6, who were delivered by him, and to many others. The midwife who delivered No. 3 carried the infection to No. 4; from No. 24 to Nos 25, 26, and successively, to every woman she delivered. The same thing is true of many others, too tedious to be enumerated (3).

Gordon extended the concept to explain why one area around Aberdeen was left unaffected.

> Now it may seem remarkable that the puerperal fever should prevail in the new town and not in the old town of Aberdeen, which is only a mile distant from the former. But the mystery is explained, when I inform the reader that the midwife, Mrs. Jeffries, who had all the practice of that town, was so very fortunate as not to fall in with the infection; otherwise, the women whom she delivered would have shared the fate of others.

Most remarkably, Gordon tracked the midwives and ventured to accurately predict who would be affected by puerperal sepsis:

> . . . upon hearing by what midwife they were to be delivered or by what nurse they were to be attended during their lying in [resting in bed for a period after delivery].

This noteworthy advance in epidemiology and theory of puerperal sepsis was published in a seven-chapter treatise (4). Given the risk of not only proposing a new theory but also suggesting that certain individuals, including himself, were responsible for the disease was a courageous act—one that cost Gordon his livelihood and reputation. Gordon took great criticism for his theory suggesting that he or others, who he named, could have been responsible for puerperal fever. Within a year or so of publication of the treatise, Gordon left Scotland. Almost 80 years before the germ theory of infectious diseases became an accepted notion, Gordon had proposed an accurate and well-reasoned theory. But the world at the time took little notice—perhaps because Gordon was not a well-known academic, perhaps because his communication style lacked sophistication, or perhaps because of the treatment he proposed. Gordon proposed bleeding and purging as treatment, holdovers from the humoral theory, for which he claimed success. Other physicians achieved no success ministering to puerperal fever with those treatments. Moreover, Gordon's theory could not be connected to any known medical framework that considered an external, transmissible agent responsible for a febrile disease. Perhaps Gordon was simply too far ahead of his time. Whatever the reasons, Gordon's treatise was largely forgotten. Unfortunately, within 4 years of publication of his treatise, Alexander Gordon died of tuberculosis and was never able to further defend his assertion.

Puerperal fever raged on in the hospitals of Europe into the 19th century. But other countries were affected, too. America had its problems with the disease, as well documented throughout the United States in the 1840s (5). To understand the prevalent concepts of the time, consider that physicians in both America and Europe used the terms contagion and infection in ways far different than we do today. In the pre-germ theory era, a contagious disease was suspected when a disease replicated itself in a predictable way, e.g., smallpox. An infectious disease was thought to occur when different diseases occurred at the same time, e.g., puerperal fever and erysipelas. (Note: erysipelas, derived from the Greek meaning "red skin," is an infection of the skin usually caused by *Streptococcus pyogenes*, the bacterium associated with puerperal fever.) How did doctors of the early 1800s explain the causes of a contagious versus an infectious disease? The answer went back to the remnants of the humoral theory. Some noxious influence in the air, miasma, was thought to be the cause of disease. There were two distinct forms—one form caused the contagious diseases, and another was responsible for so-called infectious diseases. The essential difference in the variant forms of miasma was that the form responsible for an infectious disease struck people at random and the form responsible for a contagious disease struck all, or nearly all, people in a predictable way. Smallpox threatened everyone, while puerperal fever affected only a few of the women congregated in a hospital, even during epidemics.

An example from an 1846 writing may help in demonstrating the distinction in the minds of pre-germ theory physicians between contagion and infection. Thomas Mitchell was a professor at Transylvania University in Lexington, KY, who documented the experience of a merchant who traveled to Philadelphia during an epidemic of yellow fever, avoiding contact with anyone who had yellow fever during a brief visit. When he arrived home, he became sick with yellow fever, although not one of his relatives who cared for him became ill (5). Surely, Mitchell reasoned, yellow fever was an infectious disease but not a contagious disease. This seemingly absurd distinction between infectious and contagious diseases was described in a medical journal of the time:

In applying the word [contagion] it should be made to include all cases resulting from a poison derived from a patient, and communicated either by direct contact, or through the air and in making a division of the cases into contagious and infectious, we follow this unconsciously (2).

In the thinking of the time, infectious diseases, which seem to strike randomly, often had personal and environmental influences that determined individual susceptibility to disease.

CONTAGION VERSUS INFECTION IN EARLY 19TH-CENTURY MEDICINE

While we might find the argument to hold little merit, understanding the distinction between contagion and infection in the minds of 19th-century physicians is essential to appreciate the controversy over the contagiousness of puerperal fever. Within the controversy is the evolving concept that some form of contagion was now accepted by physicians of this era, unlike 100 years earlier, when a change in the balance of humors was the explanation for all disease. The theoretical ground was maturing for the germ theory, but all the parts had not been put into place. The parts including Fracastoro's seeds of contagion and van Leeuwenhoek's animalcules could not be fit together until miasma, the humoral-theory remnant, was disproven and replaced with a specific role for microorganisms in human disease.

More than 20 years before the time when the role of microorganisms in human disease was seriously considered, two nearly simultaneous explanations of puerperal fever in the 1840s purported the contagiousness of the disease—one in America and one in Europe. Each bore the hallmarks of the men who proposed the theory, which produced important differences in their accounts. The American devised his comments largely from reading other accounts. The European used a novel numerical analysis (6). More importantly, the American did not propose and test the correct means to control the disease; the European did. The American was Oliver Wendell Holmes. The European was Ignaz Semmelweis.

OLIVER WENDELL HOLMES AND PUERPERAL FEVER IN AMERICA

Oliver Wendell Holmes graduated from Harvard Medical School in 1836. In 1843, Holmes published an article in the short-lived *New England Quarterly Journal of Medicine and Surgery* after an extensive examination into the causes of puerperal fever (7). Holmes collected a mountain of evidence that proved to his satisfaction that puerperal fever was, indeed, contagious. The evidence included accounts of physicians who had performed autopsies of patients who died of puerperal fever. Through some mishap during the autopsy, a physician became ill and died of a condition that Holmes believed was identical to the one that killed the autopsied patient. Holmes had read Gordon's report, too. He became utterly convinced of his correctness, writing,

> *The disease known as Puerperal Fever is so far contagious as to be frequently carried from patient to patient by physicians and nurses.* (7)

The italics are Holmes's. Holmes did not attempt to explain the mechanism or precise nature of the contagion, although he suggested that some sort of particle or poison was responsible.

Holmes did, however, have very strong recommendations for control of puerperal fever. He wrote that physicians should avoid autopsies of women dying of puerperal fever. If an autopsy was absolutely necessary, any physician present should change all clothing, which should be burned. The physician should wait 24 hours before delivering any other baby. If a patient under a physician's care is discovered to have puerperal fever, the physician is under moral obligation to consider his next patient in danger and take appropriate precautions. If two cases occurred within a short time frame, the physician was to suspend his practice for at least a month. He advised physicians to ensure that nurses or other assistants were not transmitting the disease. He recommended thorough washing. Holmes was crystal clear in his belief and did not mince words:

> Whatever indulgence may be granted to those who have heretofore been the ignorant causes of so much misery, the time has come when the existence of a *private pestilence* in the sphere of a single physician should be looked upon, not as a misfortune, but a crime; and in the knowledge of such occurrences the duties of the practitioner to his profession should give way to his paramount obligations to society (6).

The reaction to Holmes's eloquently written article was muted, initially. Many physicians agreed with Holmes, but he was only 34 years old when he wrote the piece. Holmes was not an obstetrician, either. He came up against powerful and vocal opponents. The most noteworthy was the 60-year-old Charles Meigs, a professor of obstetrics at Jefferson Medical College. Meigs was the recognized authority in obstetrics in the United States and criticized Holmes's paper. Meigs had no axe to grind with Holmes personally. Nor was Meigs an uncaring physician. Consider what Meigs wrote about the tragedy of puerperal fever.

> . . . there is almost no acute disease that is more terrible than this. . . . There is something so touching in the death of a woman who has recently given birth to her child; something so mournful in the disappointment of cherished hopes; something so pitiful in the deserted condition of the newborn helpless creature, forever deprived

of those tender cares and caresses that are necessary for it—that the hardest heart is sensible to the catastrophe (2).

Meigs was simply convinced that the theories of contagion in this case were incorrect.

> But while the opinion of childbed fever contagion ought, in my view of it, to find no supporters among truly educated medical men, I am well aware that you are to be left to the operations of your own judgment in adopting or rejecting it; nor have I the least desire to persuade you to think, because I think (8).

Yet, his opposition to the idea of contagion, given his stature, did exactly that. Physicians in the United States tended to discount Holmes's article simply because an authority like Meigs did.

IGNAZ SEMMELWEIS: EARLY INFLUENCES

On the other side of the Atlantic Ocean, Ignaz Semmelweis, unaware of Holmes's writings, considered the contagiousness of puerperal sepsis from an entirely different viewpoint than Holmes. More than any other person profiled in this book, the life and personality of Ignaz Semmelweis, his passions and his considerable shortcomings, color his contribution to infectious diseases and its acceptance by his contemporaries. Semmelweis was born in 1818 in the Hungarian city of Budapest. He was the fifth of nine children. His father was a successful merchant. His mother, a daughter of a wealthy Bavarian-born merchant, spoke German at home, although his father spoke a Germanic dialect called Buda-Swabian. Despite his Hungarian birth, Ignaz did not learn to speak Hungarian until he reached secondary school. His own dialect would expose Ignaz Semmelweis's country of origin for his entire life. As he proceeded through his education, he was an able student and achieved excellent examination grades. However, his skills in writing were not strong. Decades later he would point out that he had "an innate aversion to everything which can be called writing." These seemingly inconsequential features of Semmelweis, his dialect and his aversion to writing, would play a profound role in his professional life.

In 1837, Semmelweis enrolled in law school at the University of Vienna at the age of 19. Although the study of law met with his father's approval, it had little appeal to the young Semmelweis. His attendance at an anatomy lecture with a medical student friend was to change the course of Semmelweis's career. The compelling lecture by Joseph Berres caused Semmelweis to enroll in the university's medical school the following year. After one year at the University of Vienna's Medical School, Semmelweis returned to Hungary to continue his medical training at the University of Pest. After 2 years, he returned to complete his medical degree training in Vienna, the most prestigious school in the entire Austria-Hungary Empire. Semmelweis hoped that the prestige would ensure a post anywhere he wished. He was wrong. Semmelweis wanted to work under Jakob Kolletschka in the Pathology Department but was rejected. He then applied to be an assistant to Joseph Skoda, the leading physician at the University of Vienna. Again, he was rejected. Intensely fearful of another rejection, he went with his third choice: obstetrics. This department did not occupy as eminent a position as the others, so it was an easier position to obtain. Ignaz Semmelweis began his position in obstetrics on 1 July 1846.

THE UNIVERSITY OF VIENNA HOSPITAL: A SHORT HISTORY

Before one can fully understand Semmelweis's contribution to infectious disease, the control of puerperal sepsis, one must have an appreciation of the history of the University of Vienna Hospital at the time Semmelweis began working there. During the first half of the 19th century, Paris was the center of medicine. Following Morgagni's publication in 1761, physicians placed great emphasis on the correlation between a patient's symptoms and pathological findings at autopsy. Following the French Revolution, physicians of the Parisian school became increasingly expert at the physical examination, especially after the discovery of auscultation using René Laënnec's stethoscope in 1816. In Paris, the clinician did the autopsy on the patient for which they were caring. In the German-speaking countries of Europe, a new trend developed. The so-called German School of Medicine embraced the laboratory. The German School performed more detailed autopsies than the French and studied pathological physiology in the laboratory, setting the stage for the invention of new fields of biochemistry and cellular pathology.

By the 1840s, the German School had wrested the title of "The Center of Medicine" away from Paris. With the largest hospital in the world and the most prominent professor of pathological anatomy, Karl von Rokitansky, the University of Vienna Medical School and Hospital laid claim to the title itself. von Rokitansky meticulously performed or supervised every autopsy at the University of Vienna Hospital. He categorized thousands of observations on diseases, yielding distinct disease entities, later earning him the title, "the Linnaeus of pathological anatomy." His *Handbuch der pathologiischen Anatomie* became the field's preeminent publication. For the first time ever, in 1844, pathological anatomy became a compulsory subject, not an elective, at a medical school—the University of Vienna Medical School.

von Rokitansky performed over 30,000 autopsies during his career, but dissection never revealed much in the cases of childbed fever. While every detail of the pathological anatomy was known, the cause was not. von Rokitansky convinced himself and his colleagues that the solution was more autopsies. Thanks to the large number of outbreaks at the University of Vienna Hospital, there remained plenty of opportunities to do them.

When Semmelweis began his work on the obstetrics ward at the University of Vienna Hospital, he fell into the midst of a brewing revolution between the conservative senior professors and a younger generation of physicians with new ideas. Semmelweis was a conformist and did little to stand out. There was nothing in his background to suggest that he would do anything that would help create upheaval. He was well liked, of average height and stocky build. He only smiled, as his colleagues ridiculed his accent. The smiles hid a deep insecurity—feeling inferior to the Viennese elite, something that his two recent rejections for posts did not help. For now, Semmelweis would repress his insecurity and feelings of being the outsider. But later in his career, these feelings would resurface and reveal an ugly side of Semmelweis's personality when they would be expressed as volatile indignation and contempt. At the beginning of his post, Semmelweis would obediently accept his duties.

The turmoil at the University of Vienna had its origins some 23 years earlier, when Johann Klein assumed the position of Professor and Director of Obstetrics. The previous Professor and Director, Lucas Boer, had been trained in England and believed that puerperal fever was a contagious disease. Boer did not believe in medical student instruction on the women who died of puerperal sepsis, respecting the humanity of women who died so tragically. Great Britain, where Boer trained, and Ireland were exceptions to the European practice of performing autopsies on all women who died in lying-in hospitals. Boer believed in the gentleness of technique that he learned in England, restricting the number of internal examinations during labor, and limiting instrumentation with forceps. During Boer's tenure, the incidence of puerperal sepsis was generally low, around 1% maternal mortality. (Note: nearly all maternal mortality of the time was due to puerperal sepsis.)

In 1823, when Johann Klein took over as professor, the tone of the obstetric wards took a striking change. In a school dominated by the preeminent pathological anatomist of the day, Klein required students to learn in the morgue, observing autopsies and performing internal examinations on mothers who had died. Government regulations started to require autopsies on every mother who died in hospital. Klein also relaxed the restraints on internal examinations during labor. The use of forceps during delivery became more common. These changes coincided with an immediate increase in maternal mortality to about 7.5%. In 1834 a second maternity division at the University of Vienna was established. By 1839, two entirely separate Divisions existed. The First Division was operated exclusively by doctors and medical students; the Second Division was staffed solely by midwives. After only a short time, the mortality rates showed differences between the Divisions. Although fluctuations regularly occurred, the First Division generally had a threefold-higher maternal mortality rate than the Second Division. Klein was not a believer in the contagiousness of puerperal fever and insisted that epidemic infection was responsible for the elevated mortality rate (2). The conservative Klein was not searching for novel ideas about puerperal sepsis or anything else for that matter. Klein was well connected politically, an important attribute at the University of Vienna, where posts were appointed by the government. Everything was designed to maintain the status quo. In the German School, Professors/Directors ruled with an iron fist. Opposition was soundly discouraged. Klein had been Director for 20 years and had become increasingly resistant to any change when Semmelweis came on the scene.

SEMMELWEIS AND CHILDBED FEVER: A TALE OF TWO DIVISIONS

Once Semmelweis began his work on the obstetrics ward, he committed all his efforts towards his work. He enjoyed a good relationship with von Rokitansky, who helped Semmelweis dissect the bodies of women who died on the obstetric service. The sheer number of deaths from puerperal sepsis on the First Division service had a profound effect on Semmelweis after only a few months. The First Division generally had a 10%, sometimes as high as 30%, monthly mortality rate from puerperal sepsis, while the Second Division generally had a 1 to 3% mortality rate. This difference in mortality rates between the two Divisions was not lost on the public. Since admission to one or the other service changed every day at 4 p.m., there were many stories of women, in active labor, waiting to go to the University of Vienna Hospital after 4 p.m. so they might be admitted to the Second (midwives') Division. The weekends changed this alternation in admission. Sometimes the women, particularly new mothers who were unfamiliar with the routine, would miscalculate. Semmelweis wrote about it,

> Many heart-rendering scenes occurred when patients found out that they had entered the First Division by mistake. They knelt down, wrung their hands, and begged that they might be discharged. [Seeing] Lying-in patients [with childbed fever] with uncountable pulse, meteoric (swollen) abdomen, and dry tongue, only a few hours before their death, [they] would protest that they were really quite well, in order to avoid medical treatment, for they believed that the doctor's interference was always the precursor of death (2).

Semmelweis set out to determine why the First Division was so deadly and, more importantly, to find a way to end the fatalities. He worked tirelessly on the wards and continued his relationships with von Rokitansky and others in the Pathology

Department, including his friend Jakob Kolletschka. Semmelweis kept trying small alterations on the First Division when he spotted differences in practice on the Second Division. For example, midwives usually delivered women on their sides. Semmelweis tried it, without success. He became increasingly frustrated at his own failures but also the apparent apathy of his Director, Klein.

By the beginning of 1847, the situation with puerperal sepsis on the First Division had gotten so bad that even the conservative Klein could not ignore it. Klein was quick to look for the easy way out. He believed that the milk metastasis theory explained puerperal sepsis. One of the important implications for believers in this theory was that it meant that puerperal sepsis could only affect women. The key for Klein was to uncover the trigger that caused the retention of fluid (lochia) in the uterus. The trigger for the current increase in number of cases, he pronounced, was a general miasma that hung over Vienna. This elucidation did little to explain why the Second Division had one-third the number of cases as the First Division. Klein looked for other possible triggers. He believed, in part correctly as it turned out, that the number of medical students examining women in labor was too high. Foreign medical students were flocking to Vienna due to its stature. Klein reasoned that these students were the cause of the latest problems and restricted their access to the Division, with a temporary reduction in the mortality rate. The willingness to consider that medical personnel were in any way responsible for puerperal sepsis suggests some evolution in thinking towards an external cause of puerperal fever, even among the most conservative. But Klein was not one to let this kind of thinking get too close to home. Klein would not consider any theory that left him or his colleagues culpable. Klein suggested at a committee meeting that the First Division's walls, which were in poor condition, might be responsible for puerperal sepsis. Semmelweis had had enough. His volatility showed itself. Few junior physicians on the job for only a few months would rise against the Professor and Director of a service in the German School. Semmelweis did. He pointed out that there were many lying-in hospitals whose walls were in far worse condition without similar mortality rates. His comment did little to endear Semmelweis to his Director. It was as though the first shot had been fired in a simmering tumult between the old guard and the young ones. But Semmelweis realized that his Director would not be the one to solve this problem.

In early 1847, Semmelweis took on an even more zealous approach to examining the causes of puerperal sepsis. He became familiar with the books and journal articles on puerperal sepsis. He read the English contagionists' viewpoints and studied evidence both for and against contagion. He sifted through opinions and data until he began to clarify, in his own mind, essential truths. He was unerringly accurate in his critical analysis. Semmelweis had a natural ability with numbers and came up with the following observations:

The numbers of deliveries on the First Division and the Second Division were nearly the same each year, about 3,000 to 3,500.

The main difference between the services was that deliveries on the First Division were performed by male doctors and medical students and those on the Second Division were performed by female midwives.

In 1841 through 1848, the mortality rate on the First Division was nearly threefold greater than on the Second Division (Fig. 8.1).

No epidemic of puerperal fever was evident outside the walls of the University of Vienna Hospital, in home deliveries by either midwives or private doctors. Even street births rarely had puerperal fever following delivery.

Decades of record keeping showed no association between puerperal fever and changes in weather or season of the year.

Overcrowding was not associated with puerperal fever since it was usually worse on the Second Division.

Greater degrees of trauma to the mother during delivery appeared to increase the chance of developing puerperal fever.

Infants would more than occasionally die when their mothers died of puerperal sepsis. The infants had a fever similar in appearance to their mother's. Autopsy findings were frequently similar to the mother's findings as well.

None of these observations were open to debate, but taken together, they gave Semmelweis enough evidence to topple the established dogma of the day: that puerperal sepsis was a disease with miasmatic causes and that it affected only women. The lack of association with weather, season, and overcrowding was a strong argument against the general miasma in Vienna. But the low incidence outside the University Hospital and the differences between the two Divisions were nearly definitive for Semmelweis to think that miasma had nothing to do with the disease he was studying. The young physician was also struck by the fact that infants, both male and female, seemed to acquire a febrile disease like their mothers'. How could the milk metastasis theory be correct? Semmelweis had, in his own mind, discarded the established dogmas.

THE TRAGIC CLUE TO CHILDBED FEVER

For Semmelweis, the convincing evidence for a new theory of puerperal fever was an immensely tragic one. On 20 March 1847, his longtime friend, the man he originally wanted a post to assist, Jakob Kolletschka, died. Kolletschka's death occurred following a scalpel wound that he obtained during the dissection of a woman who died from puerperal sepsis. The findings at Kolletschka's autopsy seem to hold the answer for Semmelweis. Rarely in the course of a historical appraisal is a reader able to see inside the mind of a person who makes a great discovery at the moment that he or she puts all the pieces together—the "aha" moment. Even though writing was not his great strength, Semmelweis recorded the most extraordinary passage on his breakthrough:

> At once it became clear to me that childbed fever, the fatal sickness of the newborn and the disease of Professor Kolletschka were one and the same, because they all consist pathologically of the same anatomical changes. If, therefore, in the case of Professor Kolletschka general sepsis [infection in the bloodstream] arose from inoculation of cadaver particles, the puerperal fever must originate from the same source. Now it was only necessary to decide from where and by what means the putrid cadaver particles were introduced into the delivery cases. The fact of the matter is that the transmitting source of those cadaver particles was to be found in the hands of the students and attending physicians (2).

Not only had Semmelweis determined the contagiousness of puerperal fever, but also this determination had suggested the way to prevent puerperal fever: rid the students and physicians of these cadaver particles. The distinctive odor of the autopsy room remained on anyone who participated in the autopsy of a woman who had died from puerperal fever. But how to eliminate the particles? Chloride solutions, available at the time, had the ability to rid objects of vile odors. Chloride's capacity to snuff out odor, not its disinfectant properties, which were unknown at the time, drew Semmelweis to try using *chlorina liquida* in a bowl. (Note: everything sounds more medicinal in Latin.) Semmelweis required all persons entering the First Division to wash their hands in a bowl of this solution beginning in May 1847. Chloride of lime was soon substituted because of its lower cost.

PREVENTION OF CHILDBED FEVER: HAND WASHING

Like the prevention of smallpox, the means to prevent puerperal sepsis, hand washing, was determined well before the germ theory of disease became an accepted premise in medicine. Semmelweis took this one step further, what he considered to be the final step. He showed that the intervention worked! He used the same numerical analysis (Fig. 8.1). In the last seven months of 1847, only 56 women died of the 1,841 delivered in the First Division. A mortality rate of 3% was now achieved on the two divisions! In 1848, the death rate on the First Division fell below that of the Second Division for the first time, down to 1.2%. To Semmelweis, he had proven that he was correct. The hand washing was the only intervention that could account for the change, Semmelweis thought. Unfortunately for Semmelweis, Klein had recently installed a new ventilation system. Naturally, Klein was convinced that this change was responsible for the lower mortality rates on the First Division service. Klein simply could not give in to the new idea, especially one from Semmelweis.

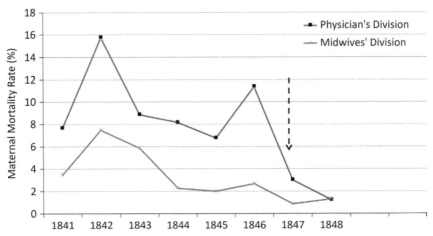

FIGURE 8.1 *Maternal mortality, which was nearly all from puerperal sepsis, at the Vienna General Hospital, by division, 1841 to 1848.*

REACTION TO HAND WASHING

Just as Jenner discovered with his means of prevention, physician acceptance of hand washing would prove astonishingly difficult, taking Ignaz Semmelweis to his breaking point. Winning physician acceptance of his method encountered two major obstacles. First, there was the growing rift between the old guard and young physicians of the University of Vienna. Semmelweis's notion got caught in this rift. Second, there was Semmelweis himself. At times, Semmelweis may have been his own worst enemy. Some historians have made it seem that the bureaucracy of the University of Vienna conspired against Semmelweis. Others have made it seem that Semmelweis alone blundered in his methods of gaining acceptance. To me, it seems that a combination of these two factors played important roles in the failure to gain acceptance of hand washing.

The climate around the University of Vienna was stifling to new ideas despite having some of the brightest researchers of the time. The great German surgeon Theodor Billroth wrote in the 1870s about the atmosphere at the University of Vienna at the time,

> A generation that had been reared in an intellectual straitjacket with dark spectacles before their eyes and cotton wool in their ears. The young people turned somersaults in the grass, and the old men, whose bodies had been hindered in their natural development by the lifelong burden of state supervision, felt their world tumbling about their ears, and believed that the end of things was at hand (9).

The two most influential professors in Semmelweis's sphere epitomized this dichotomy: Klein, the Director of Obstetrics, and von Rokitansky in Pathology.

Klein was in his sixties and owed his position to his political connections in the government. von Rokitansky, on the other hand, was only 43 years old when he took over Pathology at the University of Vienna. Kind and understanding, he fostered research and helped young assistants. Since he busily published, creating a strong reputation for himself and the school, he owed virtually no political debt. There were other youthful physicians caught up in this continued old-versus-new seesaw. Joseph Skoda, the physician with whom Semmelweis had also wished to work, was an expert diagnostician. He had mastered the stethoscope and correlated clinical findings with lessons from the autopsy. Skoda was able to quickly diagnose new clinical entities, earning considerable fame. Another physician of the time, Ferdinand Hebra, was only 2 years older than Semmelweis. Hebra had systematically dealt with the ignored problems of skin diseases and had quickly made a scientific study of them. By 1845, Hebra ran the world's first dermatology unit. He and Semmelweis had become good friends. Hebra saw the solid scientific value of Semmelweis's discovery. He encouraged Semmelweis and was a source of great support.

By the end of 1847, Semmelweis had enough information to publish. He did not. Hebra, convinced that his friend had a discovery as important as Jenner's vaccination, wrote a brief editorial in a local journal in Vienna. von Rokitansky wrote a supporting statement. Skoda spoke out in favor of Semmelweis's theory. The word was out, but not because of Semmelweis. He had done nothing. He had vocal and influential friends, but that was not enough. His aversion to writing and his insecurity, his sense of being an outsider, crippled his action. He felt unworthy to debate with the Viennese elite. Opposition to Semmelweis and his theory caught up with him. Klein, backed by members of the old guard, refused to renew his appointment on the Obstetric Service in 1849 despite repeated urging from von Rokitansky, Skoda, and Hebra. Even though unemployed, Semmelweis finally consented to publicly discuss his findings in 1850 at the Medical Society of Vienna. He handled it well. He had the support of the three young but influential academics and successfully presented his case to the physicians of the city that was at the center of medicine. Out of this apparent victory, Semmelweis managed to snatch defeat. Again, with his dislike of writing, he never published his presentation and the subsequent debate. The success of the presentation was sufficient for Semmelweis to be offered a limited clinical appointment. But for the man who felt that he had been continually treated as an outsider in Vienna, this was not enough. To Semmelweis, it was insulting. Within 5 days of the offer, Semmelweis fled the city. He left without telling his friends or supporters why he was leaving or what his plans were. Historians have debated the reasons for Semmelweis's hasty departure. Some say it was for financial or family reasons. Other historians believe that Semmelweis had a self-destructive streak where he saw himself as unworthy to live among the Viennese medical elite but with an immense ego that was bruised because of the half-hearted offer.

SEMMELWEIS'S DEPARTURE FROM VIENNA

The result of Semmelweis's departure is not in doubt. The support of his greatest allies evaporated. It was as though amid this revolution, the man with the most valuable ammunition abandoned his fellow insurgents. Skoda never forgave Semmelweis; von Rokitansky and Hebra eventually did. Without its discoverer, the theory suffered. The failure to publish was damaging, to be sure. But if Semmelweis had pressed on with his theory at the University of Vienna, there is a chance that its greatest weakness may have been overcome. There was no known medical framework on which to attach his new hypothesis about transmitted cadaver particles causing puerperal sepsis and its prevention by hand washing. What were these cadaver particles? The German School emphasized the laboratory, but Semmelweis did not utilize the one instrument that may have provided conclusive proof: the microscope. Semmelweis cannot be entirely faulted for the failure to use the microscope, since it was not even used at the University Hospital by von Rokitansky, the last of the great naked-eye pathologists. Only a few years later, Rudolf Virchow would utilize the microscope using von Rokitansky's methods to develop the new science of cellular pathology in Germany, further solidifying the German School as the center for medicine. But for Semmelweis, it did not happen. He may have been in the right place at the right time, but for Semmelweis there were ample missed opportunities.

THE RETURN TO HUNGARY

Semmelweis returned to Budapest after fleeing Vienna. His family was in financial trouble, so Semmelweis had to secure a position. As he started building a private practice, he requested a position as an unsalaried director of the lying-in ward of St. Rochus Hospital from the imperial commissioner of health. The desperate conditions of the obstetric ward of the hospital left the commissioner little choice. Semmelweis sought to mold the ward in his image with chlorine-water prophylaxis. The opposition that he saw in Vienna among physicians and students did not exist in Hungary. He was in charge and saw that everyone complied. The rates of puerperal sepsis fell. However, the zeal with which Semmelweis policed the use of hand washing did little to win him supporters in Hungary. Semmelweis slowly built up a private practice to support himself and his family. Eventually, he was appointed to the post of Professor at the University of Pest in 1855. But despite the academic post and ample data, he still did not publish his theory.

OPPOSITION TO SEMMELWEIS AND HIS THEORY

The opposition to the theory that Semmelweis proposed began to gain ground across Europe and was more powerful than ancient bureaucrats reflexively rejecting a new idea. Semmelweis's theory suggested that physicians had been responsible for killing their patients. This idea would not sit well with even the most open-minded physician. Many obstetricians simply rejected the theory on that basis alone. The English contagionists began to weigh in. They believed in miasma and not cadaver particles. Simple cleanliness, as practiced in Britain, was the answer, not hand washing with *chlorina liquida* in a bowl. Semmelweis would argue vociferously over whether puerperal sepsis was a contagious disease in later writings, pointing out that it was transmissible but not contagious. The subtleties of the distinction meant a great deal at the time but would have been quickly brought together if the nature of the cadaver particles had been discovered. But the quarrel demonstrated an unattractive side of Semmelweis. He argued with an explosive nature, convinced of his own correctness, and greeted those who disagreed with contempt. He managed to garner support of friends in Hungary. They spoke Hungarian. Semmelweis had only learned it in secondary school. Just as his dialect in Vienna left him feeling like an outsider, Semmelweis's discomfort with Hungarian left him wondering if he was accepted as a true countryman in his native country. His troubled experiences turned the well-liked, smiling gentleman into an unpleasant, contemptuous, distressed man who had the strongest sense of self-righteous certainty. Semmelweis would occasionally read a medical journal, only to find an attack on his theory. He would fire off letters to some of the leading obstetricians of Europe, soliciting their opinions on his theory in hopes of reassuring himself. The responses rarely did. Finally, he realized that the time had come to explain his own theory despite his antipathy to writing. In 1860 he completed his book, *The Etiology, the Concept, and the Prophylaxis of Childbed Fever*, which was published in 1861, some 13 years after Semmelweis's initial work in Vienna. He sent copies to physicians throughout Europe but was never satisfied with the responses to his efforts. The book was long, overly complex, and rambling. Semmelweis believed that the book's publication would be the ultimate vindication of his theory. However, there was no wholesale conversion to his way of thinking. Semmelweis often damaged his own cause. In one of his most vitriolic letters and an example of his writing style, he hurled these words at Friedrich Scanzoni, an influential professor of obstetrics at the University of Würzburg:

> If, Sir, without having refuted my doctrine, you continue to teach the students and midwives that you train that puerperal fever is an ordinary epidemic disease, I proclaim you before God and the world, to be an assassin and the history of puerperal

fever would not do you an injustice were it, on opposition to my life-saving discovery, to immortalize you as a medical Nero.

SEMMELWEIS'S LAST YEARS

By 1862, his friends began to notice Semmelweis's increasingly erratic behavior and utter disdain for those who dared to criticize this theory. By 1865, he was moody, irritable, forgetful, and argumentative. In his last faculty meeting, when called upon to give a report, he pulled a piece of paper from his trousers and read the Midwives' Oath without any understanding of where he was. His colleagues could no longer ignore the man's difficulties. Something was dreadfully wrong with Semmelweis's mind. With the help of his closest friends, his wife took Ignaz Semmelweis to an asylum in, of all places, Vienna. In Vienna, his old friend Hebra met Semmelweis and his wife on their arrival. The details of his commitment to this asylum in Vienna are not known. He was probably beaten, resulting in trauma and infection in the chest wall, evident on the records of the official autopsy report the year he died, 1865. Historians argue about what happened to Semmelweis to cause his decline, which was likely a presenile dementia or Alzheimer's disease, although it may have been syphilis or some other neurologic disorder. We will never be certain. Semmelweis was only 47 years old at the time of his death.

What happened on the maternity wards after Semmelweis abruptly left Vienna has been the subject of some debate. A common belief is that Semmelweis' requirement for hand washing was abandoned but that may not have been true (10). In 1863, Carl Mayrhofer, a physician who followed Semmelweis on the maternity ward of the University of Vienna Hospital, published an article on childbed fever in which he referred to Jakob Henle's conjecture that many infectious diseases were caused by microorganisms. Using a newly acquired microscope, Mayrhofer even described various microscopic organisms in the uterine discharge of childbed fever victims, referring to them as "vibrions." Mayrhofer never mentioned Semmelweis in his publication.

Whether the contributions to the etiology of childbed fever from Semmelweis or, to a lesser extent, Mayrhofer had any influence of the development of the germ theory of disease described in the next chapter seems doubtful. However, Semmelweis had taken on a problem that had vexed physicians for decades, if not hundreds of years, and correctly concluded that a transmissible "cadaver particle" on the hands of personnel was responsible. He reasoned this theory 20 years before the germ theory was an accepted notion. He had used a technique no one else had used, a numerical analysis (6). He showed, for the first time, that three factors were required for puerperal sepsis: the source of putrid particles, the means of transmission from the source to the victim, and an injured host, such as one with postpartum uterus, particularly one that had trauma, or an incised finger. Contagionists believed that only smallpox could produce another case of smallpox. Infectious diseases in the 1840s were thought to strike people at random via miasma. Semmelweis was clear that there was no randomness or miasma involved. Still, Semmelweis tried to distinguish childbed fever from a contagious disease: "Childbed fever is a transmissible, but not a contagious disease" (11). While he took great pains to distinguish childbed fever from a contagious disease such as smallpox, Semmelweis had bridged the gulf in 19th-century thought that distinguished between contagion and infection. Semmelweis provided an extraordinary contribution on the prevention of puerperal sepsis with hand washing using the numerical analysis. He had come so close to the germ theory, but his own foibles had gotten in the way of physician acceptance. In the next two decades, three individuals would finally establish the framework for the modern notion of the germ theory in medicine: Louis Pasteur, Robert Koch, and Joseph Lister.

REFERENCES

1. **Risse GB.** 1999. p 117–165. *In Mending Bodies, Saving Souls. A History of Hospitals.* Oxford University Press, Oxford, United Kingdom.

2. **Nuland SB.** 2004. p 31–56. *In The Doctors' Plague: Germs, Childbed Fever, and the Strange Story of Ignác Semmelweis.* W W Norton, New York, NY.

3. **Gould IM.** 2010. Alexander Gordon, puerperal sepsis, and modern theories of infection control—Semmelweis in perspective. *Lancet Infect Dis* **10:**275–278. http://dx.doi.org/10.1016/S1473-3099(09)70304-4.

4. **Gordon AA.** 1795. *A Treatise on the Epidemic Puerperal Fever of Aberdeen.* G G and J Robinson, London, United Kingdom.

5. **Parsons GP.** 1997. Puerperal fever, anticontagionists, and miasmatic infection, 1840-1860: toward a new history of puerperal fever in antebellum America. *J Hist Med Allied Sci* **52:**424–452. http://dx.doi.org/10.1093/jhmas/52.4.424.

6. **Buyse M.** 1997. A biostatistical tribute to Ignaz Philip Semmelweis. *Stat Med* **16:**2767–2772. http://dx.doi.org/10.1002/(SICI)1097-0258(19971230)16:24<2767::AID-SIM710>3.0.CO;2-N.

7. **Holmes OW.** 1843. The contagiousness of puerperal fever. *N Engl Q J Med Surg.*

8. **Meigs CD.** 1854. On the Nature, Signs and Treatment of Childbed Fever. In a series of Letters Addressed to Students of His Class. Blanchard and Lea, Philadelphia, PA. Original from Harvard University Countway Library (digitized 30 November 2007).

9. **Billroth T.** 1924. *The Medical Sciences in the German Universities.* (Welch WH, translator.) MacMillan Co, New York, NY.

10. **Carter KC.** 1985. Ignaz Semmelweis, Carl Mayrhofer, and the rise of germ theory. *Med Hist* **29:**33–53. http://dx.doi.org/10.1017/S0025727300043738.

11. **Semmelweis I.**1861. *The Etiology, the Concept, and the Prophylaxis of Childbed Fever.* Translated by F P Murphy, 1981. Classics of Medicine Library, Birmingham, AL.

12. **Morse DA.** 1866. On contagion. *Med Surg J* **15:**529–531.

9 Louis Pasteur and the Germ Theory of Disease

van Leeuwenhoek had uncovered the microscopic world using a simple microscope, one with only one lens. He kept many secrets on the lens formation and use of his microscopes. Simple microscopes with sufficient power to see bacteria were difficult to make and use. For nearly 150 years, little progress was made in microscopes. The few discoveries were largely of academic interest and not considered by physicians, who made little use of the simple microscope. While compound microscopes, i.e., those with two lenses, were capable of magnifying sufficiently to see bacteria and had been available since the 17th century, there was one major difficulty that precluded their widespread use—chromatic aberration. This aberration distorted the spectrum of colors seen through a compound microscope and produced an image with strangely colored halos around objects that would make identification of items in an image all but impossible. The problem of chromatic aberration was not solved until the 1830s, when an Englishman named Joseph Jackson Lister devised a lens system that overcame the distortion using achromatic lenses. In the mid-19th century, the compound microscope suddenly became a tool for scientific discovery. But medicine was slow to include microorganisms discovered through the microscope until Louis Pasteur opened the world of the infinitely small.

The path to the germ theory had been prepared by others before Pasteur. The Paris School of Medicine advocated that diseases were discrete entities that could be classified, first by their symptoms in life and then, by their pathologic anatomy at autopsy, in stark contrast to the holistic disease approach of Hippocratic/Galenic

Germ Theory: Medical Pioneers in Infectious Diseases, Second Edition. Robert P. Gaynes.
© 2023 American Society for Microbiology.

medicine. Disease specificity, a concept crucial to the development of the germ theory of disease, was extended by early 19th-century scientists. For example, Pierre-Fidèle Bretonneau proposed a "morbid seed" causing specific diseases such as diphtheria in the 1820s. William Gerhard in the United States provided evidence that typhus was a distinct disease from typhoid, following an 1833 epidemic in Philadelphia. William Budd in England published a book on typhoid fever, stressing the specific and unchanging nature of the disease. Claude Bernard proposed that disease had a clinical spectrum; only the end stage was seen in hospitals. He emphasized experimental medicine where the laboratory, crucial for the germ theory, could be the seat for study of disease. While others had presented a concept that microscopic living organisms were the causative agents of diseases, especially those that occurred in epidemic form (1), it was Pasteur's contributions to the germ theory of disease that ultimately led to its acceptance.

A chapter on Louis Pasteur presents a considerable challenge. Louis Pasteur is probably the most notable nonphysician in the history of medicine (Fig. 9.1). Volumes have been written about Pasteur, including a biography from his son-in-law (2). Even with all that has been already written, his importance in the development of the germ theory of disease is crucial to include and still can present a few surprises to the reader. Pasteur researched the structure of chemical crystals and described the biological basis of fermentation of wine and beer. He debunked the widely accepted concept of spontaneous generation. Pasteur solved the mysteries of silkworm diseases, chicken cholera, anthrax, and rabies, which contributed to the development of the first vaccines with attenuated microorganisms. These seemingly unrelated topics can be strung together in a surprisingly logical sequence through Pasteur's life. His work led to an astonishingly important series of contributions to science, medicine, and human health. Pasteur's personality, including his perseverance, tenacity, and thoughtful but forceful approach to problems and his critics, was responsible for much of his success and a few of the conflicts that occurred during his lifetime.

LOUIS PASTEUR: EARLY INFLUENCES

Louis Pasteur was born in 1822 to Jean Joseph Pasteur and Jeanne Etiennette Roqui Pasteur. He was the only son in his family, which included three daughters. Most of Pasteur's youth was spent in Arbois, France. His parents had a profound influence on Louis, imparting values that shaped his approach to science and his private life. His father ran a tannery and was described as hard working and frugal. Like most parents, he wished for his son to improve his position beyond his own. His father stressed education and hard work, helping young Louis with his studies in the evening. His mother was described as kindhearted, imaginative, and

FIGURE 9.1 *Louis Pasteur. Courtesy of the National Library of Medicine (NLM image ID B020574).*

enthusiastic, a marked contrast from her reserved, cautious husband. The marriage was a happy one that formed a confident base for young Louis. The senior Pasteur wanted his son to be a scholar. Louis was quiet and industrious at home. His mind worked slowly but methodically. Some teachers mistook the plodding approach that Louis followed as evidence of an inadequate mind. But the headmaster of his school in Arbois recognized Louis's potential, recommending him to the great *École Normale* in Paris. At the age of 16, Louis left home in 1838. Despite his strong will and passion for learning, Pasteur had an intense attack of homesickness. He wrote about his struggles in Paris. One day, Louis was told to head to a small café and found his father sitting at a table, waiting to take him back home. No explanation was needed. Such was the close relationship that they had. His father understood and brought Louis back to Arbois. Pasteur took up drawing and kept his mind occupied at the school in Arbois until it became clear that he needed to move on with his career and life. Fearing another attack of homesickness, Louis chose to attend a college in nearby Besançon, only 40 km away. He did well in his studies, though not exceptionally well, in Besançon. Soon, Pasteur recognized within himself the desire, perhaps the need, to go back to Paris to attend *École Normale*. After graduation from the college in Besançon, he took the required examinations for the Parisian school. His response to placing 15th among 22 candidates was indicative of Pasteur's demanding and exacting personality. Not content with his performance, he elected to give himself another year and retake the examination. His subsequent placement was 4th among the list of entrants.

In 1842, Pasteur, now more confident, returned to Paris to fulfill his father's and his ambition to become a teacher. During his first time in Paris, Pasteur came under the spell of the celebrated chemist Jean Dumas, who was noted for his chemical discoveries and his numerous books in the field. Pasteur, profoundly influenced by Dumas, had vowed to become a chemist, and plunged headfirst into his studies now that he was back in Paris, performing well. Again, his examination achievements were merely good, not exceptional. Pasteur loved learning, but while attending the school, he found a previously unidentified zeal—a passion for discovery.

PASTEUR THE CHEMIST AND THE DISCOVERY OF CRYSTALS

One of Pasteur's chemistry teachers recognized his potential and made him a laboratory assistant. During this time, Pasteur learned to use the microscope in the study of crystallography. At that time, there were two forms of tartaric acid. One, the true tartaric acid, was a component of tartar in wine fermentation vats. The other form occurred as needlelike tufts and was called paratartaric acid or racemic acid from its origin from the grape (racemus). The two forms were chemically identical and had the same overall crystal shape. The difference was in the way the two rotated the plane of light when placed in solution, as measured by a device called a polarimeter. At the time Pasteur became interested, scientists observed that a solution of tartrate could rotate the plane of light while the paratartrate was optically inactive. Pasteur was dissatisfied that two chemically identical compounds could rotate light differently. He felt that there must be a chemical difference between the two. He began a systematic evaluation of the tartrate crystals under the microscope. He quickly noticed small facets of the crystals that had escaped the attention of others studying the phenomenon. All the tartrate crystals had the small facets on one side of the crystal; a tartrate solution rotated polarized light to the right. When Pasteur examined the paratartaric acid crystals, he found that some of the crystals had facets turned to the right, similar to tartrates, but some of the crystals had facets turned to the left. Paratartaric acid was an equal mixture of right- and left-handed crystals, which explained why it was optically inert. In a painstakingly detailed effort, Pasteur separated the right-handed from the left-handed crystals and placed them into two different solutions. The right-handed crystals rotated light to the right, and the left-handed crystals rotated light to the left. This was the first scientific discovery for the young Pasteur, who became so excited that he rushed from the laboratory and exclaimed to a colleague in the hall,

> I have just made a great discovery . . . I am so happy that I am shaking all over and am unable to set my eyes again to the polarimeter (3).

This initial discovery lit a flame of interest in scientific discovery that lasted Pasteur's entire life. He never tired of telling the story of his first discovery. Pasteur was fortunate that the tartrate crystal could be observed under the microscope and separated into right- and left-handed crystals. However, Pasteur had thoroughly studied his crystals and exhibited in this and in many subsequent studies an uncanny knack in selecting the best experimental material to solve a problem. This gift would lead to one of his most famous sayings: "In the field of experimentation, chance favors only the prepared mind." For his doctorate degree in chemistry, Pasteur wrote two theses, one of which was titled "A Study of Phenomena Relative to the Rotary Polarization of Liquids."

For the next 5 years, Pasteur was consumed with his chemistry experiments. Pasteur's studies were interrupted by the Revolution of 1848 and shortly thereafter by the death of his mother. His ties to his family were strong. Louis spent weeks unable to carry on his work after she died. Eventually, Pasteur was matched with a teaching position outside Paris since the purpose of the *École Normale* was to produce teachers. In 1849 he left Paris, first for Dijon for a few months and then to

Strasbourg, where he was made Professor of Chemistry at the University of Strasbourg. Shortly after his arrival in Strasbourg, Pasteur met Marie Laurent. After only 2 weeks, Pasteur proposed marriage. Ms. Laurent hesitated for a few weeks but eventually accepted. The two were married in May 1849.

Pasteur's chemical discoveries with tartaric acid crystals led to some scientific acclaim, allowing him to travel in Europe and meet many distinguished scientists. He continued his work in stereochemistry, the branch of chemistry that involves the study of the relative spatial arrangement of atoms within molecules. He began to recognize that one of the fundamental features of living organisms was that they nearly always were asymmetrical chemically, i.e., if there were compounds that could be right- or left-handed like his tartaric acid crystals, living beings nearly always had one or the other but rarely both the right- and left-handed varieties of the chemical. He wrote,

> The universe is an asymmetrical whole. I am inclined to think that life, as manifested to us, is a function of the asymmetry of the universe and of the consequences it produces (3).

From 1850 to 1854, Pasteur continuously worked and taught chemistry in Strasbourg. He was awarded the Prize of the Society of Pharmacy in Paris and the Award of the Legion of Honor. During this period, two daughters and his son were born. In 1854, Pasteur and his family moved to Lille, France, when he was appointed Professor of Chemistry and Dean of the newly organized Faculty of Sciences of the University of Lille.

THE "DISEASES" OF FERMENTATION

Shortly after Pasteur's arrival in Lille, the father of one of his students consulted with Pasteur on the difficulties that he was experiencing with the alcoholic fermentation of beet sugar in his distillery. To understand Pasteur's contribution to the field of fermentation, one needs to consider that in the 1850s, fermentation was viewed as an entirely chemical process. There was alcoholic fermentation, where organic solutions placed together in the correct way would be converted to wine, beer, or cider. There was acetic acid fermentation, with wine being converted to vinegar. The souring of milk, where milk sugars were converted to lactic acid, was called lactic acid fermentation. It was also commonly known that natural materials such as eggs and meat could undergo a process called putrefaction. Scientists believed that putrefaction was closely related to fermentation, differing only in the products that were formed. These processes were assumed to be entirely chemical in nature, requiring a catalyst to start or speed up the chemical reactions. The catalyst was not consumed during the chemical reaction. The makeup of the catalyst was mysterious but thought to be chemical in nature and specific to the type of fermentation or putrefaction. The catalyst that was responsible for alcoholic fermentation was known as yeast, which was also the name of the leaven needed for the rising of the dough in bread making. Surprisingly, the biological nature of yeast was not known at the time even though yeast had been used for making bread and wine for centuries. A scientific examination of fermentation was only carried out in the 18th and 19th centuries and gave the impression that the fundamental nature of the process was purely chemical and had been fully discovered. That view began to change in 1835 when two scientists, Cagniard de la Tour in France and Theodor Schwann in Germany, independently published experiments suggesting that yeasts were living cellular organisms. Some controversy followed these publications. Justus von Liebig, an eminent scientist of the time, tried to incorporate de la Tour's and Schwann's work. von Liebig proposed that if yeasts contributed to alcoholic fermentation, it was not as living beings, though they might be, but because they were dying in the fermentation process and released the material that caused the reaction to proceed. But von Liebig had no experiment to back up his claim.

Pasteur was still contemplating the fermentation problem that the father of one of his students had presented to him. Pasteur realized that he had to reexamine the nature of fermentation. As famous as he was, von Liebig proved no match for Pasteur. Pasteur took on von Liebig's chemical theory of fermentation with an experiment where he obtained fermentation in a synthetic medium inoculated with minute amounts of yeast. Pasteur showed that the amount of alcohol produced ran parallel to the multiplication of the yeasts. This simple yet superbly designed experiment proved, once and for all, that organic decomposition was not necessary to start alcoholic fermentation. Indeed, Pasteur showed that yeasts grow, bud, and multiply in sugar fermentation.

> Alcoholic fermentation is an act correlated with the life and with the organization of these globules [yeast], and not with their death or their putrefaction (3).

Soon, Pasteur showed that other ferments, including lactic acid ferments, were the result of living beings producing chemicals and that these beings could be cultivated in artificial media. He began to realize that each of these microscopic beings differed by their morphology, nutritional requirements, and susceptibilities to toxic substances. The field of microbiology was born! Through publications in 1860, Pasteur contended that his findings would permit a study of the fundamental chemical nature of life. By this time, the cellular theory, i.e., that all living beings were composed of cells, was accepted. Pasteur believed that single-celled organisms would be the starting point for understanding the cellular makeup of more complex beings. He believed that the chemical and physical properties of single-celled organisms could be more easily studied in the laboratory.

The problem that the father of his pupil presented to Pasteur was that putrefaction was occurring in the distillery rather than alcoholic fermentation. After studying the biological process of fermentation and realizing that different microorganisms could cause different types of fermentation, Pasteur began to view the problem as a "disease" of fermentation. He used the microscope, the instrument that had helped him with his crystal experiments, to examine the diseases of fermentation. His pupil, the son of the man who presented Pasteur with the problem, described what Pasteur had discovered.

> He [Pasteur] had noticed by microscopic examinations that the globules were round when fermentation was healthy, that they lengthened when alteration began, and were quite long when fermentation became lactic. This very simple method allowed us to watch the process and to avoid the fermentation failures which were then so common (4).

Pasteur realized that foreign organisms competing with the yeast in the ferment caused the "diseases" of fermentation.

The leaps that Pasteur had made in his work on fermentation in Lille radically changed thinking about a process that was considered to have been fully explained—the nature of fermentation. He carefully designed experiments to refute the long-held belief in the solely chemical nature of the process. Importantly, he had moved science to the biological rather than chemical basis of fermentation. He learned to cultivate and identify microorganisms in the laboratory. Pasteur related certain microorganisms to certain fermentation processes and determined that if competing microorganisms entered the picture, the desired fermentation process would be "diseased." His logical thinking and, more importantly, his experiments had defined the germ theory of fermentation. It is no accident that Pasteur used "disease" to help define the problems encountered in the fermentation process. Unencumbered by any attachment to the humoral theory of disease as a nonphysician, his logical mind quickly saw the appearance of unwanted, competing microorganisms in the fermentation process as harbingers of disease. The time for a broader application of the germ theory to human disease would soon be at hand. Finally, Pasteur had correctly speculated that his laboratory efforts might even unlock the biochemical secrets of life. At 42 years of age, Pasteur, who had already made significant contributions to chemistry, gave birth to an entirely new field, microbiology. Amazingly, his contributions to medical science had not even begun. In 1857, Pasteur moved back to Paris when he was appointed Director of Scientific Studies at *École Normale Supérieure* (*Supérieure* was added to the school's name in 1845).

Pasteur's discoveries in fermentation and its diseases revealed an immensely practical aspect to his work. Later, in a publication from 1877, he described similar alterations in microfloras when he studied beer production. Pasteur uncovered the biological basis of vinegar production, helping the French vinegar industry adjust the production of vinegar to the demands of the market. In these efforts, Pasteur had taken scientific discoveries in the laboratory and converted them into discoveries for society. From early in his time at Lille, Pasteur became more than a laboratory scientist and oriented nearly all his efforts to practical problems of his world.

PASTEURIZATION

Since the introduction of an undesired microorganism can damage the fermentation process, some method for preventing their introduction or killing them once they have been introduced would have immediate benefit to the industry. Although meticulous controls of the technological side of the industry had been made, elimination of all undesired microorganisms could not be achieved. Pasteur set out to find a way to remove these microorganisms by adding a variety of compounds, such as bisulfites, aimed at destroying them. Unfortunately, all his attempts to find antimicrobial compounds were ineffective. Pasteur then turned to heat. Pasteur was familiar with and intensely interested in the process of making wine. He knew, for example, that wine was slightly acidic. He also knew that the antimicrobial effects of heat on microorganisms were more successful in an acidic environment. He experimented to find a temperature that would kill the unwanted microorganisms (bacteria) and still leave the process and the taste of wine unaffected. He further discovered that the heat would not affect the wine making process if applied after all the oxygen originally present in the bottle had become exhausted. We know this process as pasteurization. Pasteurization was successfully applied to wine, beer, cider, vinegar, and, of course, milk. Not content with a theoretical laboratory discovery, Pasteur took an active role in determining the effectiveness of this process in the various industries to which pasteurization was applied. Pasteur wrote,

> There are no such things as pure and applied science—there are only science, and the applications of science (4).

Pasteur's success in putting forward his theory on the biological basis of fermentation and taking on formidable opponents and critics such as von Liebig occurred by expert experimental design, careful interpretation of the results, and the force of his own personality. Perhaps just as important, pasteurization provided a practical solution that von Liebig and other critics could not offer or deny.

SPONTANEOUS GENERATION AND LOUIS PASTEUR

The role of microorganisms in fermentation brought Pasteur into an ongoing debate: what was the origin of these microorganisms? The 19th-century arguments over spontaneous generation of life were often philosophical and religious, since the idea of creation, the ultimate beginnings of life, can lie beneath the belief in spontaneous generation. Pasteur, however, would never be drawn into a philosophical or religious discussion of spontaneous generation, even though he was a devout Catholic. Pasteur acknowledged that somewhere in the universe or perhaps back in Earth's early history, life sprung from organic matter. Pasteur, instead, devoted great energy to evaluating claims that experimenters had discovered conditions where spontaneous generation was possible.

Resolution of the debate over spontaneous generation of microorganisms was an essential step in the development of the germ theory of human disease. Consider the implications if spontaneous generation of bacteria could occur. Contagion would become irrelevant, and sterile techniques in the operating room and the laboratory would be immaterial. The spontaneous formation of life could occur anywhere, even inside the human body, given the proper conditions. To Pasteur, spontaneous generation invited anarchy to what he was beginning to perceive as a specific role that microorganisms played in nature. Pasteur believed that the theory demanded challenging, even though some of his mentors from the *École Normale Supérieure* urged him to stay out of the fray, fearing that the time spent on the controversy would take too much valuable time outside of scientific inquiry. But Pasteur would challenge spontaneous generation with scientific inquiry and use his determined nature to convincingly disprove it.

Before considering Pasteur's role in the controversy of spontaneous generation in the mid-19th century, we must consider the confused situation in the debate. Scientists had ceased to believe that higher forms of life such as flies, maggots, or mice would spontaneously generate from a few grains of wheat or a piece of cheese based upon experiments from, among others, Francesco Redi, published in 1668. Microorganisms were another matter, however. Scientists recognized that broth or milk that had been boiled to destroy all life could be observed to quickly acquire numerous bacteria upon cooling, usually within 1 to 2 days. Since these organisms were so small, it was not a stretch of the imagination to believe that the simplest agents of life could form from the organic matter in solution. The question remained without definitive conclusions, since critics on both sides could find fault in experiments supporting and refuting the theory. No one had been able to find an experiment that put an end to the matter. The confused state of affairs came to a crisis when Félix Pouchet, an esteemed scientist, and Director of the Museum of Natural History in Rouen, read a paper before the Paris Academy of Science in

1858. Pouchet performed an experiment in which he took a flask of boiling water that he hermetically sealed and plunged it upside down into a tub of mercury. He then introduced a half-liter of oxygen and a small quantity of hay infusion that had previously been exposed to very high temperatures for a significant period of time. Pouchet believed that he had taken sufficient precautions to preclude the introduction of living organisms into the flask. After a few days, microbial growth occurred in the flask. Pouchet presented his findings to the Paris Academy of Sciences and published them a year later. Due to his reputation, his experiment achieved great acclaim. Pouchet's moment in the sun was short-lived. Pasteur took aim at Pouchet's work. Pasteur was not the kind of individual one wanted as an opponent. Pasteur worked with feverish energy and masterfully designed his experiments to counter any potential criticism.

The first task for Pasteur was to consider the possible flaws in Pouchet's experiment. Organic solutions, even if subjected to prolonged heat, would eventually support microbial growth when natural air was admitted. Pasteur set out to determine whether air was an essential factor for the spontaneous generation of microorganisms or whether merely the admitted air contained viable microorganisms. He examined every detail of Pouchet's experiment and the countless experiments that he considered to deal with the question at hand. Pasteur noticed that the mercury used in Pouchet's work always contained dust that could be determined to contain living microorganisms. Pasteur eliminated the mercury from the experiment since he could demonstrate that it could contaminate the fluids and the air in the flask. Next, Pasteur had to deal with the presence of air that might represent some vital component for spontaneous generation or simply contaminate a previously heated organic solution when the air is introduced. In the now-famous experiment, Pasteur used a swan-neck flask with an organic solution that had been previously heated to destroy microbial life. After it was cooled, he demonstrated that microbial life did not spontaneously generate in the organic solution that was left open to natural air. The swan neck design prevented the entry of the particles suspended in air that contained microorganisms. Pasteur convincingly demonstrated that there was no vital force in exposure to air that allowed microorganisms to spontaneously generate in the organic solution. Pasteur extended this simple yet elegant experiment by boiling organic fluid in a flask to destroy any life and drawing out the neck in a flame so that it could be easily sealed. The solutions in the sealed flask would remain sterile indefinitely. Pasteur could then break the seal under exacting conditions to avoid contamination. The admitted air could be obtained in a variety of settings and the flasks resealed and incubated. Testing with these flasks revealed that microorganisms were not in the same abundance everywhere. More microorganisms were obtained in air from low-lying places, near the earth. The numbers of microorganisms were lower when the air was still or at higher elevations. Some flasks remained sterile when the air was introduced in the atmosphere of Swiss glaciers.

Pasteur's experiments produced a sensation within the scientific community and the public. In 1862, the Paris Academy gave Pasteur the Prix Alhumbert for "Mémoire sur les corpuscules organisés qui existent dans l'atmosphère." Critics remained, however, largely due to philosophical or prejudiced beliefs. Pouchet reported that he had attempted to duplicate Pasteur's findings, but without success. Pasteur, convinced of his own work, threw down the gauntlet. He demanded that a commission from the Paris Academy of Sciences duplicate his experiments. The experiments were repeated. The official report published in 1864 in the Academy's records essentially closed the debate with Pasteur's triumph. Pasteur wrote,

> There is no known circumstance in which it can be affirmed that microscopic beings came into the world without germs, without parents similar to themselves. Those who affirm it have been duped by illusions, by ill-conducted experiments, by errors that they either did not perceive or did not know how to avoid (2).

DISEASES OF SILKWORMS AND THEIR ROLE IN THE GERM THEORY OF DISEASE

The next phase of Pasteur's career and in the development of the germ theory of disease takes us to the study of silkworms. The story of Pasteur's involvement with the economically vital silk industry in southern France played a central role in our understanding of the germ theory. In the middle of the 19th century, a strange and deadly disease began to strike the French silkworm nurseries. I never thought I would need to explain the life cycle of the silkworm to understand the germ theory of human disease. However, before one can understand the silkworm disease and how it aided in development of the germ theory of human disease, it is necessary to describe the life cycle of the silkworm. The worm begins as an egg that so closely resembles a plant seed that the eggs are often referred to as seeds. There are four molts where the worm changes its skin. The worm remains essentially motionless until the fourth molt, when, after 2 to 3 days, it begins to voraciously eat and rapidly increases its size. This stage is called the *grande gorge*. The worm then settles down on a sprig of heather and spins its cocoon. The silk cocoon is where the worm transforms into a chrysalis and then emerges as a moth. The moth does not eat and exists solely for reproduction. After mating, which takes place nearly immediately upon the moth's exit from the cocoon, the female lays generally between 600 and 800 eggs. The entire process takes about 2 months, although the eggs do not hatch until the following year in the spring, when the cycle begins again. The silk industry utilized the cocoon before the moth emerged since the silk after the moth has emerged was unfit for spinning. The cocoons were placed in a steam bath to kill the chrysalids by heat. Naturally, some cocoons were allowed to remain for the entire life cycle to be complete.

The disease that began affecting the silkworms, devastating the industry, was characterized by small black spots resembling bits of black pepper and was referred to as pébrine. The worms with pébrine could be affected at any stage. Most importantly, diseased worms would not enter the *grande gorge* stage but arrest and die before spinning the cocoon. Peculiar tiny structures designated corpuscles often were noted in diseased moths, but these corpuscles also appeared in what were thought to be perfectly healthy moths. Some investigators noted that diseased moths that mated produced diseased worms, suggesting to some that the disease might be hereditary. The relationship between the black spots on the worms and the corpuscles of the moth was unclear, but many considered them to be correlated. One of Pasteur's oldest mentors, Jean Dumas, was from one of the affected regions and convinced the Minister of Agriculture in France to appoint a commission to study the disease. It was Dumas that convinced Pasteur to take the lead. Since Pasteur knew nothing about silkworms, it was surprising that he agreed. We may never be certain about his reasons, but he was devoted to his mentor and found it hard to refuse any of his requests. However, Pasteur's work on fermentation and spontaneous generation appeared to have whetted his appetite for the study of the role of microbes in disease. Since he was not a physician, this opportunity afforded Pasteur a chance to study pathology in the field.

This study took Pasteur down another unusual path, at least for him. Because this work was government funded, Pasteur was obliged to periodically report his findings well before he was satisfied with the results. Pasteur rarely shared his views from his experiments, even with his assistants, until he felt that solid conclusions had been reached. However, the study on silkworm disease affords us an opportunity to see how Pasteur worked, how he thought, and the blind paths or erroneous attempts at solutions that he rarely allowed others to see.

Pasteur arrived at Alais, France, in early June 1865. He did not know it at the time, but Pasteur would spend the next 5 years of his life working on silkworms. Always aware of the practical impact of his work, Pasteur noted the devastating impact that the disease had on the area:

> Today the plantations are entirely abandoned; the tree of gold no longer enriches the country, and the faces of formerly happy are now downcast and sad. Where abundance once reigned, there is now poverty and distress (5).

He immediately familiarized himself with the black spots of pébrine and the corpuscles of the moth. By careful observation, he determined that the disease could be found in worms that had been feeding on leaves on which diseased worms had been crawling. Pasteur tried pricking a healthy worm with a needle that had merely scratched a diseased worm. The pricked needle induced the disease. He also noted that eggs containing the black spots gave rise to diseased worms. To Pasteur, these observations clearly suggested contagion. Pasteur used these facts to develop a method to examine the moths that produced the eggs:

> After mating, the female, set apart will lay her eggs; then one will open her, as well as the male, in order to search for the corpuscles. If they are absent from both the male and female, he [the investigator] will number their laying which shall be preserved as eggs absolutely pure and bred the following year with particular care (5).

This approach to egg selection, using his trusty microscope to identify corpuscles in moths, proved practical and immensely helpful. The method was not infallible, however. Pasteur found diseased progeny from moths that exhibited no corpuscles. The worms sickened and died without showing the typical black spots or corpuscles. He also tried to cultivate the corpuscles to produce bacteria or yeasts with techniques that he developed during his fermentation studies, but to no avail. While Pasteur had developed a method of egg selection that had shown good success, he was dissatisfied and frustrated. His correspondence showed that he was more and more disturbed until he realized why there were diseased worms from moths that had not exhibited corpuscles. One day Pasteur finally collapsed in a chair, nearly in tears, and exclaimed,

> Nothing has been accomplished. There are two diseases.

The disease that Pasteur recognized as distinct from pébrine was known as *flacherie*. Like pébrine, flacherie was a contagious disease that was fatal to the worms. Unlike pébrine, it was primarily an intestinal infection, with enormous numbers of the disease-causing bacteria forming in the gastrointestinal canal of the worm and did not produce any black spots on the surface of the insect. Pasteur's disappointment was short-lived. The discovery of two diseases resolved the failures and frustrations that Pasteur had experienced in the method of egg selection for pébrine. Pasteur eventually developed a method for determining the presence in the silkworms of not only pébrine but also flacherie. He examined the adult moths for both the corpuscles (ultimately determined to be produced due to a protozoan) and for the bacteria in the alimentary tracts for flacherie. From his methods, he accurately predicted lots of eggs to the immense satisfaction of the commission and the silkworm industry.

Pasteur had succeeded where others had failed because he had identified the causes of the two infections affecting the silkworms. From there, he found a practical method for selection of eggs that saved the industry. He published his findings in a paper, "The Diseases of Silk Worms." But the years that Pasteur had spent in Alais, France, were some of the most personally distressing of his life. In 1865, he lost his father and one of his daughters, Camille, only 2 years old. In May 1866, another daughter, Cécile, died from typhoid fever. Pasteur tried to console himself from these tragedies through work. But the intense effort he made brought on a stroke in 1868. Pasteur slowly recovered, but his left side remained somewhat paralyzed for the rest of his life.

Unlike his controlled laboratory experimentations, Pasteur had entered the unpredictable world of biological life. He was initiated into problems of infectious diseases. He witnessed the variable susceptibilities of individual organisms of the same species to an outcome of an infectious disease. In particular, Pasteur appreciated that these variations could be from an assortment of environmental factors, such as excessive heat or humidity or inadequate ventilation. His work on fermentation, spontaneous generation, and now the diseases of silkworms had been like an apprenticeship for the study of pathological processes of microorganisms. Pasteur was keenly aware of the direction that his career was taking. He would tell those who came to work with him, as he later wrote,

> Read the studies on the silkworms; it will be, I believe, a good preparation for the investigations that we are about to undertake (6).

When Pasteur returned to Paris after his work on silkworms, the country was rife with rumors of an approaching war with Prussia. When war broke out in 1870, many individuals at the *École Normale Supérieure* volunteered for military service. Because of his stroke, Pasteur was rejected for service. Pasteur remained forever bitter towards the Germans after shells bombarded *École Normale Supérieure*. In 1868, the University of Bonn had awarded Pasteur an honorary doctorate to acknowledge his work in the role of microorganisms. During the Franco-Prussian war, Pasteur sent back the degree with an acerbic letter explaining his reasons. His dislike of Germany would play a role in a professional dispute with Robert Koch some years later.

THE GERM THEORY OF DISEASE, PASTEUR, AND MEDICINE IN THE 19TH CENTURY

Between 1857 and 1878, Pasteur contemplated and sought further evidence for the germ theory of disease, but the concept had not sprung from his mind alone. There were others in the 19th century who had come before Pasteur and deserve some consideration for their efforts. Agostino Bassi, sometimes called de Lodi, was an entomologist who had discovered a disease of silkworms in Italy caused by a fungus that now bears his name, *Beauveria bassiana*. In 1844, he theorized that microorganisms caused human diseases.

In England, livestock diseases in the mid-19th century were subject to laboratory investigation to identify their agents of disease, including studying living disease-causing organisms (7). Perhaps the most prominent 19th-century individual whose work preceded Pasteur was a pathologist named Jacob Henle. Henle wrote in his *Pathologische Untersuchungen* that:

> ... material of contagion is not only organic but living, endowed with individual life, and standing to the diseased body in relation of a parasite organism (8).

The difficulty that scientists had in proving this theory of disease to a skeptical physician audience was that there was little means to differentiate one microorganism from another despite the extraordinary differences in symptoms and signs of various infectious diseases. Henle even predicted that specific microscopic agents would be consistently found to cause certain diseases but did not have the methods at his disposal to prove his assertion. Curiously, Henle taught Robert Koch, who, along with Pasteur, would provide the scientific evidence for the germ theory. One other name worth considering for his pre-Pasteur work is John Snow. In 1846, a cholera epidemic spread through the cities of Europe. The cause of the disease continued to be in question. Even midway through the 19th century remnants of the humoral theory remained. Many scientists postulated miasma from vegetable decomposition or elements in the air as the cause of cholera. Snow believed that cholera was an infection of the gastrointestinal tract, and that water was the vehicle for transmission. Few paid attention to Snow's theory until his spectacular demonstration of a cholera outbreak that he traced to a water pump on Broad Street in London. While Snow did little to determine the microorganism causing the disease, his groundbreaking use of epidemiological methods convincingly showed that the agent of cholera was waterborne and could survive outside the human body.

As the concept of the germ theory of human disease was taking hold among many academic scientists, physicians had a difficult time believing that a living being so small could cause profound pathological damage to a human. The theory received a hostile reaction from a surprising source within the medical field. In the mid-19th century, Rudolf Virchow had revolutionized medicine with the cellular basis of disease. He was a staunch opponent of spontaneous generation, putting forth the concept that every cell comes from another cell. His concept of cellular pathology took Morgagni's anatomic basis of disease a step further—to the level of the cell that he identified as the functional seat of human disease. For example,

Virchow was able to show that cancer cells originated from normal cells within specific tissues. This theory of cellular pathology forever put to rest any notion of humors or miasma. But Rudolf Virchow stated in his later years,

> If I could live my life over again, I would devote it to proving that germs seek their natural habitat—diseased tissues—rather than causing disease.

Virchow viewed Semmelweis's work unfavorably. For a pathologist who essentially invented cellular pathology, his opposition to viewing microbes as the cause of human disease was a formidable obstacle to the acceptance of the germ theory in medical circles.

In 1873, Pasteur was elected as an Associate Member of the Académie de Médecine in France as a nonphysician. However, associate membership was not sufficient to give Pasteur standing in the medical community. Two physicians would soon step into the debate on Pasteur's side, Joseph Lister in Great Britain and Robert Koch in Germany. Each provided key elements that fostered support for the theory that will be detailed in the next two chapters.

PASTEUR'S WORK ON ANTHRAX

Pasteur took the medical community head on in his studies on anthrax. Anthrax was a disease largely of sheep that had an enormous economic impact on French agriculture. In 1850 Casmir Davaine had successfully isolated a small rod-shaped microorganism from the blood of diseased sheep. After the sheep died, however, the bacterium was often difficult to isolate, leading some investigators to question whether something else was required to produce anthrax. Despite Davaine's discovery, the significance of these organisms was unclear. Nor was it clear how the disease was transmitted. For example, the anthrax affected animals that grazed on certain fields years after the diseased animals had died from anthrax. How could this difficult-to-isolate bacterium be responsible so many years later? As we shall see in the next chapter, Robert Koch, early in his career, determined that the organism responsible for anthrax had various life stages, including a spore-forming stage that would allow the bacterium to resist drying in the environment and remain viable for years. Doubters remained, suggesting that the bacterium, which could not be uniformly isolated from deceased animals, might not be the cause of anthrax. The source of this doubt was an experiment by fellow Frenchmen, Claude Leplat and Pierre-François Jaillard, who obtained blood from a rabbit that unquestionably died of anthrax and injected the blood into other rabbits. The inoculated rabbits died, but the rods that Davaine had proposed as the cause of the disease could not be found. Still, blood from these rabbits, apparently without the microscopic rods, could be inoculated into other rabbits that died. Pasteur became intrigued with the dispute because of his earlier studies on spontaneous generation. He knew from those studies that blood from a healthy animal that was removed with aseptic care and cultured with nutrient fluids would not spontaneously generate microorganisms. Pasteur reasoned that blood from an animal with anthrax could also be taken aseptically and could be added to nutrient broth. Using his experiences on the diseases of silkworms, Pasteur realized if any microorganisms appeared, the cultures would yield the causative organism of the disease. Pasteur proved to be correct. Pasteur saw rapid and abundant growth of the rods of anthrax by cultivating a small amount of blood from the diseased animal in neutral urine. He could maintain the cultures through countless generations by transferring a small amount of the material to sterile neutral urine. He kept diluting the original culture that contained about 50 ml of the animal's blood (about 2 oz) so that after 100 generations of culture, he calculated that hardly a molecule of the original animal's blood was in the culture. Yet, if Pasteur injected the bacteria into rabbits, anthrax occurred. Pasteur reasoned that it was the bacteria that produced anthrax, not some other component in the animal's blood. Proof of the germ theory of disease was at hand! During the next 20 years, numerous bacterial causes of infectious diseases were discovered. But it was the work of two men, Pasteur and Koch, on the disease of anthrax that broke the backs of the doubters.

THE DISCOVERY OF TOXIN PRODUCTION FROM ANTHRAX BACILLI

One of Pasteur's lesser-known discoveries also involved his work on anthrax. Pasteur passed blood from an anthrax-infected animal through a series of plaster filters that removed the anthrax bacillus. He injected the filtered material into normal fresh blood. When blood cells in an animal are exposed to natural infection with anthrax, they immediately clump or agglutinate. The filtrate induced identical agglutination of fresh red blood cells, suggesting that a soluble toxin from anthrax bacilli could produce disease. Pasteur was the first to demonstrate the importance of toxin-producing bacteria in the production of disease and even death.

While Pasteur and Koch met on the same ground, namely, the proof that a bacterium, eventually determined to be *Bacillus anthracis*, was the causative agent of anthrax, their views of infectious diseases began a remarkable divergence. Koch and his colleagues in Germany identified and described innumerable bacteriological causes of disease. Pasteur and the French became interested in the attenuation of pathogenic microbes and the role of this attenuation in the immunity to microorganisms.

CHICKEN CHOLERA AND ATTENUATION OF MICROORGANISMS

In July 1878, Pasteur began to study chicken cholera, a disease that bears no relationship to human cholera but the name. The disease could wipe out an entire barnyard full of chickens in as little as 3 days. Rabbits are also susceptible, but adult guinea pigs are strangely resistant to the illness. The guinea pigs may develop an abscess, but it really does not affect their general well-being. Pasteur became intrigued:

> Chickens or rabbits living in contact with a guinea pig suffering from such abscesses might suddenly become sick without any apparent change in the health of the guinea pig itself. It would be sufficient that the abscesses open and spread some of their contents onto the food of the chickens or rabbits. Anyone observing these facts and ignorant of the relationship that I have just described would be astounded to see the chickens and rabbits decimated without any apparent cause and might conclude that the disease is spontaneous . . . How many mysteries pertaining to contagion might some day be explained in such simple terms (9)?

Pasteur's study on chicken cholera led to discoveries about the nature of infection, contagion, and immunity. One discovery was the concept that one species of animal could serve as a reservoir of infection for another. Pasteur also noted that after recovering from the disease, chickens could carry the causative organism and be the source of infection for other chickens—the so-called carrier state. To Pasteur, the observations meant that something happened to the microorganism or the host because the microorganism did not invariably produce infection. He considered his earlier work to determine if the microorganism might change. During his studies on fermentation, he observed a mold called *Mucor mucedo* that grew in a filamentous form when in the presence of air but was a round, yeast form under anaerobic conditions. Depending upon the conditions, a microorganism was capable of transformation, but no one had any proof that any change in a microorganism could produce immunity in an animal. Luck then intervened. But, as Pasteur was fond of saying, "Chance favors the prepared mind."

In 1879, some cultures had to be set aside for several weeks. When Pasteur attempted to inoculate the bacteria from the old cultures onto new media, he found that growth of the microorganisms was unusually slow. He then inoculated them into fowl that were apparently unaffected by the inoculation. No disease occurred. Many of us would simply have started over with fresh cultures and fresh chickens. Not Pasteur. He had already been thinking about possible changes to microorganisms from his earlier work on mold in fermentation studies. He inoculated the chickens that had been given this slow-growing microorganism, a facet that suggested some transformation of the microorganism, with a fresh culture of the chicken cholera bacterium. To his surprise, nearly all the chickens survived, whereas newly purchased fowl inoculated with the same fresh cultures perished. The observation immediately suggested that the first lot of chickens had been protected against chicken cholera because the causative microorganism had been attenuated. The determination that the virulence of a microorganism could be moderated and used to protect an animal from the disease-causing form was one of Pasteur's greatest discoveries. Jenner stumbled onto a virus that conferred immunity to smallpox, but Pasteur realized that the virulence of microorganisms was not fixed and could be modified to help produce immunity. The trick, of course, was determining the method of transformation that would produce attenuation suitable for vaccination (Pasteur used the term in deference to Jenner) that was both effective and safe. Pasteur determined the attenuation of the chicken cholera bacillus to be from the harmful effect of air, specifically oxygen, on an aging culture. He showed that cultures from vials that were sealed maintained their virulence to chickens for months, whereas cultures from vials that had only cotton plugs lost their disease-producing activity rapidly. Within the astoundingly short period of 4 years, Pasteur developed vaccines for chicken cholera, swine erysipelas, anthrax, and rabies.

PASTEUR AND THE ANTHRAX VACCINE

Armed with the success of chicken cholera vaccination, Pasteur turned his attention to the production of a veterinary anthrax vaccine in the early 1880s. Some biological observations suggested to Pasteur that an attenuated anthrax vaccine might be possible to produce. He noticed that of a group of eight sheep that had been grazing in a pasture where an animal that died from anthrax had been buried, several animals survived after inoculation with virulent anthrax bacilli. Unfortunately, efforts to attenuate the anthrax bacillus proved difficult because anthrax spores do not undergo any significant changes. Pasteur determined that he needed to prevent the formation of spores but keep the bacilli alive. He did this by adding a certain compound, potassium dichromate, to the cultures and then keeping the cultures in a shallow layer at 42 to 43°C. After 8 days, the bacilli were harmless to laboratory animals. But for large animals, Pasteur used a culture of very low virulence followed 12 days later by a second, more virulent vaccine. Pasteur kept this method of attenuation secret (10).

Scientists, particularly physicians, were sometimes openly skeptical of and even hostile to Pasteur's vaccines and his comparing them to Jenner's vaccine. The editor of Veterinary Press, a surgeon named M. Rossignol, typified some of the medical community's attitude toward Pasteur in a statement dripping with sarcasm:

> Will you have some microbe? There is some everywhere. Microbiolatry is the fashion, it reigns undisputed; it is a doctrine which must not even be discussed, especially when its Pontiff, the learned M. Pasteur, has pronounced the sacramental words, *I have spoken*. The microbe alone is and shall be the characteristic of a disease; that is understood and settled; henceforth the germ theory must have precedence of pure clinics; the Microbe alone is true, and Pasteur is its prophet (11).

Rossignol challenged Pasteur to perform a public experiment of his vaccine for anthrax. Pasteur tried hard but was often annoyed at the medical community's response to his work. He readily accepted the challenge to silence his critics. Pasteur conducted a very public experiment of his anthrax vaccine at Pouilly-le-Fort, near Melun, in May 1881. Since anthrax was responsible for the loss of many millions of animals every year in France during this period, the economic impact of a preventative vaccine was enormous. Pasteur vaccinated 24 sheep and a few other animals with his anthrax vaccine. He used 24 other sheep and some other farm animals as controls. The sheep were then inoculated with a virulent culture of the anthrax organism. The results were amazing to all.

> When the visitors arrived on June 2, they were astounded. The twenty-four sheep, the goat, and the six cows that had received the vaccinations of the attenuated anthrax, all appeared healthy. In contrast, twenty-one sheep and the goat which had not been vaccinated had already died of anthrax; two other unvaccinated sheep died in front of the viewers, and the one remaining sheep died at the end of the day (10).

Pasteur concluded,

> ... the development of a vaccination against anthrax constitutes significant progress beyond the first vaccine developed by Jenner, since the latter had never been obtained experimentally (10).

The economic impact of the vaccine was great enough to the agricultural community that it helped France pay reparations to Germany from the treaty ending the Franco-Prussian War.

THE RABIES VACCINE

One of the greatest achievements attributed to Pasteur was the discovery of a vaccine against rabies. With the near disappearance of rabies in developed countries, we no longer recognize the fear that the disease produced, which was well described by a journalist in Pasteur's day:

> Every year at the same season, a terror strikes Paris. The sun burns ever and ever hotter. Rabies draws near, and gains force. Every dog becomes the object of suspicion— the poor dog, good as he is (12).

In the 1880s, humans usually contracted rabies from the bite of rabid dogs. Pasteur set out to find the causative agent. Since viruses were not discovered until the very end of the 19th century, finding the agent proved impossible for Pasteur, even though he believed it to be a short, twisted rod that he described from saliva of a child infected with rabies using his microscope. He soon realized that saliva from uninfected children contained the same rod.

The study of rabies presented great difficulties for Pasteur and his colleagues. The agent remained undetectable through the available 19th-century techniques for cultivating microorganisms. Additionally, the bite of a rabid dog did not always give the bitten person rabies. If it did, rabies symptoms would occur after an extended incubation period, often weeks. These features of rabies did not lend themselves to Pasteur's previous laboratory studies. The symptoms of rabies suggested that the nervous system was involved in the disease. After much discussion, Pasteur devised an ingenious method for cultivating an invisible agent of disease in the laboratory. Using a surgical technique, called trephination, that involves making a circular hole in the skull of an animal, one of Pasteur's colleagues, Émile Roux, inoculated dogs' central nervous systems with saliva from rabies-infected animals. It was not fear of contracting rabies that kept Pasteur from performing the trephination. Pasteur was courageous enough to extract saliva from a child suffering from rabies. Pasteur had repugnance toward vivisection, according to Roux. Once Roux perfected the technique, he could infect a dog and take the spinal cord from the deceased animal, which would be preserved in a flask. An emulsion of the infected cord would be used to infect other dogs. This cultivation did not yield the actual infectious agent, which remained unidentified. The technique provided scientists with a new approach for cultivating unknown agents of infectious diseases, even though Koch's postulates (detailed in the next chapter) were not satisfied. Work on rabies progressed and turned to a method of attenuation of the virulent virus. Pasteur saw Roux's specimens of rabies-infected spinal cords in flasks. He came upon the idea of suspending them in dry, sterile air. He used caustic potash to prevent putrefaction and allowed oxygen to attenuate the virus over a period of about 2 weeks. Into the dog, Pasteur inoculated emulsions from these spinal cords of progressively less attenuated virus. Dogs received emulsions from cords dried for 14 days, then 13 days, and so on. Fresh cords were eventually used on these dogs, and immunity to rabies was established.

Since a human bitten by a rabid animal may not show symptoms for weeks, even a month or more, Pasteur considered the possibility of vaccination after exposure. He experimented on dogs bitten by rabid animals. Using his rabies vaccine, he found encouraging results. But Pasteur and Roux were unsettled when it came to trying the vaccine on humans. Roux, in particular, felt that the vaccine had been insufficiently studied. Unfortunately, fate interfered. Joseph Meister, then age 9, was brought to Pasteur from Alsace 60 h after suffering bites from a rabid dog. Pasteur agonized over the possibility of treating the boy with his untested vaccine. Vaccines had only been used as preventatives, not treatment. Medical dogma of the day said that once an infectious agent had settled in the body, vaccines could not and should not be used. The medical community was largely antagonistic towards Pasteur using his vaccine. Even Roux refused to participate. But two trusted colleagues, one a physician, assured Pasteur that the nature and size of the bites which were over the boy's hands, legs, and thighs made it quite likely that he would develop a fatal case of rabies. On 7 July 1885, Meister was treated with the rabies vaccine. The physician that Pasteur consulted gave the boy injections with progressively less attenuated cord emulsions until 16 July, when he received an inoculation of the virulent cord emulsion. Meister recovered and returned to Alsace. A second case, a 15-year-old boy bitten by a rabid dog, was treated, and also survived. A flood of requests for the vaccine followed these successes. Fifteen months after Meister's treatment, some 2,490 persons had received the vaccine.

REACTION TO THE RABIES VACCINE

The medical community viciously attacked Pasteur. His colleague, Roux, finally put aside his own objections when he saw how savage and personal the attacks became. In November 1885, a tragic failure of the rabies vaccine occurred when Pasteur attempted to treat a little girl who had been bitten by a rabid dog. Despite the vaccine, the girl died of rabies. Pasteur's adversaries dramatically used the failure against him. But the father of the little girl wrote a letter that touchingly describes how history now remembers Louis Pasteur:

> Among great men whose life I am acquainted with . . . I do not see any other capable of sacrificing, as in the case of our dear little girl, long years of work, of endangering a great fame, and of accepting willingly, a painful failure, simply for humanity's sake (13).

Pasteur's health suffered from the constant strain from his work on the rabies vaccine. He suffered from a cardiac condition and his previous stroke. He took some time away from his work in 1885 on the advice of a physician. Renewed with vigor, Pasteur returned to Paris and was warmly received at the Academies of Sciences and of Medicine. Perhaps his colleagues realized that their attacks had had an impact on a great scientist or what a great loss Pasteur's absence would be. In July of 1887, an English Commission issued a report from a 14-month study of Pasteur's rabies vaccine and its use in human treatment and fully supported Pasteur and his findings.

PASTEUR'S LAST YEARS

In 1887, the Academy of Sciences asked Pasteur to become its Life Secretary. He only maintained his duties for a short time, since his health began to slow him. He suffered another small stroke in October 1887 and resigned from the Academy in January 1888. During 1888, Pasteur would watch the construction of the Pasteur Institute as it was erected in Paris. On 14 November 1888, many of Pasteur's friends and colleagues assembled in the Institute's library to honor the man and the Institute. Pasteur was overcome but was unable to read his own speech, asking his son to do this for him.

> Keep your early enthusiasm, dear collaborators, but let it ever be regulated by rigorous examinations and tests. Never advance anything which cannot be proved in a simple and decisive fashion . . . But when, after so many efforts, you have at last arrived at a certainty, your joy is one of the greatest which can be felt by a human soul, and the thought that you will have contributed to the honour of your country renders that joy still deeper (14).

In the last few years of his life, Pasteur was weak and ill. Some other tributes came his way, but Pasteur made no further contributions to science. He died on 28 September 1895. He was given the rare honor for a scientist of a state funeral. Thousands lined the streets of Paris to pay their respects. His remains were placed in a crypt in the Institute that bears his name. A widely circulated story suggests that Joseph Meister, the boy whom Pasteur first treated with his rabies vaccine and who was so grateful to Pasteur, eventually became the gatekeeper of the Institut de Pasteur. In 1940, during the Nazi occupation, Meister was about to be compelled to open the crypt where Pasteur's remains were kept so that the remains could be removed. He committed suicide rather than agree to do so. However, recent investigations show this story to be a myth (15). Meister did commit suicide—not because of the requirement to move Pasteur's body but under the mistaken belief that his family had been killed in a bombing raid after he had sent them away from Paris as the Nazis were approaching the city. Today, Pasteur's grave remains undisturbed below the Institute, where great work continues.

On the 100th anniversary of his death in 1995, tributes to a man whose body of scientific work has few rivals became clouded and, in some publications, decidedly negative. One book, by G. L. Geison, attacked Pasteur's character and ethics, citing that he misrepresented facts to marginalize opponents, lied about his research, and stole ideas (16). One account found Geison's criticisms unfounded or overdramatized (17). Like any successful man, he had (and has) detractors. But Pasteur's body of work remains spectacular regardless of his faults. His influence on medical science, especially the germ theory of disease, is immense. There were pieces of a germ theory before Pasteur. But Pasteur provided a proven context for the germ theory of disease. The impact of Louis Pasteur's work goes beyond the germ theory of disease. Luc Montagnier, co-winner of the 2008 Nobel Prize in Medicine and Physiology for his discovery of human immunodeficiency virus, wrote that Pasteur's legacy provides three lessons for science. First, Pasteur introduced modern research methodology in science. Second, Pasteur linked basic research and applications in a multidisciplinary approach, what today might be called translational research. Third, Pasteur introduced the concept of global humanism in science:

> He considered that his discoveries and those of his institute should be shared and used by all people in the world, regardless of borders, that knowledge is the patrimony of humanity (18).

Pasteur continues to inspire. Françoise Barré-Sinoussi, co-winner of the 2008 Nobel Prize in Medicine and Physiology together with Montagnier, worked at the Institut Pasteur. She echoed her colleague's view:

> It was his vision to encourage scientific collaboration among disciplines and countries for the benefit of humanity worldwide. He is what inspired me to work here [at the Institut Pasteur].

REFERENCES

1. **Codell Carter K.** 2003. *The Rise of Causal Concepts of Disease: Case Histories.* Ashgate, Aldershot, United Kingdom.
2. **Vallery-Radot R.** 1923. p 257–296. *In The Life of Pasteur.* Doubleday, Page & Co, New York, NY.
3. **Dubos RJ.** 1950. p 90–115. *In Louis Pasteur: Free Lance of Science.* Little, Brown & Co, Boston, MA.
4. **Dubos RJ.** 1950. p 116–158. *In Louis Pasteur: Free Lance of Science.* Little, Brown & Co, Boston, MA.
5. **Holmes SJ.** 1924. p 125–142. *In Louis Pasteur.* Harcourt, Brace and Company, New York, NY. http://dx.doi.org/10.5962/bhl.title.4574.
6. **Dubos RJ.** 1950. p 209–232. *In Louis Pasteur: Free Lance of Science.* Little, Brown & Co, Boston, MA.
7. **Worboys M.** 2000. p 51–60. *In Spreading Germs: Disease Theories and Medical Practice in Britain, 1865–1900.* Cambridge University Press, Cambridge, United Kingdom.
8. **Dubos RJ.** 1950. p 233–266. *In Louis Pasteur: Free Lance of Science.* Little, Brown & Co, Boston, MA.
9. **Dubos RJ.** 1950. p 267–291. *In Louis Pasteur: Free Lance of Science.* Little, Brown & Co, Boston, MA.
10. **Pasteur L.** 2002. (1881.) Summary report of the experiments conducted at Pouilly-le-Fort, near Melun, on the anthrax vaccination. *Yale J Biol Med* **75:**59–62. (Original publication. *C R Acad Sci* **92:**1378–1383.)
11. **Vallery-Radot R.** 1923. p 297–340. *In The Life of Pasteur.* Doubleday, Page & Co, New York, NY.

12. **Pasteur L**, **Illo J.** 1996. Pasteur and rabies: an interview of 1882. *Med Hist* **40:**373–377. http://dx.doi.org/10.1017/S0025727300061354.
13. **Dubos RJ.** 1950. p 317–358. *In Louis Pasteur: Free Lance of Science.* Little, Brown & Co, Boston, MA.
14. **Vallery-Radot R.** 1923. p 413–444. *In The Life of Pasteur.* Doubleday, Page & Co, New York, NY.
15. **Dufour HD**, **Carroll SB.** 2013. History: great myths die hard. *Nature* **502:**32–33. http://dx.doi.org/10.1038/502032a.
16. **Geison GL.** 1995. *The Private Science of Louis Pasteur.* Princeton University Press, Princeton, NJ.
17. **Martínez-Palomo A.** 2001. The science of Louis Pasteur: a reconsideration. *Q Rev Biol* **76:**37–45. http://dx.doi.org/10.1086/393744.
18. **Montagnier L.** 1995. Pasteur's legacy. *Am J Med* **99**(6A)**:**4S–5S. http://dx.doi.org/10.1016/S0002-9343(99)80277-0.

10 Robert Koch and the Rise of Bacteriology

As Pasteur introduced the concept of the germ theory of disease, the medical community was slow to embrace it. Some doctors had trouble believing that these invisible creatures could take down human beings millions of times their size. But even the most astute physicians had difficulty recognizing how microscopic organisms could cause different clinical syndromes, since there was little means to distinguish one microorganism from another. But that changed with Robert Koch (Fig. 10.1).

Robert Koch's life can be measured in extremes. Koch was a kind country doctor who did not perform formal medical research until he was 37 years old, but in the subsequent 9 years he made spectacular contributions that earned him an appointment as a professor at the University of Berlin. To strangers he was suspicious and aloof, but to his friends he was described as warm and friendly. He appeared to many to be stereotypically German in character, militaristically ordering many assistants who were at his bidding. But Koch regularly savored performing his own complex research tasks, even in old age. He was meticulous in his research, but after one of his most valuable discoveries, that of the tuberculosis bacillus, he released information suggesting that he had found the cure for tuberculosis before carefully seeing it through his usual rigor of testing. Koch is noted for his postulates that still serve as a guide for determining if a microorganism is the cause of a disease. But some of his most important contributions were in public health, where he receives little credit. Koch made one of the most important contributions in the history of immunology, the discovery of the tuberculin

Germ Theory: Medical Pioneers in Infectious Diseases, Second Edition. Robert P. Gaynes.
© 2023 American Society for Microbiology.

FIGURE 10.1 *Robert Koch. Courtesy of the National Library of Medicine (NLM image ID B016691).*

reaction, but never fully understood the significance of his discovery. For his work, Koch reached dazzling fame, yet a few years later, he was forced to resign his position in disgrace. Koch's life and achievements are both inspirational and cautionary.

ROBERT KOCH: EARLY INFLUENCES

Heinrich Herrmann Robert Koch was born on 11 December 1843 in Clausthal-Zellerfeld, a village in Germany. He was the third oldest of 13 children born to Herrmann and Mathilde Koch. His father was a mining official who had an understandable problem supporting such a large family. Young Robert showed sharp intellect at an early age, astonishing his parents with his self-taught reading ability at age 5. His first toy, a magnifying glass, may have been a predictor of his future path. As Robert grew, he began to show a strong interest in nature and biology. In school, he showed a strong aptitude for mathematics and sciences. He did not do as well in his study of languages, though he did better with English (he eventually became fluent) than French. Koch attended the University of Göttingen when he was 19 in 1862. The university had several well-known professors, but one of the most influential to Koch was Jakob Henle. Henle was a pathologist, but 18 years earlier had predicted that specific microscopic agents would be consistently found to cause certain diseases. Even though Henle did not have the methods at his disposal to prove his assertion, he influenced the very individual who would provide these methods to medical science. While at university, Koch carried out an anatomical study of the nerves in the ganglia of the uterus for which he won a monetary prize. He used the money to travel to Hannover, Germany, to attend a conference where he had a chance to interact with Rudolf Virchow, at that point the most famous physician in Germany. Koch published an additional paper before he left the university. By age 23 he passed all his examinations and graduated with his medical degree. He spent an extra 3 months of study in Berlin to attend lectures by Virchow. Despite his two publications, Koch's road to research would have to wait an additional 14 years.

In March 1866, Koch passed the state examinations and was "licensed" to practice medicine. His initial desire was to travel as a ship's doctor, but this penchant for travel was curtailed when he became engaged to Emmy Fraatz. His first position was in Hamburg General Hospital as a medical assistant, where he carried out microscopic examinations of pathological tissues. In Hamburg, Koch first became acquainted with cholera patients during an outbreak of the disease in the city. However, his impending marriage to Emmy forced Koch to choose a post that would provide enough income to support his wife and himself. In October 1866, Koch took a position in the tiny village of Langenhagen, near Hannover. Koch worked at an institute that educated and cared for retarded children but also opened his own private practice. He quickly became successful and popular. Unfortunately, after 2 years of a comfortable position at the institution in Langenhagen, the budgetary problems at the institute forced Koch out of that job. Robert Koch and his wife, now pregnant, moved near her home in Clausthal,

Germany. Koch had difficulties setting up a medical practice there. After his daughter, Gertrud, was born, Koch eventually found a village called Rakwitz, in eastern Germany, for a private practice. With the outbreak of the Franco-Prussian War in 1870, Koch volunteered to be a physician in a battlefield hospital, where he became acquainted with wound infections and typhoid fever. The exposures to infectious diseases, patients with cholera in Hamburg, and typhoid fever and wound infections during the war fascinated Koch, enriching the mundane experience in his private practice.

After the war, Koch returned to Rakwitz. To supplement his income from private practice he took an exam to be a district medical officer, passing it in 1872. Koch and his family moved to Wöllstein, Germany, when he was 29 years old. The village of Wöllstein had a population of 3,000 and was surrounded by forests and farms in what is now Poland. Koch's research interests began to show themselves when, despite a meager income, he purchased a microscope and created a laboratory in his home. He became a popular doctor among the inhabitants of the town, with a flourishing practice, but he had responsibilities in his role as district medical officer, too. He administered smallpox vaccinations, gave public health advice, and oversaw the local hospital. The lure of research crept into Koch's busy life during his years in Wöllstein.

THE DISCOVERY OF ANTHRAX SPORES

Koch began to study anthrax, a public health problem among livestock in the area around Wöllstein, just as it was in France. As early as 1873, Koch had made initial microscopic examinations of the blood of infected sheep. Anthrax bacteria are large and were relatively easy for Koch to see with his microscope. In 1874, Koch wrote his first observations that the anthrax bacteria might form spores:

> The bacteria swell up, become thicker, and much longer. Slight bends develop. Gradually, a thick felt develops. Within the long cells, cross walls appear, and small transparent points develop at regular intervals (1).

These are the first recorded words on endospores in history. After a trip to the research centers in Germany in 1875, Koch worked tirelessly, putting his practice, his work as district medical officer, and even his family behind efforts to cultivate the bacteria. On 23 December 1875, Koch was called to examine an animal that had died of anthrax. He removed some blood from the animal and inoculated it into a rabbit. The rabbit quickly exhibited illness and died on Christmas Eve. After quickly saving some tissue specimens from the rabbit, he returned to work without the interruption of patients on Christmas Day. He examined the tissues and found them teeming with bacteria. Koch took samples of the tissues and injected them into surgically made slits in the cornea of another rabbit. After several days, the rabbit died. Koch noted bacteria present in the rabbit's blood, spleen, and the fluid behind the cornea, or aqueous humor. An idea popped into Koch's mind—use the aqueous humor fluid from a rabbit's eye to artificially culture anthrax bacteria outside the animal. It worked! Koch rapidly determined the best growing conditions for the bacteria—a temperature of 30 to 35°C and the requirement for oxygen. He began using slide cultures, which could be sealed with paraffin to keep them from drying. As Koch began examining his slide cultures using aqueous humor of the rabbit's eye as a growth medium, he noted rows of long filaments containing spheres that refracted light. Even after drying when the filaments disappeared, Koch noted that the spheres remained in rows. In a matter of a few weeks, working in his house, Koch had (i) developed an artificial means to cultivate the anthrax bacterium; (ii) passed the bacterium through successive cycles *in vitro* and *in vivo*, strongly suggesting that this was the causative organism of anthrax; (iii) determined the optimal growing conditions for the bacterium; and (iv) determined that it was capable of forming spores under more adverse conditions. The French researcher Casmir Davaine had shown that animals could contract anthrax without contact with other diseased animals but from the soil. Koch had supplied the critical finding, spore formation, to explain the phenomenon. In Koch's own words,

> The bacteria form spores which possess the property of growing into new bacteria after longer or shorting resting states. All my further experiments were directed to discovering this suspected developmental stage of the anthrax bacillus. After some fruitless experiments, it was possible finally to achieve this goal and therefore to determine the true etiology of anthrax (2).

Koch's first paper was titled "The Etiology of Anthrax, Based on the Life Cycle of *Bacillus anthracis*." Note the use of the words, etiology and life cycle. Koch had done much to establish that the disease, anthrax, had a specific bacterial cause and showed that the life cycle of the bacteria had much to do with the observed phenomenon that animals could acquire the disease from the soil. Koch eventually developed an animal model for anthrax using mice. He injected them with the anthrax bacterium at the base of their tail. He found that their spleens would become markedly swollen with large numbers of the bacteria that were not spore formers. The disease was reproducible for multiple series of mice, and the pathological state and the bacteria never seemed to change. Koch was not the first to develop animal models for infectious diseases, but he was the first to use animal models to incorporate other experimental methodologies, as we shall see.

The remarkable findings on anthrax were obtained in nearly total isolation from the scientific community. Koch began to have doubt about his own findings. Fortunately for Koch, one of the period's leading researchers on bacteria, Ferdinand Cohn, was close by at the Institute of Plant Physiology at the University of Breslau. Koch wrote to Cohn, asking to demonstrate his findings at the university. Soon, Koch traveled to the university with his equipment and cultures. He set up an experiment with a mouse that had died from anthrax and used his slide cultures to demonstrate the filamentous bacilli and the formation of spores. Cohn had recently described spores for another bacillus, *Bacillus subtilis*, and was quite excited by Koch's demonstration. Cohn recognized the medical significance of the work and called the director of the Institute of Pathology at the University of Breslau, Julius Cohnheim, a disciple of Virchow. Both Cohn and Cohnheim were most impressed with Koch and his work and were to become Koch's ardent supporters as he and his work were introduced to the scientific community. Cohnheim later told his assistants,

> He has done everything himself and with absolute completeness. There is nothing more to be done. I regard this as the greatest discovery ever made with bacteria and I believe that this is not the last time that this young Robert Koch will surprise and shame us by the brilliance of his investigations (3).

After Koch returned home, he continued his correspondence with Cohn, who invited Koch to publish in Cohn's journal, *Beiträge zur Biologie der Pflanzen*. Koch excitedly agreed. Cohn even helped Koch with his figure for "The Etiology of Anthrax, Based on the Life Cycle of *Bacillus anthracis*." The paper was published in 1876 and brought Koch some acclaim, although it took a few years.

IMPROVEMENTS IN MICROSCOPY

A vexing problem for Koch when he worked in his meager laboratory involved visualization of bacteria. Anthrax bacteria were large and relatively easy to see. Koch wanted to obtain better clarity of his microscopic organisms and wanted to take photographs of the bacteria. Koch's desire to view bacteria came at the beginning of a revolution in microscopy. From 1876 to 1879, Koch helped to adapt the light microscope for studying bacteria. The innovations included the use of the oil immersion lens and the Abbe condenser. Koch did not invent these advances, but his continual curiosity pushed him to seek out these improvements. The oil immersion lens increased the resolution of microscopic images. Ernst Abbe was a scientist and consultant to the Carl Zeiss Microscope Company. That company was instrumental in developing a condenser, known as the Abbe condenser, to provide optimal illumination of images such as those of bacteria. Koch obtained one of the first oil immersion lenses and an Abbe condenser for his microscope. He also adapted his microscope to obtain photographs of bacteria. During this period, Koch perfected methods for preparing, staining, imaging, and photographing bacteria. He published his methods in an 1877 paper in Cohn's journal. The images from that article are remarkably clear and would be considered for publication by contemporary journals. Koch was moving to the forefront in a new field of science, bacteriology.

Scientists of the 1870s were still at odds as to whether microorganisms seen under the microscope were one species or many species. It is difficult for the modern-day scientist to fathom that notion, but Koch and Cohn were on one side of the controversy. Other scientists, such as Carl von Nägeli, fell on the other side of the controversy: von Nägeli, a Swiss botanist, published a claim that he had examined thousands of different organisms and was completely unable to distinguish one from another. Koch was infuriated by von Nägeli's publication and vowed to set the story straight. Koch thoroughly examined blood and tissues of normal animals for the presence of bacteria and could not find them. But diseased tissues, particularly traumatized tissue from wounds such as those he saw from battlefields that carried pus, were different. Koch was familiar with Joseph Lister's work that implicated bacteria in wound infections. But Lister had not tried to cultivate and determine the species of bacteria present. Koch examined diseased tissues for bacteria, but the methods that he used for anthrax were not appropriate for many bacteria. Koch's connection to the University of Breslau proved valuable. Karl Weigert had discovered that bacteria could be seen with aniline dyes in the tissues of animals. Bacteria took up the dye where artifacts and normal animal tissue did not. Koch used that information together with his improved microscope. In July 1878, Koch described what he saw:

> [The oil-immersion lens] completely altered the pictures. In the same slides, which had previously shown nothing, the smallest bacteria are now visible with such clarity and definition that they are very easy to see and to distinguish from other colored objects . . . now we can see bacteria which can be differentiated by size and shape (4).

Koch set up animal experiments with mice, injecting them with small doses of anthrax-diseased animal material. He made successive passages from one animal to another, carefully noting that death would come about 50 h after the injection, always with the same symptoms. Koch then turned to cultivating specific bacteria, attempting to prove that each pathogenic bacterium represented a distinct species. Koch had to rely on animals for bacterial cultivation, as he had not perfected any plate or slide cultures, but he was able to cultivate them using the animals. He also recognized an important aspect of animal cultivation—the "law" of increasing virulence. This law suggested that animal passage might select the more pathogenic bacteria with each passage and that the selected ones might be more virulent than the original.

KOCH'S MOVE TO BERLIN

As time in Wöllstein passed, Koch realized that working as a lone researcher was not going to remain satisfying. Cohn had pressed the faculty at the University of Breslau to create a position for Koch. Koch briefly took a position at the University of Breslau, but the position did not materialize in a manner that suited Koch, and he returned to Wöllstein. Following the unification of the German states by Otto von Bismark in 1871, the country needed a central office for public health. The Imperial Health Office was established in Berlin a few years later. The Office was under the guidance of an advisory council. Ferdinand Cohn, a council member, helped Koch with an appointment to the council while he was in Wöllstein. By March 1880, the Director of the Imperial Health Office began to work for a position for Koch in Berlin. Finally, in July 1880, Koch had a salaried position in the Imperial Health Office in Germany's capital city. Koch was 37 years old. Koch's title in the Imperial Health Office was essentially a Government Councilor. The Office had little room for Koch, but he set out to work in a small laboratory with two assistants, Georg Gaffky and Friedrich Loeffler. Koch performed some of the most important research of his career in this laboratory, working tirelessly and inspiring his assistants and coworkers with his work ethic and enthusiasm.

Koch's early work in Berlin may seem mundane and simple to us today, but Koch's premise, "the pure culture is the foundation of all research in infectious disease," remains true and was the basis of his famous postulates. This simple premise constitutes one of the greatest contributions of anyone in the history of infectious diseases. As great as Pasteur's innovative work was, his methods of cultivation of bacteria were in animals or in liquids and broths. The only method to determine if the broth was a pure culture was to examine a small portion under the microscope. Pasteur's methods enriched the bacteria of interest but could not ensure that one was dealing with a pure culture. This uncertainty became a point of contention when Koch commented on Pasteur's conclusions on anthrax.

THE DEVELOPMENT OF PURE BACTERIAL CULTURES

Koch developed the means to definitively cultivate pure cultures without animal hosts. The approach was another instance of serendipity. One day, Koch entered the laboratory and found a portion of a boiled potato that someone had left in the lab. As he was about to throw it out, Koch noticed differently colored spots on the surface of the potato. Curious about the nature of these spots, he examined them under a microscope. He found that the differently colored spots represented different types of bacteria but that each spot was entirely pure (3). Koch and his colleagues took the concept (not all bacteria will grow on a potato) and succeeded in developing a gelatin beef broth semisolid medium that could be used to grow all types of bacteria. One of his assistants, R. J. Petri, even lent his name to the petri dish, which is still used today. In 1882, another Koch assistant, Walther Hesse, introduced agar as a solidifying agent that would not melt at temperatures needed for incubation of bacteria. The 1881 paper "Methods for the Study of Pathogenic Organisms," published in *Mittheilungen aus dem Kaiserlichen Gesundheitsamte*, remains one of the most influential papers, if not the most influential, in the history of bacteriology (5). Koch's approach was so simple, so reproducible, and so effective in isolating pure cultures that over the next 20 years, Koch and/or one of his colleagues isolated the causative microorganisms of most of the major bacterial diseases affecting humankind. Koch traveled to London in the summer of 1881 to demonstrate his plate technique for obtaining pure cultures. Little record is left to us of a meeting that included Joseph Lister, Louis Pasteur at the height of his fame, and Robert Koch. Koch did not present a formal paper but demonstrated the plate technique at King's College in London, where Pasteur is said to have taken Koch's hand and said, "C'est un grand progrès, Monsieur [This is great progress, sir]" (6). This meeting would be the most civil of all the face-to-face meetings between these two great men. Within a year, the two men would be in open conflict. Koch was 38 years old and a relative unknown at the time, but his fame was about to skyrocket with the discovery of the tubercle bacillus.

THE DISCOVERY OF THE TUBERCLE BACILLUS

The discovery of the causative agent of tuberculosis was, as the *New York Times* would later describe, "one of the great scientific discoveries of the age" (7). Tuberculosis was one of the greatest scourges of humankind. Evidence of its effects can be traced back to antiquity. During the mid-1800s, tuberculosis was the cause of one-seventh of all deaths and nearly one in three deaths among adults. Pulmonary tuberculosis, also called phthisis, was well described by the Hippocratic physicians. Through the centuries, another form of tuberculosis, miliary tuberculosis, was recognized. Pulmonary tuberculosis affects the lungs, causing a chronic, productive cough and a wasting or consumptive picture of the affected person, whereas miliary tuberculosis is a condition where many organs of the body are affected, with small lesions the size of millet seeds. There was great controversy over whether these two clinical conditions were part of the same spectrum of illness or two distinct diseases. Rudolf Virchow was convinced that these were two distinct diseases. René Laennec believed that these two clinical entities were part of the same disease process. The difficulty in settling the debate was that no one had demonstrated the causative organism, although many investigators had ardently tried. As Koch was about to discover, the causative organism, *Mycobacterium tuberculosis*, was (and still is) difficult to grow and stain, requiring demanding patience. Such patience, which must have required more faith from Koch rather than simply patience, made its discovery by Koch the crowning achievement of his lifetime. Here is how he did it.

About 2 weeks after returning from London in 1881, Koch began his work to isolate the causative organism of tuberculosis, but he was not starting from scratch. Jean Villemin, a French physician, had shown that tuberculosis could be transmitted to experimental animals in 1865, a finding confirmed by others. No one seemed to be able to isolate or even visualize a bacterium from tissues. Koch began his search by inoculating guinea pigs with tuberculous material from several different sources. Guinea pigs developed tuberculosis experimentally, permitting investigators to remove affected tissues that had cheesy, so-called caseous, changes that were typical of tuberculosis in humans, in whom it was so often fatal. Koch took some of this material and streaked it on slides to look for microorganisms. He used various dyes, including methylene blue, which demonstrated very tiny thin rods. Koch found these rods only in tuberculous materials, not in control specimens. Koch remained unsure that these rods were responsible for the disease. He next tried counterstaining the tissues using a brown dye called vesuvin. Koch could now clearly see blue rods in the preparations where he used methylene blue staining followed by vesuvin counterstaining, where everything else was brown. Other bacteria did not hold the blue stain like the rods from tuberculosis-infected guinea pigs. Koch experimented with this procedure, eventually determining that the methylene blue dye was taken up more avidly when the solution was alkaline, either by adding sodium or potassium hydroxide or aniline. The definitive staining of this organism from tuberculous tissues began to unlock the mystery of tuberculosis. Koch had quickly recognized the unusual properties of the bacilli that made them stain differently than other bacteria.

> It seems likely that the tubercle bacillus is surrounded with a special wall of unusual properties, and that the penetration of a dye through this wall can only occur when alkali, aniline, or a similar substance is present (8).

Koch also noted that while these bacteria stained differently than most other bacteria, the staining of the tubercle bacteria was like the staining of the bacteria that were associated with leprosy. Koch made extensive observations of both animal and human materials and was able to show that the tubercle bacillus was constantly present in infected tissues, usually in large numbers. The presence of the tubercle bacillus was not enough for Koch to conclude that it was the causative agent.

In his 1882 paper "The Etiology of Tuberculosis," Koch reasoned,

> In order to prove that tuberculosis is brought about through the penetration of the bacilli, and is a definite parasitic disease brought about by the growth and reproduction of those same bacilli, the bacilli must be isolated from the body, and cultured so long in pure culture, that they are freed from any diseased production of the animal organism which may still be adhering to the bacilli. After this, the isolated bacilli must bring about the transfer of the disease to other animals and cause the same disease picture which can be brought about through the inoculation of healthy animals with naturally developing tubercle materials (9).

This passage is Koch's first presentation of what would become "Koch's postulates." The postulates would be published 2 years later as a formalization of Koch's approach to determining the causative agent of a disease. The difficulties in working with the tubercle bacillus forced Koch to be deliberate in his methods and conclusions. So, Koch's next step was to culture the organism. Koch had developed the plate technique, but the tubercle bacillus was a fastidious organism. He tried a variety of media, eventually settling on coagulated blood serum. Instead of plates, he placed the nutrient material in test tubes placed on a slant to increase the surface area for growth of the bacilli. After removing tuberculous tissues from animals, Koch inoculated the slants. Koch must have been very certain that he was going to find the microscopic culprit, since he kept incubating and examining daily the slants even though nothing was seen following 1 week of incubation. Finally, after 2 weeks, tiny colonies appeared. He stained them, finding the characteristic blue bacilli using the staining technique he developed. Now for the test of virulence of these pure cultures. Were they going to produce disease? They did. Guinea pigs inoculated with these pure cultures of the tubercle bacilli developed the same symptoms and pathology as animals that had been inoculated with tissue from human cases. Koch also isolated the tubercle bacillus from these guinea pigs.

After only 8 months of investigation, Koch was ready to present his findings to the world on the etiology of one of the most important diseases of humankind, tuberculosis. He chose the Berlin Physiological Society to give a lecture, vaguely titled "On Tuberculosis," in March 1882. Koch had misgivings about the reception that he would receive, stating that he believed that it might take a year for the medical profession to accept his findings. He was wrong. After the lecture there was a stunned silence. Paul Ehrlich was quoted as saying, "I hold that evening to be the most important experience of my scientific life" (8). The enthusiastic reception after his lecture generated excitement in both the public and the scientific community. News of Koch's discovery quickly spread to England and then to the United States and the rest of the world. Within 2 months, not a year, the world knew the name Robert Koch.

Even Rudolf Virchow, who was skeptical of the germ theory in general, realized that Koch had found something of great significance. With the staining technique that Koch described, a unified model for tuberculosis, including miliary tuberculosis, could be constructed, since the organism could now be seen in all of the affected tissues. Two German physicians, Franz Ziehl and Friedrich Neelsen, improved on the basic staining technique that Koch began. The Ziehl-Neelsen staining procedure, developed 1 year after Koch's lecture, is still used to identify mycobacteria today. Within 3 months of its discovery, Koch began to receive recognition, first from Germany, including a salary increase and more research support, and then from the rest of world in the form of fame. Visitors traveled from all over the globe to his laboratory to learn and consult. In May 1883 the German Exposition of Hygiene and Public Health was held in Berlin. Koch presented his laboratory techniques for plating pure cultures, disinfection, and staining. The Exposition carried Koch's name and, more importantly, his techniques across the globe. Koch's techniques were a break with the past. His methods were simple and transparent. Importantly, they could be easily repeated in other laboratories, in marked contrast to those of Pasteur, who often kept his methods secret, even from his laboratory assistants.

KOCH'S POSTULATES

An 1884 Koch publication titled "The Etiology of Tuberculosis" contained a more expansive explanation of his work on isolating the tubercle bacillus but is best remembered as the formal presentation of what we now call Koch's postulates (10). The postulates are known to most students who study microbiology as the guide to determination that a microorganism is the cause of a disease.

> The putative organism must be constantly present in diseased tissue.
> The organism must be isolated in pure culture.
> The pure culture must induce disease when injected into experimental animals.
> The same organism must be isolated from these diseased animals.

The postulates are credited to Koch, but history shows that he was not the first person to develop the concepts (11). Koch's teacher, Jakob Henle, first proposed the guiding hypothesis but never had the means to actually test them. Later, Edwin Klebs, in 1877 and again in 1878, lectured on the criteria for causality from microorganisms (12). Oddly, Koch did not rigorously adhere to the postulates in his anthrax work and could not adhere to them in his next triumph, human cholera, since humans are the only hosts, and no experimental animal model exists.

THE DISCOVERY OF THE CAUSATIVE AGENT OF CHOLERA

Fresh from his achievement on the etiology of tuberculosis and with an improved salary, laboratory, and prestige, Koch was eager to test his newfound success. The chance came quickly. In 1883, a cholera epidemic hit Egypt. Concern across the continent of Europe flared; the worry was that it might begin in European cities. Koch remembered the Hamburg epidemic in 1866. He believed that he had developed the techniques to contain the devastating illness and, quite possibly, find the etiologic agent.

A small port city, Damiette, in Egypt became the first city to announce the outbreak in 1883, but soon cholera was spreading throughout the country. Egyptian officials contacted France and Germany for help. Both countries sent delegations. In a way it was a national competition. Pasteur directed the French delegation in absentia. His young but trusted assistant, Louis Thuillier, and Émile Roux were on the ground in Egypt 1 day earlier than the German delegation. Koch, Georg Gaffky, Bernhard Fischer, and Hermann Treskow formed the German Commission, as it was known. The German team came well equipped to study cholera. A number of inoculations to animals failed to produce an animal model. Cultures of patients' blood, liver, spleen, or liver also failed to produce any bacterial colonies. Koch had dissected a number of victims of the disease and found a characteristic organism only in the intestines. As Koch described the curious microorganism, it was not as a straight rod, but a little bent, like a comma. The bending could be so great that the organisms resembled half-circles. But Koch could not be certain that this strangely shaped bacterium was the cause of the horrible disease. There were masses of bacteria in the intestine. Although this comma-shaped bacillus was noted only in cholera victims, for Koch, the answer was still not clear. Unfortunately for Koch, the outbreak disappeared in Alexandria, Egypt, where the German Commission was working. Koch had no fresh material for culture. He did, however, track the disease through epidemiological studies to ascertain the effectiveness of quarantine measures. He correlated the disease incidence with the rise and fall of the Nile waters and made studies on the relationship of cholera to the water supply and with meteorological conditions.

While the German Commission had some success in Egypt, the French delegation met with disaster. About the time the epidemic died down, the French began to study another disease affecting cattle, rinderpest, which was ultimately determined to be caused by a virus. Two weeks into the investigation and away from any known cases of cholera, Louis Thuillier became ill and died from cholera. The French, including Pasteur, were devastated. Koch and his German colleagues went to the funeral. Koch even helped to carry the coffin. The French team, disillusioned and unsuccessful, returned to France.

The German Commission headed to India, where cholera was still present. Koch, Gaffky, and Fischer (Treskow had returned to Berlin) arrived in Calcutta, India, on 11 December 1883—Robert Koch's 40th birthday. With the help of the British, the German Commission set up their laboratory in the Medical College Hospital. Because the German Commission was a government function, Koch had to write a number of reports to Berlin, so we have extensive documentation of the German Commission. Fresh samples from cholera victims were the key to Koch's success in cultivating the curved bacteria he had seen in Egypt. In less than 2 months in India, Koch described his findings:

> It can now be taken as conclusive that the bacillus found in the intestine of cholera patients is indeed the cholera pathogen. . . . In pure culture these bent rods may even be S-shaped . . . They are very actively motile, a property which can best be seen when examining a drop of liquid culture attached to a cover slip. . . . Another important characteristic is the behavior of the bacteria in nutrient gelatin. Colonies are formed which at first appear compact but gradually spread out as the gelatin is liquefied. In gelatin cultures, colonies of the cholera bacillus can therefore be readily distinguished from colonies of other bacteria, making isolation into pure culture easy (13).

Koch concluded that after finding only the comma-shaped bacillus in the cholera victims, together with his Egyptian findings, he had found the pathogen responsible for cholera. The German Commission did further epidemiological investigations that implicated drinking water as the vehicle for cholera. These findings had important public health implications, even though Koch was unaware of a similar conclusion drawn by John Snow several decades earlier. While there was some skepticism about their findings, including from Pasteur, the vast majority of scientists accepted Koch's findings. After 8 months away, Koch, Gaffky, and Fischer returned to Germany in April 1884 to a heroes' welcome. Koch received a medal from the Kaiser. Even Rudolf Virchow, who had been cool to Koch for most of his career up until the cholera expedition, accepted his findings, giving Koch full credit for the work.

THE RIVALRY BETWEEN KOCH AND PASTEUR

The efforts of Pasteur and Koch defined the germ theory of disease in medicine. But their relationship was not one simply of scientific colleagues, or even rivals. For about a decade, the word that can best describe the relationship between Pasteur and Koch is hate. For reasons on every possible level, they detested each other. The two did not hide their contempt of each other. In fact, their conflict was carried on in the open with letters and publications for all the world to read. After a cordial meeting in London in 1881, how did things turn so acrimonious? The bitterness began after Pasteur presented some of his work on vaccination with the anthrax bacillus. Pasteur even made a passing favorable comment on Koch's discovery of the spore formation of the anthrax bacillus but generally ignored Koch's work. No one knows for certain, but Koch may have been somewhat offended by Pasteur's lack of comment on his work. In the issue that included Koch's paper "Methods for the Study of Pathogenic Organisms" in *Mittheilungen aus dem Kaiserlichen Gesundheitsamte*, Koch, Gaffky, and Loeffler had other papers that assailed Pasteur and his methods. Koch was particularly ferocious, accusing Pasteur of making errors and especially for having impure cultures, since Pasteur did not use Koch's plating technique. Koch went on,

> Of these conclusions of Pasteur on the etiology of anthrax there is little which is new, and that which is new is erroneous . . . Up to now, Pasteur's work on anthrax has led to nothing (14)

There was some scientific basis for disagreeing with Pasteur's method of attenuation of the anthrax bacilli. First, Koch argued that Pasteur had no way of knowing if he was dealing with a pure culture of anthrax. Second, Pasteur recognized that anthrax bacilli that formed spores would be deleterious to his attenuation since they not only would not attenuate with his methods but also, once introduced into the animal, could actually produce the anthrax disease. Pasteur contended that with his method, growing the organism at 43°C, spore formation did not occur. Koch found that spores were evident at that temperature. Since Pasteur did not use plate cultures, the significance of Koch's findings is uncertain but surely did not warrant Koch's tirade. Pasteur was 20 years older than Koch, who was just beginning his career. Pasteur could easily have ignored Koch, but he did not. Pasteur was used to forcefully silencing critics with his experiments. At first, he attempted to respond to Koch by sending his assistant, Louis Thuillier, to Germany to run some experiments. Thuillier wrote to Pasteur in 1882 before the cholera expedition, making it clear to his superior that Koch viewed the entire affair as a direct competition. Pasteur decided to respond to Koch's paper at a meeting that was held in Geneva, Switzerland, in September 1882. Pasteur was scheduled to deliver a paper at the meeting, the IVth International Congress of Hygiene and Demography. Pasteur spoke only French. Koch knew some French, but his language skills were less than perfect. Koch attended Pasteur's speech. One should remember that Pasteur was near the height of his fame. All of Europe was listening to his every word. Koch, however, was fresh from his notoriety for the discovery of the tubercle bacilli. As Pasteur attempted to defend his experimental practices, Koch did the unthinkable. In the midst of Pasteur's speech, Koch stood and attempted to

interrupt him. Pasteur, who did not understand and was incensed by the disturbance, angrily silenced Koch. A shocked audience buzzed. When Pasteur was fin-
ished, Koch was asked to respond. The response became the subject of record in
the proceedings of the conference.

> Professor R. Koch, of Berlin, took the podium and made the following speech, in
> German, which was immediately translated into French by M. Haltenhoff:
> "When I saw in the program of the Congress that M. Pasteur was to speak today
> on the attenuation of virus [note: the 19th-century use of the word virus was differ-
> ent than its contemporary meaning; it denoted any pathogenic microorganism],
> I attended the meeting eagerly, hoping to learn something new about this very inter-
> esting subject. I must confess that I have been disappointed, as there is nothing new
> in the speech, which M. Pasteur has just made. I do not believe it would be useful to
> respond here to the attacks which M. Pasteur has made on me, for two reasons: first,
> because the points of disagreement between Pasteur and myself relate only indi-
> rectly to the subject of hygiene, and second because I do not speak French well and
> M. Pasteur does not speak German at all, so that we are unable to engage in a fruitful
> discussion. I will reserve my response for the pages of the medical journals."
> [Applause].
> M. Pasteur responded to M. Koch that if he had been able to follow the lecture,
> he would have easily understood that new material was presented today. M. Pasteur
> awaits confidently the reply of M. Koch and will reserve the right to reply to him
> further at that time." (15)

THE MISTRANSLATION OF A WORD

The proceedings politely recorded the exchange but hid the profound animosity
that both scientists felt and exhibited. Why did Koch behave, as Émile Roux,
Pasteur's colleague, suggested, "like a fool"? The behavior was set off by a mistranslation, according to Mollaret (14). In 1925, Charles Ruel, an administrative assistant for the 1882 Congress, wrote a letter to Émile Roux, then Director of Institut
Pasteur. Ruel wrote,

> In the course of the remarkable and conscientious presentation, when he [Pasteur]
> listed and commented on appropriately and properly to the work of Koch and his
> School, he referred several times to the *German collected works (Recueil allemand)*.
> Now Koch and his friend, Prof. Lichtheim, were sitting side by side: they knew
> French only imperfectly and both mistook the word *pride (orgueil)* for *collection
> (recueil)*. They felt their self-respect profoundly wounded and interpreted the words
> *German pride* as a grave insult. Immediately, Doctor Koch at the instigation of his
> compatriot got up and tried to interrupt the orator in order to protest the term which

he regarded disrespectful. The assembly, ill at ease and amazed, witnessed this quarrel but without understanding. I have this explanation from Professor Lichtheim himself (14).

The word *orgueil* is generously translated as "pride." Actually, it is more closely translated as "arrogance," which better explains Koch's reaction to Pasteur's words. Neither Koch nor Pasteur was aware of this miscue at the time. The exchanges set off a lengthy volley of published insults from Koch to Pasteur and back again. Three months after the Congress, Koch published a vicious attack on Pasteur. Koch criticized Pasteur's inclusion of what he thought to be irrelevant topics in his speech at a congress on hygiene. Koch went on to write,

> In particular, the less as Pasteur's polemic was not directed to defeat me by real proof, but by general phrases and to a major part personally direct against me in an angry tone . . . As a result of his poor methods, Pasteur drifted off course immediately the moment he started to answer a new question on the contagion of rabies (14).

Koch became even more personal, chiding Pasteur for not being a physician. In his conclusion, he could not resist one final swipe:

> When Pasteur was celebrated as the second Jenner at the Congress in Geneva, this occurred slightly prematurely. Obviously, in the desire to be enthusiastic, it was forgotten that Jenner's beneficial discovery was not in sheep but in humans (14).

Pasteur wasted little time responding in a tone just as acerbic as Koch's. In January 1882, Pasteur published the following:

> This is another mistake on your part . . . The day you would like to be informed on this point and on all the preceding points, I will be to your disposition during a congress or a commission where you can designate the members. If you accept . . . You may not be able to sustain the tone of assurance reflected in your brochure . . . You, Sir, who entered in Science in 1876 only after all the famous names that I just mentioned, can recognize without derogation that you are a debtor of French Science . . . There are in your brochure numerous sections where the impertinence or mistake, the way Pascal would say it, "is really too much." (15)

The rancorous disagreement between these two giants can be understood not only by considering the mistranslation of words but also by their differences on nearly every level imaginable. The 19th century was a time of intense nationalism. No two countries exhibited nationalism more passionately than France and Germany.

For example, the two countries disagreed over anthrax findings that actually extended back to a time before Koch and Pasteur were involved. In the 1850s, German scientists like Franz Pollender disputed French discoveries on anthrax by Davaine and Pierre Rayer. Pasteur detested Germany, especially after the 1870 Franco-Prussian War. Pasteur returned his honorary degree from the University of Bonn in 1871, describing the sight of it as "odious." Koch was fiercely anti-French going back to his days volunteering as a physician to help in the 1870 war with France. There were scientific differences between the men, too. The initial shot across the bow in the debate was over whether the anthrax bacillus could form spores at a certain temperature. On the subject of attenuation of the anthrax, the matter was of little scientific significance. But the French school and the German school had serious and important differences regarding the issue of attenuation. Koch's postulates themselves demand a rigid conservation of bacterial form. If the bacterium or its virulence factor changes in the process, Koch's postulates will not be satisfied. Pasteur's continual attempts to find vaccines demanded that the investigator find some way for the microorganism to change to produce immunity in a host but not disease. We know today that both aspects of microbiology and immunology hold truth: bacteria tend to conserve their form but also can attenuate. When the field was at its birth, it is understandable how these scientists could believe that the two features of microorganisms could not simultaneously be present. On a personal level, both men seemed destined for dispute. Pasteur was harsh on his critics, with a need to be proven correct. He could be prone to fits of anger. Koch was authoritarian, aloof, and easily offended. Neither man could tolerate being told they were wrong, and each would go to great lengths to prove their positions. Finally, the language barrier may have added fuel to this burning fire. Neither man could understand the other's publications, including the nuances contained within the prose.

After Koch returned from the cholera expedition, where the German Commission returned in triumph and the French came back in disastrous failure, Koch visited France during a French outbreak of the disease around Toulon in 1884. Since Koch was afraid to bring the cholera pathogen back from India in fear that it might cause disease in Europe, the isolation of cholera in France was the first time it had been cultivated in Europe. Pasteur rigidly opposed Koch's presence in his country and discounted his efforts. It is likely that the feud between the two men caused a delay in French applications of Koch's plate culture method. In Germany, Koch initially opposed the use of Pasteur's rabies vaccine, although public opinion forced Koch's hand. The dispute lessened over the years. Pasteur sent a telegram to Koch congratulating him for his discovery of tuberculin in 1890. But the ill will never entirely disappeared. On the celebration of Pasteur's 70th birthday in 1892, Koch was noticeably absent from the esteemed collection of European scientists that attended the event. The feud may not have had any lasting effects on microbiology and medicine, but the nationalistic, scientific, and personal views of these two pioneers give insight into the roles that personality and ego can play in science.

THE INSTITUTE OF HYGIENE

In 1885, the University of Berlin established a new Institute of Hygiene, appointing Koch as its first director. Koch also received the title Professor of Hygiene. Together with a friend, Carl von Flügge of Göttingen, Koch helped to establish a new journal, *Zeitschrift für Hygiene* (*Journal for Hygiene*), where he and others could publish papers on the developing new science of bacteriology. Koch became frightfully busy with his professorial duties, teaching courses to the near exclusion of any research until around 1889. During this period, his daughter, to whom he was quite close, became engaged. Koch and his wife grew apart.

In mid-1889, Koch threw himself back into the laboratory and did not publicly emerge until August 1890 to make an announcement at the Tenth International Congress of Medicine in Berlin that would stun the world but would bring disgrace to Koch.

THE TUBERCULIN FIASCO

The months leading up to the Tenth International Congress of Medicine showed a secretive side of Robert Koch. He worked in the laboratory performing his own experiments, with only a laboratory helper to carry out dead guinea pigs. No one knew what he was doing. He was invited to give a talk at the conference simply titled "On Bacteriologic Research" that appeared to be intended to summarize the achievements of the last decade or so. But in front of a crowd of 8,000 scientists in Berlin on 3 August 1890, Koch made the following statement:

> I ultimately found substances that halted the growth of tuberculosis bacilli not only in test tubes but also in animal bodies. As everyone who experiments with tuberculosis finds, investigations of the disease are very slow; mine are no exception. Thus, although I have been occupied with these attempts for nearly one year, my study of these substances is not yet complete. *I can only communicate that guinea pigs, which are known to be particularly susceptible to tuberculosis, if subjected to the operation of such substances, no longer eat when infected with tuberculosis bacilli, and that in guinea pigs in which tuberculosis has already reached an advanced stage, the disease can be completely halted without otherwise harming the body.* At this time, I conclude that it is possible to render harmless the pathogenic bacteria that are found in a living body and to do this without disadvantage to the body (16).

A cure for tuberculosis?! What was this substance? Koch was not saying. He stated that more work was needed, but that was not what was reported. The entire knowledge that the scientific community had for this "tuberculosis cure" was contained in the brief statement above. Koch's prominence is the main reason that the announcement sent the scientific world, and soon the general public, into an uproar.

In the months following the uncharacteristically vague August 1890 announcement, Koch began testing the substance on humans. The source of patients was the Charité Hospital in Berlin, but the "treatments" were hardly undertaken as part of any clinical trial. Today, any claims of "cure" would be highly suspect. What did occur during these secretive months in Berlin? Much later Koch revealed that the substance, which became known as tuberculin, was actually an extract of virulent tubercle bacilli that were kept in a glycerin solution. Subcutaneous injection of tuberculin produced (and still produces) a type of immunologic reaction termed delayed-type hypersensitivity reaction. Contemporary immunologists know that delayed-type hypersensitivity reactions are among the most complex of all immune phenomena, elucidated only within the past 30 years. Koch had no idea what he was getting into but observed many reactions that he fervently believed were cures.

An injection of tuberculin (material from the tubercle bacillus) caused the host to exhibit a reaction after 24 to 48 hours. The reaction could be local or systemic, depending upon the amount of material injected and the immune status of the host. If a person had never encountered the tubercle bacillus before the injection, virtually no reaction was seen. If the person had been infected with the bacillus previously, the tuberculin would cause the joint near the site of injection to swell and become red and tender. Fevers would also occur. If the reaction was strong enough, the skin at the site of injection would die and slough off. Some patients would even become prostrate for several days. This reaction, which became known as the Koch phenomenon, has important diagnostic value, and is still used in skin testing of people to see if they have been previously exposed to the tubercle bacillus. However, several of Koch's colleagues presented information that suggested that it was curative (17).

Koch remained silent until November 1890, when he published a three-page paper in *Deutsche Medicinishe Wochenschrift*, which was reprinted in the *British Medical Journal* (18). In the paper, Koch clearly stated that the remedy did not kill the tubercle bacteria but, instead, caused the death of the tissue in which the tubercle bacteria lived. He stated that the remedy could not be taken orally but required injection under the skin, i.e., subcutaneously. Koch detailed the effect of injections in healthy individuals and the diagnostic value of the material. Koch described the dose for the commonly diagnosed tuberculous syndromes: tuberculosis of the bones and joints, pulmonary tuberculosis (also called phthisis), and advanced cases of phthisis. Koch described the human effects of tuberculin injection on patients who had a form of tuberculosis called lupus (not to be confused with the rheumatologic condition) in which a patient's skin was disfigured by tuberculosis, usually around the nose and ears:

> The effect of the injection upon the lupus tissue is to destroy it more or less thoroughly and cause it to disappear. In some parts, the dose may suffice to cause this directly, whereas in others the tissue rather melts or wastes away, requiring repeated injection of the remedy to complete the process (18).

Koch did not include any information on the identity, source, or preparation of the substance. He only described the physical character of the remedy as a brownish transparent liquid. Despite detailing information on human use of tuberculin, strongly suggesting that it cured tuberculosis or at least melted away the tissues containing the bacteria, Koch refused to divulge how tuberculin was prepared. Koch received severe criticism for not divulging the nature of the remedy. There is little evidence that Koch planned to receive or received monetary gain from tuberculin. Why was Koch so secretive? No one is sure, although Koch may simply have wanted to protect his remedy from fraudulent scientists or doctors. There was little control from government authorities on treatments in Germany or elsewhere. If the nature of the treatment were known, anyone could have made it. In the case of tuberculin, if it was unsuccessful, the blame would have fallen on Koch. Koch gave an interview with Sir Arthur Conan Doyle in which he claimed that his secrecy was out of a sense of duty and the "purest unselfishness" (19).

PUBLIC REACTION TO KOCH'S ANNOUNCEMENT

Like all good intentions, announcement of Koch's remedy had unanticipated effects. Patients and doctors flocked to Berlin in search of this treatment. As the *British Medical Journal* described,

> Medical men are flocking to Berlin from all parts of Europe; at the present about 1,500 have arrived, and it will be easily believed that consumptive [tuberculosis] patients of all classes clamor for treatment. Dr. Cornet, one of Koch's co-workers, has no fewer than eight temporary consulting rooms in various parts of the city, which are crowded night and day by patients, rich and poor, old and young, from such as have to be carried upstairs to those with only a slight cough (20).

The pressures exerted on Koch were tremendous. He asked to be relieved of his administrative duties at the Institute of Hygiene and resigned his post as professor at the University of Berlin. He requested that a new institute be created by the

German government to administer and control the remedy, similar to the manner in which Institut Pasteur, established in 1886, handled requests for the rabies vaccine. Koch would head the new institute, which would have a clinical department and a scientific research department. Such was the origin of the Institut für Infektionskrankheiten (Institute for Infectious Diseases). As the weeks went by, concern rose over the numbers of tuberculosis patients descending on Berlin. Emergency measures were soon needed, including regulations for disinfection not only for hospitals but also for hotels and boarding houses.

Enthusiasm for tuberculin was initially extreme. Notable physicians from around the world, including Joseph Lister, came back from Berlin thoroughly impressed. Even Pasteur sent a telegram of congratulations to Koch. The lay press heaped praise on the discovery. The *New York Herald* even warned American consumptives (as those with tuberculosis were often called) about traveling to Berlin due to the crowds and limited availability of the remedy. Koch provided small amounts of tuberculin to respected physicians in Germany and throughout the world. The amounts were not enough to carry out trials, at least outside Germany. Some critics contended that the distribution of tuberculin was more of a marketing ploy than a well-intentioned allocation of a precious treatment.

At the end of 1890 and into early 1891, reports began to circulate that tuberculin was not the panacea for tuberculosis. In January 1891, Koch began to release information about the nature of tuberculin, "a solution of glycerin and extracts from tubercle bacilli cultures" (16). The release of this information took the secrecy away from tuberculin but may have also taken away some of its allure. The breaking point came in early 1891 when a report in *Klinisches Jahrbuch* summarized the clinical trials of tuberculin in Germany from September 1890 to January 1891. The publication included 55 clinical studies with 2,172 patients (2). Only 1,769 patients were evaluable with sufficient information to assess the effect of the treatment. The results are summarized in Table 10.1. Overall, more patients died than were cured. With no control group, i.e., tuberculosis patients who received a sham injection with no tuberculin, it is difficult to determine what was meant by "cured" or "substantially improved." The concept of double-blinded clinical trials was years away in medicine. Clearly, tuberculin was not a booming success;

TABLE 10.1 Results of treatment with tuberculin[a]

Type of tuberculosis	No. of cases				
	Cured	Substantially improved	Improved	Unimproved	Died
Extrapulmonary	15	148	237	298	9
Pulmonary	13	171	194	586	46

[a] Adapted from reference (2).

less than 20% of all patients treated were considered to be substantially improved. Not only was tuberculin not a cure for tuberculosis but also physicians began to consider tuberculin use as dangerous. Patients with tuberculosis of the larynx and in the meninges were especially at risk for bad reactions to the treatment, with increased inflammation and even death (21). Koch conceded that he could not delineate the toxic from the therapeutic components of tuberculin. By mid-1891 the furor over tuberculin died down. Koch continued to experiment with the material for the remainder of his life. Tuberculin continued to be used in Europe and the United States well into the 20th century as a treatment for tuberculosis until antituberculous antibiotics were introduced. Even as late as 1950, tuberculin was mentioned as a therapeutic alternative (22). Tuberculin is still used as a diagnostic for exposure to tuberculosis, but few contemporary physicians even realize that it was ever used as a treatment.

THE EFFECTS OF TUBERCULIN FAILURE ON KOCH

The intense pressure from the tuberculin controversy took a toll on Koch. Some historians believe that the strain caused him to resign his posts as professor at the University of Berlin and as Director of the Institute for Hygiene in disgrace. But Koch had tired of his work in hygiene. He used the initial enthusiasm for tuberculin to leverage a promise from the German government to create the Institute for Infectious Diseases, which he was to head. At the end of 1890, he wrote his longtime colleague Georg Gaffky, who was then at the University of Giessen, to say that he was going on an extended vacation, returning to work when the Institute was ready. Koch headed off to Egypt with a friend in the winter of 1891.

Koch's personal life unraveled during this period. After several years of estrangement from his wife, Koch became involved with a young art student, Hedwig Freiberg, who was 17 in 1889 when they met. By the time Koch left for Egypt he was romantically involved with her. When the press got wind of the relationship, it caused quite a scandal in Berlin society. Koch divorced his wife, Emmy, in 1893. Several months later, Koch, now 50 years old, married Hedwig, who was 20. Despite the gossip in the press, the marriage seemed to be solid. As his wife, Hedwig accompanied Koch on most of his foreign travels through the remainder of his life.

THE INSTITUTE FOR INFECTIOUS DISEASES IN BERLIN

After a month or so in Egypt, Koch was refreshed and ready to come back to his main passion, the field of infectious diseases. Koch returned to Berlin. In the spring of 1891, the Institute for Infectious Diseases opened in a triangular building in Berlin and quickly became known as the Triangle. With Koch as Director, the Triangle became a beehive of activity, with a staff that included Emil von Behring,

Bernhard Proskauer, Shibasaburo Kitasato, Richard Pfeiffer, August van Wassermann, and Paul Ehrlich. Each of these individuals contributed greatly to the field. van Wassermann discovered the test for syphilis that bears his name; Pfeiffer discovered immune lysis, the "Pfeiffer phenomenon"; von Behring discovered an antitoxin for diphtheria; Kitasato isolated the causative agent for tetanus, *Clostridium tetani*, and the agent of virulence, a tetanus toxin; Proskauer helped to develop the Voges-Proskauer test, one of the main diagnostic tests for enteric bacteria; and Ehrlich made a number of contributions to infectious diseases that are detailed in chapter 12. Koch remained busy with work on tuberculin. In 1892 a cholera epidemic hit Hamburg, the first German city with the disease since Koch's discovery of the causative agent. Koch dropped everything and went to the Hamburg site.

CHOLERA IN GERMANY: A PUBLIC HEALTH TRIUMPH FOR KOCH

Despite the efforts of John Snow and earlier work by Koch, the transmission of cholera when outbreaks occurred in cities was still the subject of controversy. The 1892 cholera outbreak proved to be an ideal "experiment" for Koch to prove that cholera was indeed waterborne. Two cities, Hamburg and nearby Altona, obtained their drinking water from the Elbe River. Hamburg obtained its water upstream from the city, whereas Altona obtained its water farther downstream, below the city of Hamburg. Because of the possibility of sewage contamination from Hamburg, Altona had installed sand filtration for its municipal water supply. Hamburg officials deemed this precaution unnecessary. In Hamburg, 17,000 cases of cholera occurred, with 8,000 deaths in several weeks. Altona was virtually free of disease. Koch required little time to discern the reason for the observed difference in disease.

> For a bacteriologist, nothing is easier than to explain why cholera is restricted to Hamburg. The cholera bacteria are brought into the Hamburg water either from the Hamburg sewers, or from the dejecta of persons living on the boats anchored near where the water is taken . . . Altona takes its water from a source which is much worse than Hamburg's, but careful filtration renders it completely, or nearly completely, free of cholera bacteria (23).

Koch's plating techniques provided definitive proof for the effectiveness of the filtration of the water supply. Quantitative bacterial counts of the water showed the presence of large numbers of cholera bacteria before filtration and virtually none after filtration. These efforts were utterly convincing to the German government and set the standards for regulations in Germany and throughout the increasingly urbanizing world. Koch's careful efforts in isolating *Vibrio* organisms from the waters around Hamburg led to the observation that many vibrios are nonpathogenic, i.e., noncholera vibrios. Koch's coworker Richard Pfeiffer used the phenomenon of immune lysis, where he showed rapid lysis of vibrios by specific antisera that provided a means to distinguish pathogenic from nonpathogenic types.

Even though Koch had tired of his public health work, his work on the Hamburg cholera outbreak ranked as one of the most important public health contributions of the last century. Not only did the installation of water filtration plants help control cholera but also typhoid fever was controlled with water filtration. Water filtration plants in cities throughout the world owe a great debt to Robert Koch.

THE ROBERT KOCH INSTITUTE

From 1892 until 1896, Koch worked on tuberculin and consulted with government officials on his findings from the cholera outbreak. During this period, the Charité Hospital expanded and needed the space occupied by the Institute for Infectious Diseases. The Robert Koch Institute, the fourth institute with which Koch was associated, was planned, and still exists in Berlin. Koch, who always loved to travel, embarked on a trip to Africa. He studied a variety of tropical diseases, including some of veterinary interest. Koch's work, while important, served more to confirm previous findings. For example, in 1897, Ronald Ross carried out work in Calcutta proving that mosquitoes transmitted the malarial parasite, efforts that were to win him the 1902 Nobel Prize in Medicine. Koch confirmed these findings in Africa and Italy. Koch made trips to India, New Guinea, and the Dutch East Indies during the period from 1896 to 1900.

In 1900, the new Robert Koch Institute opened in Berlin. Koch returned to Berlin and worked at the new institute. He became involved in another controversy in 1901, when he switched a previously held belief, now stating that human and bovine tuberculosis were two distinct diseases (21). Koch also described the phenomenon of the carrier state for typhoid fever. He noted that water filtration, which nearly eliminated cholera in European cities, decreased the incidence of typhoid fever but did not eliminate it. Koch's careful analysis of a village near Trier, Germany, led him to conclude that typhoid infections were not from contaminated water but from other people who were perfectly healthy but could be shown to be carriers of the organism.

THE 1905 NOBEL PRIZE IN MEDICINE

In 1902, Koch returned to Africa to continue research efforts in tropical medicine, although he never made any further significant contributions to science. In 1905 he was called back from Africa to receive the Nobel Prize for Medicine for his discovery of the tuberculosis bacterium. Koch had been passed over for the prize, which was first awarded in 1901 to Emil von Behring for his work on diphtheria antitoxin. In 1902 Ronald Ross received the Nobel Prize in Medicine for his efforts on malaria. The 1903 Prize went to Niels Finsen for his work on phototherapy in the treatment of skin diseases. In 1904 the Nobel Prize in Medicine went to Ivan Petrovich Pavlov for his fundamental work on condition reflexes. Robert Koch had been passed over four times! Koch's nomination came from an unlikely source, Elie Metchnikoff, then Director of Institut Pasteur. He wrote the Nobel Prize Committee in April 1905,

> I have nominated Koch for the prize for years and as long as Koch has not received the Prize, I can on principle support no other candidate. It is my opinion that Robert Koch's service to medicine has far surpassed that of all other possible candidates (24).

Koch's controversies and enemies may have delayed his Nobel Prize, but his career and extraordinary discoveries could not be ignored. Koch returned to Africa briefly and retired in 1907. He toured the United States and Japan after his retirement. During his U.S. visit the German Medical Society of New York City held a festive dinner in Koch's honor. William Henry Welch, the first dean of Johns Hopkins Medical School, described Koch's career in detail and concluded with words of great tribute:

> Is there a period in the history of medicine where such discoveries issued from one man, and from those working under his supervision; or, in the whole history of medicine, is there a like period where such discoveries are found as the laying of the foundations of modern bacteriology, forging the instruments with which we work today, exploring these newly discovered fields, demonstrating the specific microorganism of tuberculosis, the greatest discovery in the whole field of bacteriology, the entire field of which has not yet been fully reached (25)?

From 1876, when Koch published his first paper on anthrax, to 1906, most of the main bacterial pathogens that affect humans were discovered. These pathogens included *Bacillus anthracis* (Koch discovered), *Staphylococcus* (Koch discovered), *Neisseria gonorrhoeae*, *Salmonella typhi* (now called *Salmonella enterica* serovar Typhi), *Streptococcus*, *Mycobacterium tuberculosis* (Koch discovered), *Vibrio cholerae* (Koch discovered), *Corynebacterium diphtheriae*, *Clostridium tetani*, *Escherichia coli*, *Streptococcus pneumoniae*, *Neisseria meningitidis*, *Salmonella enteritidis* (now called *Salmonella enterica* serovar Enteritidis), *Clostridium perfringens*, *Yersinia pestis*, *Clostridium botulinum*, *Shigella dysenteriae*, *Salmonella paratyphi* (now called *Salmonella enterica* serovar Paratyphi), *Treponema pallidum*, and *Bordetella pertussis*. These pathogens were discovered not just in Koch's lifetime but because the discoverer had a direct connection with Robert Koch. Perhaps no other person profiled in this book witnessed and was largely responsible for the complete transformation of a field, in this case, bacteriology.

After traveling to Japan, Koch returned to Berlin in October 1908. His health began to fail, though he continued to work. He gave his last paper on tuberculosis before the Prussian Academy of Sciences on 7 April 1910, 2 days before having a heart attack. He traveled to Baden-Baden to recover, but he died on 27 May 1910. His body was cremated. His remains are housed in a mausoleum at the Robert Koch Institute.

During his lifetime, Koch had recognized that the burgeoning field of infectious disease still had a long way to go.

> In many respects, and where we would not have expected it, bacteriology has failed us. We have no knowledge of the causes of diseases like measles, scarlet fever, and smallpox. Of the germs of influenza, whooping cough, yellow fever, pleuro-pneumonia, and many other undoubtedly infectious diseases, we also know nothing, although skillful work and patient study have not been lacking. I am inclined to think that here the causal agents are not bacteria, but organisms of a far different character (23).

Koch made this extraordinarily perceptive statement in 1890 at the congress where his statement on tuberculin caught everyone's attention. Most of the pathogens noted above turned out to be viruses, a class of microorganisms for which even today there is limited therapy. Koch's major failing, the tuberculin debacle, demonstrated that despite the explosion in knowledge of causative agents, therapies for these agents and their diseases were largely unavailable during Koch's lifetime. Like the Cnidian approach from ancient Greece, Koch's efforts and those of his colleagues who discovered causative agents suffered since one could not have offered much after the effort in diagnosis was undertaken. As the Cnidian physicians discovered, diagnosis is of little use unless it can be followed by a specific treatment. But the advent of therapy for infectious diseases was at hand.

REFERENCES

1. **Brock TD.** 1999. p 27–37. *In Robert Koch: a Life in Medicine and Bacteriology.* ASM Press, Washington, DC.
2. **Anonymous.** 1891. Official report on the results of Koch's treatment in Prussia. *JAMA* **11:**526–529.
3. **Ligon BL.** 2002. Robert Koch: nobel laureate and controversial figure in tuberculin research. *Semin Pediatr Infect Dis* **13:**289–299. http://dx.doi.org/10.1053/spid.2002.127205.
4. **Brock TD.** 1999. p 70–83. *In Robert Koch: a Life in Medicine and Bacteriology.* ASM Press, Washington, DC.
5. **Brock TD.** 1961. p 101–108. *In Milestones in Microbiology.* Prentice-Hall, Inc, Englewood Cliffs, NJ.
6. **Godlee RJ.** 1924. p 446. *In Lord Lister.* Oxford University Press, Oxford, United Kingdom.
7. **New York Times.** 3 May 1882. Tyndall on Koch's work; parasites found to transmit tubercular disease, p 2. New York Times, New York, NY.

8. **Brock TD.** 1999. p 117–139. *In Robert Koch: a Life in Medicine and Bacteriology.* ASM Press, Washington, DC.

9. **Brock TD.** 1961. p 109–115. *In Milestones in Microbiology.* Prentice-Hall, Inc, Englewood Cliffs, NJ.

10. **Brock TD.** 1961. p 116–118. *In Milestones in Microbiology.* Prentice-Hall, Inc, Englewood Cliffs, NJ.

11. **Evans AS.** 1976. Causation and disease: the Henle-Koch postulates revisited. *Yale J Biol Med* **49:**175–195.

12. **Carter KC.** 1987. p ix–xxv. *In Essays of Robert Koch.* Greenwood Press, New York, NY.

13. **Brock TD.** 1999. p 140–168. *In Robert Koch: a Life in Medicine and Bacteriology.* ASM Press, Washington, DC.

14. **Koch R.** 1881. Zur Aetiologie des Milzbrandes. *Mitt Kais Gesundheitsamte* **1:**49–79.

15. **Mollaret HH.** 1983. Contribution to the knowledge of relations between Koch and Pasteur. NTM-Schriftenr. Gesh. Naturwiss. Technik. *Med Leipzig* **20:**S57–S65.

16. **Burke DS.** 1993. Of postulates and peccadilloes: Robert Koch and vaccine (tuberculin) therapy for tuberculosis. *Vaccine* **11:**795–804. http://dx.doi.org/10.1016/0264-410X(93)90354-Z.

17. **Von Bergeman P.** 1890. Demonstration of cases treated by Koch's anti-tubercular liquid. *Lancet* **ii:**1120–1122. http://dx.doi.org/10.1016/S0140-6736(01)85847-5.

18. **Koch R.** 1890. Further communication regarding a cure for tuberculosis. *BMJ* **2:**1193–1199. http://dx.doi.org/10.1136/bmj.2.1560.1193.

19. **Doyle AC.** 1890. Dr. Koch and his cure. *Rev Rev* **1:**556.

20. **British Medical Journal.** 1890. General notes from Berlin. *BMJ* **22:**1327.

21. **Anonymous.** 1891. The dangers of tuberculin. *JAMA* **2:**634–636.

22. **Stewart CA, Wilson JL.** 1950. Tuberculosis, p 456. *In* Roscoe I (ed), *Communicable Diseases.* Pullen Lea & Febiger, Philadelphia, PA.

23. **Brock TD.** 1999. p 214–236. *In Robert Koch: a Life in Medicine and Bacteriology.* ASM Press, Washington, DC.

24. **Brock TD.** 1999. p 237–266. *In Robert Koch: a Life in Medicine and Bacteriology.* ASM Press, Washington, DC.

25. **Brock TD.** 1999. p 267–285. *In Robert Koch: a Life in Medicine and Bacteriology.* ASM Press, Washington, DC.

11 Joseph Lister, the Man Who Made Surgery Safe

Modern operating rooms are positioned at the center of contemporary medical centers. Their location drives the location of intensive care units, radiology suites, and emergency rooms. In the 18th and early 19th centuries, this centralized location would have been unthinkable. The operating rooms in that era were set as far away from other patients as possible. First, there was the smell. This odor was the result of the unclean conditions where surgery was performed. Cleanliness was far from the mind of the early-19th-century surgeon. There were no surgical scrub suits or filtered air. Those surgeons never used a surgical scrub. Surgeons performed surgery either in their street clothes or in a surgical smock that had dried blood and pus caked on from years of work. Then, there was the screaming. For centuries, medicine searched for chemicals or medicines that would make surgery painless. Until the mid-19th century, opiates and alcohol were the best medicine had to offer. Neither worked well enough to deaden the pain from an amputation, the most common operation of the 18th century in Europe's developing hospitals. Often, half a dozen assistants had to hold down the poor individual who was undergoing the procedure. Charles Darwin began his career in medicine but was forever troubled by the only two operations he attended, as he describes in his autobiography:

> I attended on two occasions the operating theater in the hospital in Edinburgh, and saw two very bad operations, one on a child, but I rushed away before they were completed. Nor did I ever attend again, for hardly any inducement would have been strong enough to make me do so; this being long before the blessed days of chloroform. The two cases fairly haunted me for many a long year (1).

Germ Theory: Medical Pioneers in Infectious Diseases, Second Edition. Robert P. Gaynes.
© 2023 American Society for Microbiology.

Not surprisingly, surgeons had to be quick. No human could withstand the pain from surgery for more than 4 minutes. An 18th-century surgeon performed an amputation in less than 2 minutes; a talented surgeon took only 30 seconds. How the surgeon could become accustomed to the scene is difficult for us to comprehend.

THE DISCOVERY OF ANESTHESIA

The first great American contribution to the field of medicine ended the pain of surgery. A dentist, not a physician, formally introduced surgical anesthesia. In 1846, William Morton coaxed a surgeon at Massachusetts General Hospital to allow him to use his anesthetic chemical on a patient who had a tumor at the angle of his jaw. The operation required careful, slow dissection. The surgeon, John Collins Warren, allowed Morton to use his chemical, which he called letheron, but which was later determined to be ether. It worked. Modern surgery was born. Or was it? Crawford Long in Georgia had used ether 4 years earlier to remove a tumor on a patient's neck but did not publish his experience until 1849. An acrimonious controversy over the discoverer of anesthesia ensued. More importantly, the discovery of anesthesia did not make surgery safe, only painless. In the 20 years between the introduction of anesthesia and Lister's innovations, the problem of surgical infection and its associated mortality remained.

SURGERY BEFORE JOSEPH LISTER

To fully understand Joseph Lister's contribution, one must consider the state of surgery immediately before his efforts. Infection was the anticipated result of surgery. You read that sentence correctly. When someone had surgery, an infection was the expected outcome. Certainly, there were patients who escaped infection entirely, but they were the exception. Postoperative infections of this era were not all the same. If a patient was lucky, golden, creamy pus would develop about 7 days following surgery but would remain largely confined to the area of the wound but blowing open the incision. The drainage of this type of pus would leave a sizable scar, but its appearance usually ensured that the patient would survive. This postoperative mess was the most desirable result of mid-19th-century surgery and was termed "laudable pus." Today, we would identify this outcome as an infection from *Staphylococcus aureus* and believe it was anything but laudable. During the 19th century, just as now, patients with this type of result could easily succumb to such an infection, but it was never as nasty or fatal as other infectious outcomes of the time. Sir William Watson Cheyne was a student and then a colleague of Lister.

He was honored to give the first Lister Memorial Lecture to the Royal College of Surgeons on 14 May 1925. He chose as his topic "Lister and His Achievement" to set the record straight on the man's life and achievements. Cheyne described these more horrific postoperative outcomes before Lister's accomplishments:

> But a large proportion of cases, varying no doubt with the surroundings, the treatment, and the resisting power of the patient, the nature of the injury and other factors, instead of improving, developed after a few days, a number of very serious ailments, e.g., septicemia, pyemia, erysipelas, various forms of gangrene, and tetanus, and ultimately the patient died after much suffering. This series of events was so frequent in some hospitals that no operations were performed except such as were immediately required to prevent otherwise certain death (2).

These outcomes were described as syndromes that were observed in patients following surgery, since no methods of culturing bacteria were yet available. For septicemia, we would now recognize the outcome as a severe infection of any kind that results in the presence of bacteria in the bloodstream; for pyemia, a patient would have fever (pyemia from the Greek, meaning "to have fever") from any cause; for erysipelas (from the Greek, meaning "red skin"), we now know that streptococci cause this rapidly fatal infection, which spreads its effects by the toxins the bacteria produce. Among other effects, these toxins destroy the human body's white cells, which are normally sent into a wound or an area of inflammation to protect us from microorganisms. This syndrome produces high fevers, teeth-chattering chills, and a deep red color to the surrounding skin that rapidly spreads like a flame. Erysipelas was usually fatal. Even today, a patient with erysipelas may die if the toxins have been produced in sufficient amounts, even if antibiotics are initiated. Fortunately, today, erysipelas is rare. Before Lister's time, the syndrome was not unusual and inspired several colorful names: Ignis sacer, Holy fire, and St. Anthony's fire. For gangrene, which could occur in various forms, this outcome was invariably and horribly fatal. Bacteria that we know as anaerobes caused the gangrenous infections following surgery. These organisms grow in the absence of oxygen. The tissues involved are devoid of a blood supply and develop, with the help of these bacteria, into a foul-smelling mess, the odor of which would permeate the clothes and wards of pre-Lister hospitals. This syndrome was the most terrifying form of postoperative infection. No one survived hospital gangrene. Tetanus was not unusual following surgery in the 19th century, a fact not well appreciated by physicians today. Taken together, this myriad of syndromes following surgery produced a staggering mortality rate, above 50% in nearly all hospitals in Europe and the Americas in the mid-19th century.

HISTORY OF THE TREATMENT OF WOUNDS

Prior to the time of Hippocrates, the cause of disease was so mysterious that divine involvement was assumed; treatment of disease was thought to require divine intervention. Treating wounds goes back further than treating disease; it was attempted thousands of years before the ancient Greeks tried treating disease. Advances in wound care improved, but haphazardly and by trial and error, often on a battlefield. The Hippocratic Corpus described the use of tar and wine in the wounds of patients. During Avicenna's time, wine was also touted for surgical wounds, but cauterization of wounds, with either a red-hot iron or boiling oil, became more common, a practice that persisted until the Renaissance. One can only imagine the thoughts of a wounded soldier being carried back after a spear or sword had pierced his flesh, knowing that the military surgeon was about to pour boiling oil into his wound!

After the invention of gunpowder, wounds on the battlefield became even more severe. The great Renaissance surgeon Ambroise Paré found himself treating scores of severe wounds following a battle in 1537. Running out of oil, he had to abandon the standard practice of placing boiling oil in a wound and improvise. Using a concoction of egg yolk, oil of rose, and turpentine, Paré found, to his surprise, that the wounds healed better than those treated with boiling oil. He vowed never again to so cruelly burn poor men wounded with gunshot. The first scientific study of wound inflammation came in 1794 from John Hunter. Hunter, you will recall, was a surgeon and friend to Edward Jenner. Hunter's greatest publication, *A Treatise on Blood, Inflammation, and Gunshot Wounds* (3), was completed just before and published just after his death. In the treatise, Hunter asserted that inflammation was a "salutory" mechanism of the body, although it can sometimes become a damaging process, too. But Hunter was unaware of the role of microorganisms in setting off the process of inflammation in a wound. Until Lister, medicine lacked a complete understanding of how wound healing occurred.

JOSEPH LISTER: EARLY INFLUENCES

Upbringing influences all of us. We may need to overcome its influences. More often, the experiences of early life shape us toward the adults that we become. We are often unaware of those early influences on our personality or even our career paths. In the case of Joseph Lister, perhaps more so than anyone profiled in this book, his early influences exerted a most powerful and lasting effect on his life and career.

Joseph Lister was born in 1827 into the Quaker community in the countryside outside London, England. Lister owed much of his serene manner to the upbringing in this Society of Friends, as the Quakers were often called. The Quakers were founded in England around 1654. The term Quaker was originally a derisive term referring to the manner in which followers "trembled in the way of the Lord." Since the official religion in England was the Church of England, the first few decades of the sect saw widespread persecution and imprisonment of its leaders and followers. Thanks to the Toleration Act of 1689, Friends in Great Britain were no longer outlaws, but persecution continued. Quakers were not allowed to earn academic degrees for a time. Most Quakers became businessmen and manufacturers. A reputation for integrity served the Quakers well. People learned to trust them. The community was close-knit. The Friends lived a life without dance, song, or frolic. The Quakers maintained an intellectual direction and honesty in their life, devoting themselves to business and education. Many became quite successful. Among them was Joseph Jackson Lister, the father of Joseph Lister.

Joseph Jackson Lister followed his own father into the wine importing business. He achieved considerable success, allowing him to purchase a sizable Queen Anne-style mansion called Upton House outside London. His success allowed him to devote time to his favorite hobby—microscopy. This pastime became much more than a hobby. Joseph Jackson Lister learned to grind lenses and eventually solved the problem of chromatic aberration of compound microscopes. This problem had so bedeviled scientists that even the inquisitive John Hunter believed that microscopic descriptions of tissue or organisms before the achromatic microscope were so distorted that they could not be trusted at all. Such a distrust of microscopy was widespread in Britain. But Joseph Jackson Lister's discovery meant that his son would grow up in a house without such a prejudice. Joseph Jackson Lister was a man of high achievements. He was a self-made scientist whose achromatic microscope earned him the award of the Fellowship of the Royal Society in 1832. He was also an excellent artist and a good linguist—a skill imparted to his son, Joseph, who became fluent in French and German. Joseph was the fourth child and second son in a prosperous, loving family. His mother was described as loving and kind, but most biographers credit Lister's father with the greater influence on the lad.

Joseph Lister attended the Quaker School at Tottenham. Before he left the school, he had already made up his mind to become a surgeon. No one is quite certain why Joseph made such an announcement since no one in his family worked in the medical field. By age 16, Joseph Lister was ready to attend university. For a Quaker, there was little choice. Despite his keen curiosity and intellect, Oxford and Cambridge were closed to him. Lister attended the University College in London, popularly named "the godless college" because of its acceptance of all religious faiths. Lister struggled initially in college. After acquiring smallpox, he returned too early and suffered a breakdown in mental health. He recovered for a month in Ireland. Refreshed, Joseph Lister returned to University College and immersed himself in medical studies. William Sharpey particularly influenced him. Sharpey had trained in medicine in Edinburgh, where he had come to know James Syme, a distinguished Edinburgh surgeon of the day. Sharpey took a keen interest in Lister. During his medical studies, Lister published two articles; both studies utilized the microscope. Lister, unlike most of his colleagues, came to school with one of the best microscopes and was thoroughly expert and comfortable in its use.

True to his professed desire, Lister distinguished himself in surgery as a student. During the early stages of his career as a student, Lister was fortunate enough to see the first major operation to be performed in Britain under ether anesthesia by Robert Liston in University College Hospital on 21 December 1846, only 2 months after the first report of ether use from Massachusetts General Hospital. Although there are few records on the event, the experience seemed to have deeply impressed Lister, far differently than Darwin's initial experience in operating rooms. As a student, Lister received his introduction to septic diseases following surgery, including hospital gangrene. He never forgot them, feeling that no matter was more urgent than to find the cause, treatment, and prevention of these devastating illnesses (2). Lister went on to receive high marks in his surgery classes. Soon after graduation, Lister became a Fellow of the Royal College of Surgeons of England. He eventually became house surgeon to Sir John Erichsen, a leading London surgeon. After 9 years at University College in London, Lister's mentor, Sharpey, recommended that he work with James Syme in Edinburgh, Scotland. It would be a transforming experience for Lister.

JAMES SYME IN EDINBURGH

In September 1853, Lister headed north to Edinburgh to work with one of the most distinguished surgeons of the period. No two personalities could have been more different than Joseph Lister and James Syme. Syme was 54 years old at the time. He was somewhat short. His plain, almost homely face masked a furious temper. Syme was outspoken, self-assured, and combative, making him a formidable opponent in any argument. His dazzling technical expertise in surgery and razor-sharp mind made up for his frequent abrasive manner. He challenged the students who were brave enough to submit themselves to his tutelage, but many came. Syme was generally considered to be the best surgeon in the British Isles. Joseph Lister engendered kind and warm superlatives from anyone who ever worked with him. He was a tall man, over 6 feet. John Rudd Leeson, a British physician who worked with Lister, attempted to describe the man:

> He was always carefully and neatly dressed, and the Quaker spirit of substantiality was unobtrusively evident ... His face, once seen, could hardly be forgotten. There was something so distinctive about it, such a blending of sweetness and power and purpose (4).

In 1922, J. A. Erskine, an acquaintance of Lister, tried his hand at describing a man who seemed to have no detractors.

> Lister was a great figure in my day. His personality was a charming one, so modest, and yet so brilliant, and pains taking and conscientious. He had the highest ideal of what a doctor should be of any man I ever met, and he not only promulgated this counsel of perfection, but he lived it to the letter . . . I shall never forget his judgment and pity when a child wept whose wound was going to be dressed. He stooped down and kissed the little sufferer. Smiles took the place of tears, and all went well. He was a most lovable man. There was a genuine look about him, a naturalness, an entire lack of affectation (5).

Despite their personality differences, the two men struck an immediate and life-long friendship. Lister never tired of talking about "Mr. Syme." Lister would end-lessly describe Syme's improvements in operative procedures and instruments he had invented. Lister often repeated Syme's maxims and aphorisms. Syme, too, was delighted by Lister's enthusiasm and sense of purpose. Lister came to Edinburgh initially for a post of a few months, taking what was closer to a student position, simply to work with Syme. But when an official post opened, Lister leapt at the opportunity. He had found his life's calling and his surrogate father in surgery. He wrote of his experience to his own father,

> If the love of surgery is a proof of a person's being adapted for it, then certainly I am fitted to be a surgeon: for thou canst hardly conceive what a high degree of enjoy-ment I am from day to day experiencing in this bloody and butcherly department of the healing art. I am more and more delighted with my profession and sometimes almost question whether it is possible such a delightful pursuit can continue. My only wonder is that persons who really love Surgery for its own sake are rare (6).

MARRIAGE OF JOSEPH LISTER AND AGNES SYME

The relationship between Syme and Lister had an additional benefit for the young surgeon. The frequent visits to Syme's home put Lister in contact with Syme's eld-est daughter, Agnes. The relationship between Lister and Agnes grew rapidly. Agnes became smitten quickly. Syme was much in favor of marriage between Lister and his daughter. Lister was the hesitant one, since marriage outside the Quaker faith meant severing his ties to the Friends community. On 23 April 1856, the two were married.

Following their marriage, the couple took a most unusual honeymoon. For 3 months, they toured the continent of Europe, visiting many of the major hospitals and clinics in their travels. They visited Padua and Bologna in Italy.

The couple stopped at the University of Vienna, where they dined with Karl von Rokitansky since von Rokitansky was an acquaintance of Lister's father. This dinner, in 1856, occurred more than 6 years after Semmelweis's discovery of the importance of hand washing, the cadaver particles, and puerperal sepsis. The event has been the subject of some historical controversy. The question has been posed whether Semmelweis's work was discussed and had influenced Lister. There is no record on the matter, so speculation remains. However, after Semmelweis's abrupt departure from Vienna in 1850, von Rokitansky rarely mentioned him or his work. Lister later wrote that he was unaware of Semmelweis's findings on the importance of hand washing until long after his own discovery. There is no reason to think otherwise. After stopping in Prague, Berlin, Würzburg, and other German cities (Lister spoke fluent German), the couple returned to Edinburgh. To anyone that knew Lister, it was clear that his marriage to Agnes was not one of convenience to gain favor with Agnes's father. Agnes was an ideal partner for Joseph Lister. She acted as assistant in experiments, secretary to Lister when he dictated (there were no typewriters), hostess, and devoted companion. The couple spent 3 very happy years together in Edinburgh. Lister, together with his wife, set up a laboratory of sorts in their home. He began a series of experiments on a wide range of topics. One of those topics was the formation of blood clots. Physicians of Lister's time had difficulty explaining why blood was normally fluid within blood vessels but clotted when it was outside the vessels. Lister's early research suggested to him that when blood comes in contact with foreign material outside the body's vessels, the process of coagulation, or blood clotting, begins. The idea that something foreign to blood outside vessels initiated coagulation gave Lister an idea. Perhaps other alterations in body physiology might be affected by the influence of foreign or external materials. He used his microscope in all his studies. Lister began to look at the inflammation in wounds with his microscope.

THE MOVE TO GLASGOW

When a position opened as Chair of Surgery in Glasgow, Scotland, Lister was chosen (Fig. 11.1). The couple moved to Glasgow in 1860. The move was an important, life-changing one for the couple. For Agnes, she left home and family. For Lister, he would become chief of a surgical service at one of the great institutions in Great Britain. He quickly became a favorite of the students. At the end of his first year in Glasgow, 161 students took the unusual step to give Lister a signed petition proclaiming, "your eminent ability as a teacher of Surgery."

In Glasgow, as in any major European or American city at the time, postoperative infection was rampant. One of the leading surgeons in Paris in 1870 said,

FIGURE 11.1 *Joseph Lister. Courtesy of the National Library of Medicine (NLM image ID B017429).*

> . . . when amputation seems necessary, think ten times about it, for too often, when
> we decide upon operation, we sign the patient's death-warrant (7).

Lister had been distressed over the problem of postoperative infection for his 10 years in the field. Since infection in the wound was so common, how did the mid-19th-century surgeon explain the phenomenon? The explanation was that oxygen that came in contact with the tissues in the wound caused all the trouble, oxidizing in the tissues and creating the postoperative chaos of infection. This hypothesis was attractive to surgeons of the day for two reasons. First, oxygen was everywhere. No mystical ingredient had to be found. Once a wound was made, the chemical process of oxygen combining with the tissues made for the mess that was observed following surgery. Second, the oxygen thesis kept the surgeon from having to look inward. There was no causation involved. It was not the surgeon's fault that postoperative infection occurred. The hypothesis smacks of the last holdover from the humoral theory and miasma. Oxygen was now the culprit, not miasma. For the mid-19th-century scientific mind, the chemical explanation may have been more appealing than the mystical miasma.

Lister refused to accept the notion that every wound was destined to end in a putrid jumble. To Lister, there were cracks in this oxygen argument. First, blood carried oxygen to the tissues, but there was no spontaneous infection in the body's tissues. Lister had also noted cases of trauma in which broken ribs had punctured lungs. No external wound was evident, but surely oxygen had entered the wound made by the broken rib between the lung and chest wall. Infection was not the invariable result. He also carefully pondered fractures. A closed fracture, severe as it might be, rarely, if ever, was complicated by infection. But if the bone pierced the skin, creating an opening, sometimes called a compound fracture, infection nearly always occurred, and death followed in half of those cases. In those who survived, amputation of the affected limb was usually required. Why was there such a difference between closed and open fractures? Lister pondered the idea that, like in coagulation, something foreign was introduced into wounds to cause the pus, redness, or, worse, hospital gangrene to form.

THE CLUE TO WOUND INFECTIONS

The clue came when a colleague, Thomas Anderson, a professor of chemistry, drew Lister's attention to the work of Louis Pasteur, who had acquired some repute as a chemist. Pasteur's work on the problems of fermentation fascinated Lister. Fermentation was a process that heretofore had been considered to be entirely chemical in nature. Pasteur, you will recall, realized that foreign microorganisms had been competing with the yeast to cause what he termed the diseases of fermentation. Lister, who was fluent in French, read Pasteur's papers over and over. Lister recognized in Pasteur's experiments the nature of the foreign material that was entering wounds. To Lister, it was not the air or its oxygen that oxidized wound tissues, ending in putrefaction. It was something in the air, microbes that produced the problems of putrefaction of wounds. First, Lister would have to convince himself and his colleagues that Pasteur's discoveries had application to medicine. He used his microscope to demonstrate the presence of these bacteria in wounds. Then, Lister repeated Pasteur's swan-neck flask experiments, only using boiled urine. Some of his flasks had straight necks; some had swan necks. Dust settled in the straight-neck flasks. The fluid rapidly decomposed. But in the swan-neck flasks, air had access to the boiled urine, but the liquid remained germfree. Lister used these flasks in lectures to introduce the germ theory to skeptical colleagues and students. Lister concluded there had been:

> . . . a flood of light thrown upon this important subject by the philosophic researches of Monsieur Pasteur (8).

Lister was greatly impressed by the work of Pasteur and acknowledged the debt that he owed Pasteur throughout his life. In the 1870s, Lister and Pasteur corresponded in a gracious and complimentary fashion. Each man recognized the debt owed to the other. Most physicians of the day could not see any link between diseases of fermentation and human disease. As a physician, Lister did something that the chemist-turned-biologist Pasteur could not do. Lister transferred the discovery of diseases of fermentation into medical terms—diseases of wounds, a practical application that the French scientist often attempted in his own work. In a 1902 letter to G. H. Edington, Lister recalled his discovery:

> It was not until after I went to Glasgow that bacteria and other microbes claimed my attention. All efforts to combat decomposition of blood in open wounds were in vain until Pasteur's researches opened a new way, by combating microbes (7).

ANTISEPTIC SURGERY

The new concept of combating microbes in wounds meant finding a means to destroy them. Clearly, the use of heat, as Pasteur had done in the process of pasteurization, could not be applied to human wounds. Lister turned to chemicals. He tried several substances, without success. Then he noticed in the newspapers that the nearby town of Carlisle had successfully used carbolic acid as a chemical means to rid the town of a sewage smell. What caught Lister's attention to this chemical was not the value in controlling the stench but that the substance destroyed the parasites that affected cattle grazing on the sewage-contaminated grass. Lister obtained a sample of carbolic acid and experimented with it. The initial samples of carbolic acid that Lister obtained had a disagreeable odor, were irritating to the skin, and were insoluble in water. He eventually acquired a purer sample that was soluble in water but retained its odor and irritating properties. Lister demonstrated experimentally that the use of carbolic acid would kill microbes in his flask experiment. But he needed to show its value conclusively on humans. Lister reasoned that the difference in infection risk from a closed (simple) compared to an open (compound) fracture provided the best opportunity to test his idea. To Lister, it was the human application of the simple flask experiment. In the simple fracture, microbes could not gain access to the injury. In the compound or open fracture, the injury was open to the germ-laden air. Lister believed that if he could kill the microbes that enter the wound caused by the bone piercing the skin, it would heal just as the simple or closed fracture healed.

THE FIRST SUCCESS WITH ANTISEPSIS

Armed with a rational, albeit revolutionary, hypothesis, most researchers might rush to find an individual to try out the experiment. Not Lister. Other investigators may have attempted to test carbolic acid on the first available patient and been met with disaster. Lister waited 10 months for an appropriate test subject. In March 1865, 1 month before the end of the American Civil War, the last battlefields before the germ theory, his first test case ended in failure due to what Lister described as "improper management." Some months later, on 12 August 1865, James Greenlee, an 11-year-old boy, came to Glasgow Royal Infirmary with a compound fracture of his left leg only a few hours after an empty cart passed over the limb. The break in the skin was about $1^1/2$ inches and had very little blood. Treatment of the wound consisted of a thorough application of undiluted carbolic acid and dressing with lint soaked in the same antiseptic. Tin foil covered the wound to prevent evaporation. After 4 days, Lister personally redressed the wound using bandages soaked in carbolic acid. Under Lister's careful watch, a scab formed. Greenlee did not die; he did not require amputation; he did not even suffer infection. After 6 weeks, the boy walked out of the hospital! Lister treated 10 more patients with compound fractures. All but 2 recovered. One patient, whose leg had been fractured by the kick of a horse, came down with an infection while Lister was away. He required an amputation but survived. The other patient, who had fractured his femur near the hip, died of hemorrhage when a sharp fragment of bone pierced a nearby artery. Lister related one case in this extraordinary series of successful recoveries to his father:

> There is one of my cases at the Infirmary which I am sure will interest thee. It is one of compound fracture of the leg, with a wound of considerable size and accompanied by great bruising, and great effusion of blood into the substance of the limb, causing great swelling. Though hardly expecting success, I tried the application of carbolic acid to the wound to prevent decomposition of the blood and so avoid the fearful mischief of suppuration throughout the limb. Well, it is now eight days since the accident, and the patient has been going on exactly as if there was no external wound, that is, as if the fracture were a simple one (7).

With his success in treating compound fractures, Lister turned his attention to the treatment of abscesses. After a number of experiments, he developed a mixture of carbolic acid, linseed oil, and common whitening, which he called antiseptic putty. The putty was layered onto tin foil and placed on the wound of an opened abscess, changed daily. Lister published his experience in 1867 under the title "On a New Method of Treating Compound Fractures, Abscesses, etc." in *Lancet*. As described in the paper, he clearly had altered his thinking on the pathogenesis of infections in wounds.

> If we could see with the naked eye a few only of the septic organisms that people every cubic inch of the atmosphere of a hospital ward, we should rather wonder that the antiseptic treatment is ever successful than omit any precautions in conducting it (9).

ANTISEPSIS AND SURGICAL WOUNDS

Lister's next step was to apply his antiseptic techniques to ordinary surgical wounds. Lister described the essential conditions for success.

> The first thing that has to be done is to destroy the germs on the patient's skin, on the surgeon's hands, on the instruments which are to be used and on everything surrounding the area of operation.
>
> The second is to prevent living germs from entering the wound from the air of the surrounding objects during the performance of the operation.
>
> And the third is to prevent germs from spreading into the wound after the operation (2).

These principles hold true today. Lister had completely revolutionized thinking on how postoperative wound infection occurred. It was not oxygen but microbes that caused the problem. Just as important, these microbes could be killed when they entered the wound. Lister focused on antisepsis—killing microorganisms when they got into the wound, initially. But soon, Lister would recognize the importance of not allowing the microorganisms access to the wound at all.

PROBLEMS WITH CARBOLIC ACID

There were significant practical problems with wide adoption of Lister's approach from the beginning. At first, carbolic acid was placed in linseed oil. While the concoction had antiseptic effects, its yellow color and oily consistency obscured the field of vision of the wound. Once purer carbolic acid became available, obtaining watery (aqueous) solutions led to more rapid extension of his methods. Lister also insisted that the skin of the patient and the surgeon's hands be washed in this irritating 1:20 carbolic acid solution. Interestingly, he never removed his coat when performing operations. He only rolled up the sleeves. Instruments were immersed in carbolic acid. Many of today's disinfectants still contain carbolic acid or its derivatives, now termed phenolics. Phenolic disinfectants are still widely used on inanimate objects. However, we have found effective but gentler agents than phenolic solutions for antiseptics—antimicrobial chemicals for use on humans.

Lister was so concerned that dust carrying microorganisms would fall from the air that he eventually developed an antiseptic sprayer that filled the operating theater with the pungent smell of carbolic acid, unpleasant to be sure. He meticulously dressed wounds postoperatively with carbolic acid-soaked bandages. Unfortunately, the postoperative dressing changes that Lister demanded were so complicated and troublesome that they did not appeal to the surgical profession on a practical basis alone. But Lister's most difficult task was convincing his colleagues that the underlying principle—microorganisms are responsible for the infections—was true and made all the fuss worthwhile. An additional four papers appeared in *Lancet* in 1867 that described his ideas, methods, and some results. Although the successful results today would be considered anecdotes, Lister described several cases and then, an unanticipated benefit in a September 1867 *Lancet* article:

> Since the antiseptic treatment has been brought into full operation, and wounds and abscesses no longer poison the atmosphere with putrid exhalations, my wards, though in other respects under precisely the same circumstances as before, have completely changed their character; so that during the last nine months not a single instance of pyremia, hospital gangrene, or erysipelas has occurred in them (10).

The Chief of Surgery in Glasgow pressed ahead with his radical methods until he believed that he had sufficient experience with one procedure, amputations, to publish in *Lancet*. He summarized them in an 8 January 1870 article:

> Before the antiseptic period, 16 deaths in 35 cases; or 1 death in every 2½ cases.
>   During the antiseptic period, 6 deaths in 40 cases; or 1 death in every 6½ cases (11).

Lister concluded in his article,

> The antiseptic system is continually attracting more and more attention in various parts of the world; and, whether in the form which it has now reached, or in some other and more perfect shape, its universal adoption can be only a question of time (11).

Lister was correct but a bit hasty in his conclusion. Members of the surgical profession in his own country and in America would prove to be among the slowest to adopt his antiseptic system. But he would not use the Glasgow Infirmary as the pulpit from which he would preach his new gospel.

MOVE BACK TO EDINBURGH

Lister's father-in-law suffered a stroke in 1869 just as his 10-year appointment in Glasgow was ending. The couple moved back to Edinburgh when Lister was 42 years old. The position in Edinburgh held promise for Lister but began with troubling events. Within the year, both Syme and Joseph Jackson Lister, Lister's father, passed away. Shortly after his return to the city, Lister was summoned to Balmoral Castle to attend to Queen Victoria. The Queen had an abscess that formed in her axilla. Lister opened the abscess while the Queen's personal physician worked the carbolic sprayer. The operation was successfully performed. The Queen congratulated Lister, which was not the last royal honor for him.

REACTION TO SURGICAL ANTISEPSIS

British surgeons outside Glasgow had not taken well to Lister's antiseptic surgery. Lister's first step after his move back to Edinburgh was to convince his colleagues, a process that would take years. While the concentration of carbolic acid that Lister used decreased and lessened irritation of the skin of patients and surgeons alike, his approach to antisepsis and to dressing wounds became more and more complex. The most contentious aspect of the process was the carbolic acid sprayer. Leeson describes the typical early Listerian surgery.

> The whole scene of an operation was enveloped in its spray, which dispersed its globules into every nook and cranny of the wound, and our faces, and coat sleeves often dripped with it. . . . The spray was devised to counter the idea that sepsis was largely due to air-borne organisms . . . Subsequent experience convinced Lister that they might be disregarded, but the truth dawned but slowly, and it was years before he learnt that "sepsis is a question of dosage," that the living tissues can deal with a few organisms but are overwhelmed by the many (12).

ACCEPTANCE ON THE EUROPEAN CONTINENT

While British surgeons bickered over the value of antiseptic surgery, the 1870–1871 Franco-Prussian war produced, literally, an army of believers. The antiseptic techniques were available but promulgated too late to help most of the wounded. During the war, the French amputated about 13,200 limbs, with a mortality rate of 76%. The grim war-related figures kept both French and German surgeons open to anything that would improve outcome. Due to inquiries from both the French and German sides, Lister produced a pamphlet in 1870 titled *A Method of Antiseptic Treatment Applicable to Wounded Soldiers in the Present War* (13). The methods were a practical adaptation of his antiseptic methods for the battlefield. Faced with the enormity of casualties, the German surgeons von Nussbaum, von Langenbeck, and von Volkmann enthusiastically adopted Lister's methods, with excellent results. Ritter von Nussbaum wrote Lister,

> We experienced one surprise after another . . . Not another case of hospital gangrene appeared . . . Our results became better and better, the time of healing shorter, and the pyemia and erysipelas completely disappeared . . . I hold that next to that of chloroform-narcosis your discovery is the greatest and most blessed in our Science. God reward you for it and grant you a long and happy life (6).

As the German military surgeons continued antiseptic surgery in their civilian practices, adoption of the antiseptic approach spread to others in the German-speaking medical establishment. von Nussbaum wrote a short book on antisepsis that was translated into Italian, Greek, and French. Surgeons like Theodor Billroth used antiseptic surgery, which enabled him to dare more invasive surgeries in parts of the body previously considered untouchable. The French, too, began to adopt the antiseptic principles in their surgeries. After the war, both the Germans and the French sent their surgeons to Edinburgh to learn from the great Lister.

By 1875, the Germans held the German Surgical Congress where von Nussbaum, among others, extolled the virtues of Lister's antiseptic system. Only a few months later, Lister, with his wife, toured the German hospitals to determine the progress made. The trip was more of a triumphant march for Lister. Banquets were held in his honor. Poems and songs were composed for him. In the 19 June 1875 issue of *Lancet,* an article appeared extolling the heights that the German hospitals had reached in surgery.

> Nowhere on the continent is Lister's treatment of wounds so thoroughly followed out, and nowhere are its excellent effects better seen than in the wards of Prof. Bardeleben. Amputations, resections, and even the more trifling operations, as opening abscesses, are conducted under the carbolic spray, and subsequently treated with the eight coverings of carbolized gauze, protective covering, gauze bandages, and carbolized cotton-wool, exactly as practiced by Prof. Lister himself. That wounds treated in this way do absolutely heal without suppuration may be seen constantly (14).

LISTERISM IN THE UNITED STATES

Two countries were slow in changing their surgical practice, the United States and England, especially in London. In America, the tradition of laboratory-based and evidence-based medicine did not yet exist. The germ theory was the stuff of untrustworthy foreign influences. Few articles in the American medical literature seriously discussed the possibility that microorganisms were to blame for postoperative putrefaction. The leading textbook of surgery of the time, written by Samuel Gross of Philadelphia, denounced the theory. "Little, if any faith, is placed by any enlightened or experienced surgeon on this side of the Atlantic in the so-called (antiseptic) treatment of Professor Lister" (15). Gross, who was made more famous in an 1875 Thomas Eakins painting, *The Gross Clinic,* refused to follow Lister's principles. In the painting, he is shown about to perform surgery in his street clothes, holding his glasses. The victim's, or rather the patient's, mother is cringing in the background. No antisepsis for Gross. His authoritarian influence helped slow the progress of surgery in the United States. Even Lister's appearance in Philadelphia in 1876 at the Centennial Medical Commission to celebrate America's hundredth anniversary of independence failed to sway the American surgical community. Lister was well received and politely greeted. However, Gross wrote a review for the meeting and stated that surgeons in the United States simply did not believe in Listerism. Lister, with all the energy of an evangelical priest, set out on a transcontinental journey by train to persuade American surgeons about his methods. He visited San Francisco, Salt Lake City, Chicago, Boston, and New York. As he sailed back to England, he quietly and correctly doubted that his lectures had the desired effect on the surgical practices in the United States (16). American medical education in the latter part of the 19th century generally ignored basic sciences. The focus was largely on hasty clinical treatment. The lack of understanding of underlying human physiology and pathology left American physicians with little or no understanding of the new science of bacteriology. Americans viewed Lister's theory with skepticism. American surgeons found his methods to be simply tedious and unnecessary. As late as the early1880s, American surgeons remained slow on the uptake of this theory. The grandson of J. Collins Warren, the surgeon that allowed William Morton to use ether for the first time at Massachusetts General Hospital, had visited Lister in Glasgow. He later wrote that his attempts to bring Lister's principles to that hospital failed. An improperly performed trial did not successfully demonstrate carbolic acid's utility. Warren was "coldly informed that the carbolic acid treatment had been discarded" (6).

THE DEATH OF PRESIDENT JAMES A. GARFIELD

The backwardness of the American surgical community came to a head in the summer of 1881. On 2 July 1881, a would-be assassin shot U.S. President James A. Garfield at a Washington, DC, railroad depot. One of the shots lodged in his back; the other hit his right flank. A team of well-meaning but arrogantly ignorant surgeons tended to Garfield. There was not an iota of antisepsis in any of the multiple procedures performed on the leader. The wound was probed with unwashed hands contaminated with manure. Garfield suffered for 80 more days with terrible infections, ultimately leading to his death. Garfield's boyhood friend and former Civil War surgeon Willard Bliss headed the surgical team. Bliss repeatedly probed the tracks of the gunshot wounds with dirty hands and unwashed probes. Garfield had daily fever spikes and shaking chills. Abscesses developed across Garfield's back, requiring drainage using the same unclean instruments. Another physician, Silas Boynton, was appointed to the team but relegated to nothing more than a nursing role. Boynton later wrote,

> I think the President had a reasonable chance for recovery, but it was thrown away by
> the bad management of the case. Pus had through carelessness and neglect been
> allowed to be in the wound till it rotted and pyremia had done its perfect work (17).

Bliss was pompous enough to have predicted full recovery, ignored the multiple
pus cavities and drop in Garfield's weight from 230 lb. on the day of the shooting
to 130 lb. just before he died, on 19 September 1881. An autopsy showed that no
vital organs had been damaged. Rather, Garfield had died of sepsis at the hands of
his doctors. This tragedy helped to fuel an American interest in Lister's methods.
The change in surgical practice awaited the influence of William Stewart Halstead
at Johns Hopkins Medical School and Hospital. His German training and meticulous approach to dissecting tissues led to the phenomenal success of the so-called
Halstead Surgery of Safety in the last part of the 19th century (18).

ANTISEPTIC SURGERY IN ENGLAND

Back in Britain, particularly in London, Lister had to deal with the skepticism of
surgeons in his own country. After seven or so years in Edinburgh, an opportunity
presented itself in 1877 for Lister to deal with the solid opposition more effectively to his methods in his own country. Lister was offered the Chair of Surgery
at the Medical School of King's College in London. To the horror of students and
surgeons of Scotland, Lister and his wife headed down to London. The decision
to leave the prestigious post in Edinburgh to head to a lesser position in an environment of cynicism or even hostility towards antisepsis in London must have
seemed unfathomable to some. Behind Lister's serene personality lay a determined sense of purpose, the result of his Quaker upbringing. The move for Lister
was more of a mission. The first year or two were frustrating for Lister at King's
College. English surgeons and students sparsely attended his lectures. Lister was
bolstered by the attendance of European surgeons, mostly from France and
Germany. Often, Lister delivered half of the lectures in French or in German,
depending upon the national origin of those attending. Even the signs for no
smoking were translated into French and German. Lister persevered, permitting
only momentary lapses in his own enthusiasm when met by opposition to his
theory in London.

As late as 1879, over 10 years after Lister's first antisepsis publication in *Lancet*,
London surgeons held a meeting at St. Thomas's Hospital to debate the value of
Lister's methods. The meeting opened with a full description of a Listerian surgery
by William MacCormac, who later wrote about the event (19). The ensuing debate
was stunning in its obstinate comments. One surgeon called for statistical proof.
Another claimed that he believed in neither the germ theory of disease nor the

value of Listerism. One idea put forth was a trial at King's College Hospital, where one surgical firm would follow the antiseptic method, and another would not. However, this trial was dismissed by the senior surgeon of King's College Hospital, John Wood, who decried that he would not permit patients at his hospital to be the subjects of an adverse experiment. But Wood had visited Lister a year earlier in Edinburgh and came away impressed with what he saw on his wards. While too old to change his own surgical habits, Wood began to accept Lister's theory. Lister had staunch supporters at the meeting, including Spencer Wells, who gave figures that showed how he had reduced mortality rates following ovariotomy using a variation on Lister's antiseptic techniques. Delving deep into a body cavity to remove an ovary was inconceivable only a few years earlier. Lister himself was concerned about the operation and wrote about the use of scrupulous cleanliness by Wells and his friend Thomas Keith:

> Mr. Spencer Wells and Dr. Thomas Keith achieved results which astonished the world before strict antiseptic treatment was thought of; and when several years ago Dr. Keith expressed to me an intention of performing ovariotomy antiseptically I strongly dissuaded him from his purpose. I knew his already brilliant success; I felt that our spray apparatus was as yet inadequate for the production of a cloud sufficiently large to cover the whole field of operation and sufficiently fine to avoid needless irritation. I was also aware that such operations are often both very protracted and very anxious, while in proportion to the duration and the anxiety of an operation is the chance of the neglect of some apparently trivial yet important element of the procedure. And if the antiseptic treatment were attempted in ovariotomy and failed in its immediate object, I felt that it would be not only nugatory but injurious (2).

With anesthesia and Lister's antiseptic methods, surgeons could now venture into uncharted waters with success. Lister addressed the crowd, presenting some of his figures. Problems with the older surgeons in London lingered, but the Congress represented something of a turning point.

The reasons for the English animosity towards Lister's theory seemed to be similar to the American explanation—a lack of understanding of pathology and physiology among the English surgeons. In particular, the long-standing skepticism towards microscopy among British physicians and surgeons may have hampered their belief in the germ theory and their interest in antiseptic methods. But the tide was turning. After 2 years in London, Lister began to win over the younger generation of surgeons. The days of the dazzling speed of the surgeon began to disappear. Older surgeons were giving way to a younger breed of talent. With the advent of anesthesia and antisepsis, surgeons could use a careful, scientific approach to dissecting human tissues.

ANTISEPSIS AND ASEPSIS IN SURGERY

The irony of Lister's methods is that he paved the way to make his own approach obsolete. The first part of Lister's process to disappear was the carbolic acid spray in 1887. With Koch's development of solid media in bacteriology, Lister recognized that careful examination of the bacterial dosage in the air showed, to Lister's satisfaction, that the spray, which had been the most contentious part of his process, was not needed. Lister never really abandoned the remaining antiseptic process he invented, but others looked to improve the process. Antisepsis is the technique of killing microorganisms after they have entered the surgical wound. Asepsis is the process that does not allow microorganisms access to the wound in the first place. The processes are complementary, not antagonistic. The initial architects of asepsis were German. Ernst von Bergman of Berlin first used aseptic dressings in the Franco-Prussian war. von Bergman, while crediting Lister, began to use heat to sterilize instruments and dressings, by boiling or steaming them. He developed operation rooms that were germfree with cleansable metal surfaces and steam-sterilized gowns and towels. His assistant, Schimmelbusch, published a book, *The Aseptic Treatment of Wounds,* in 1892 that won wide support in the German-speaking medical world. Spectators were limited in the operating room, no longer called a "theater." The surgeons wore caps and gowns beginning in 1883. William Halstead was the first to use rubber gloves in surgery, which were developed when a nurse, then his fiancée, complained that the antiseptics used were harming her skin. Surgical masks were first introduced in 1897. Surgical scrub suits were introduced at the beginning of the 20th century. Surgery transformed from a generation earlier, depicted in Eakins's *The Gross Clinic,* to the modern, germfree version, heaped in ritual, which remains largely unchanged today. The brilliance of Joseph Lister was not in his antiseptic methods but in how he opened the world's eyes to the real cause of putrefaction and suppuration of wounds. Surgery and humanity were forever changed by Lister's lifework.

LISTER'S OTHER ACCOMPLISHMENTS

Antiseptic surgery was by no means the only achievement by Lister. When Lister began his career as a surgeon and postoperative infections were raging, the surgical practice included tying off blood vessels with wires or threads that could not be absorbed. The suture thread would be pulled from the filthy frock of the surgeon, as needed. The ends of these sutures would be left long, sticking out of the wound. Once the tissues began decomposing from infection, the material could be pulled out of the patient. Unfortunately, the act of removal often resulted in catastrophic hemorrhage, sometimes the cause of death. When Lister began using his antiseptic methods, wounds would not suppurate. The only way to remove the suture material was to reopen the wound. Lister recalled the use of catgut, made from the intestinal linings of sheep and other animals. Galen and Ambroise Paré used catgut to tie off blood vessels. Its primary value was that the human body could absorb the material, so it could be cut short and not protrude through the incision. After soaking the catgut suture material in carbolic acid, Lister experimented with it in surgery. He eventually found that the catgut material would dissolve in the body but that he could make the suture hold longer by soaking it in salts of chromic acid before the carbolic acid. Chromic catgut remains a mainstay of every operating room today, another achievement we owe to Lister.

At King's College, London, Lister became the second person to operate on a patient with a brain tumor. He developed a method to repair the cartilage in a damaged knee joint and improved the technique of mastectomy.

HONORS AND ACCOLADES

Lister spent 15 years at King's College. In 1881, Lister had the pleasure of introducing Louis Pasteur to Robert Koch during the Seventh International Medical Congress in London. In 1883, Queen Victoria knighted Lister for his accomplishments. As his mission in London eventually became successful and antiseptic surgery was accepted, Lister slowed his work schedule. He and Agnes took well-deserved vacations. They developed a love of bird watching and fly-fishing. Agnes was his most beloved companion, laboratory assistant, and confidant. They were nearly inseparable.

Lister received a host of honors later in his life but was modest and gracious in their receipt. Lister seemed more comfortable giving awards than receiving them. Perhaps this aspect of his personality was best exemplified in 1892 at the 70th birthday celebration of Louis Pasteur. Lister had just recently retired from his position at King's College at the mandatory retirement age of 65. In late December, he came to France to pay tribute to Pasteur as a representative of his country but more as a man who felt personally indebted to the great French scientist. Pasteur owed Lister a debt of gratitude since Lister was the doctor who recognized and disseminated Pasteur's discovery of the germ theory. Lister gave his moving speech in French, congratulating Pasteur and paying homage to the man from a grateful British nation. Lister told the audience that thanks to Pasteur's work—a debt Lister acknowledged repeatedly—surgery had been revolutionized and stripped of its horrors. Looking right at Pasteur, Lister concluded,

> You can well understand, Monsieur Pasteur, that Medicine and Surgery welcome this solemn occasion of doing you honour and of expressing their admiration (20).

Pasteur, frail from a series of strokes, was overcome with emotion. Lister moved to Pasteur and embraced him. The touching scene was captured in a solemn painting by Jean Andre Rixens titled *The Jubilee of Louis Pasteur*. Nearly every important scientist of Europe, with the notable exception of Robert Koch, attended the event. But Rixens chose to show the tribute from Lister since these two men's discoveries so changed the way we view the world.

DEATH OF LADY AGNES SYMES LISTER

In the following spring, Sir Joseph and Lady Lister headed for a vacation in Italy. Agnes developed a shiver. Six days later, she died from pneumonia. Lister was never the same man after losing his close, intimate companion of 37 years. The Listers had no children, so Joseph endured both loss and loneliness.

LISTER'S LATER YEARS

Lister lived an additional 19 years, but his life and career felt hollow without Agnes. In 1895, Lister was elected president of the Royal Society. In 1897, he became the first medical doctor to be elevated in British peer society and bear the title of Baron. In 1903, Lister's health began to deteriorate. While he remained somewhat active, he never had vigor after that year. The occasion of Lister's 80th birthday, 5 April 1907, found widespread celebration, with telegrams and flowers sent to his home. A "Lister Meeting" was held in Vienna at the Surgical Institute. The audience of 500 broke into a loud, sustained ovation when a portrait of Lister was projected above the platform. The viewers knew Lister's health was too poor for him to attend and speculated that he would be on Earth only a little while longer. Joseph Lister died quietly in his home on 12 February 1912. A huge public funeral was held in Westminster Abbey. Leeson attended and described the event:

> It was a never-to-be-forgotten scene. Representatives from universities and learned societies filled the choir; but what was particularly noticeable was the number of nurses who were present, and the many who from their appearance, looked as though they might have been former hospital patients. The vast assembly waited in solemn silence, wrapt in the memory of his sweet personality, and of the priceless benefits he had bestowed upon mankind (21).

Lister was not buried in Westminster Abbey, though he had every reason to have been included there. He left expressed wishes to be placed next to his beloved, Agnes, in the West Hampstead Cemetery. I cannot improve words about Joseph Lister from those written by the Royal College of Surgeons on Lister's passing in 1912:

By the death of Lord Lister, P.C. O.M., F.R.S., F.R.C.S., etc. surgery has lost her most brilliant student and her greatest master, England one of her most famous sons, and the world one of its most illustrious citizens.

He raised surgery from a dangerous and precarious practice to a precise, safe, and beneficent art, and in doing so his name became renowned throughout the civilized world.

His methods have been adopted in every clime and country, and the benefits, which flow from his discoveries are blessings conferred upon every race of mankind.

His perspicacity, natural insight, fertility of resource, power of close and discriminating observation, philosophical reasoning, inflexible pursuit of truth, steadfastness of purpose, capacity for taking pains, unwearied patience, and undaunted efforts to triumph over difficulties stamped him as a great example of a true and scientific genius.

The human sympathy which caused him to deplore the great mortality due to infective surgical disease, his solicitude to prevent suffering and premature death, his patience and unceasing labour to overcome them, and his gratification at the ultimate success of his efforts to ameliorate pain and prolong life eminently distinguish him as a great philanthropist.

His gentle nature, deep compassion, imperturbable temper, resolute will, indifference to ridicule and tolerance to hostile criticism combined to make him one of the noblest of men.

His work will last for all time; its good results will continue throughout all ages; humanity will bless him evermore, and his fame will be immortal. (21)

And so, it has.

REFERENCES

1. **Darwin C.** 1958. *In* Barlow N (ed), *The Autobiography of Charles Darwin*. W W Norton, New York, NY.
2. **Cheyne WW.** 1925. *Lister and His Achievement*. Longmans, Green and Co, London, United Kingdom.
3. **Hunter J.** 1796. *A Treatise on Blood, Inflammation, and Gunshot Wounds*. Thomas Bradford, Philadelphia, PA.
4. **Leeson JR.** 1927. p 48–55. *In Lister as I Knew Him*. William Wood and Company, New York, NY.
5. **Leeson JR.** 1927. p 135–164. *In Lister as I Knew Him*. William Wood and Company, New York, NY.
6. **Nuland SB.** 1988. p 343–385. *In Doctors: The Biography of Medicine*. Random House, Inc, New York, NY.
7. **Guthrie D.** 1949. p 42–53. *In Lord Lister: His Life and Doctrine*. Williams and Wilkins Co, Baltimore, MD.
8. **Guthrie D.** 1949. p 54–66. *In Lord Lister: His Life and Doctrine*. Williams and Wilkins Co, Baltimore, MD.
9. **Lister J.** 1867. New method of treating compound fracture, abscess, etc. *Lancet* **90:**95–96. http://dx.doi.org/10.1016/S0140-6736(02)51390-8.
10. **Lister J.** 1867. On the antiseptic principle in the practice of surgery. *BMJ* **2:**246–248. http://dx.doi.org/10.1136/bmj.2.351.246.
11. **Lister J.** 1870. Effects of the antiseptic system of treatment upon the salubrity of a surgical hospital. *Lancet* **95:**40–42. http://dx.doi.org/10.1016/S0140-6736(02)31303-5.
12. **Leeson JR.** 1927. p 105–120. *In Lister as I Knew Him*. William Wood and Company, New York, NY.

13. **Lister J.** 1870. A method of antiseptic treatment applicable to wounded soldiers in the present war. *BMJ* **2**:243–244. http://dx.doi.org/10.1136/bmj.2.505.243.

14. **Anonymous.** 1875. Sketches of continental hospitals. *Lancet* **108**:875.

15. **Clarke EH, Bigelow HJ, Gross SD, Billings JS.** 1876. p 213. *In A Century of American Medicine, 1776–1876.* Henry C Lea, Philadelphia, PA.

16. **Herr HW.** 2007. Ignorance is bliss: the Listerian revolution and education of American surgeons. *J Urol* **177**:457–460. http://dx.doi.org/10.1016/j.juro.2006.09.066.

17. **Boynton S.** 1881. President Garfield's case. *Am Obs Med Mon* **18**:493.

18. **Nuland SB.** 1988. p 386–421. *In Doctors: The Biography of Medicine.* Random House, Inc, New York, NY.

19. **Gilmore OJA.** 1977. 150 years after. A tribute to Joseph Lister. *Ann R Coll Surg Engl* **59**:199–204.

20. **Leeson JR.** 1927. p 203–223. *In Lister as I Knew Him.* William Wood and Company, New York, NY.

21. **Leeson JR.** 1927. p 182–202. *In Lister as I Knew Him.* William Wood and Company, New York, NY.

12 Paul Ehrlich and the Magic Bullet

Despite staggering progress in the fight against infectious diseases during the latter half of the 19th century, there were still few advances in treatment. Vaccinations could only prevent disease. Even Lister's marvelous developments in surgery prevented infection; they could not treat infection once it developed.

By the last decade of the 19th century, the only effective medical treatment for an infectious disease was quinine, a plant derivative used to treat malaria even today. The malady is an ancient disease with clinical descriptions at least as far back as the time of Hippocrates. Malaria, derived from the Latin, *mala aria*, or bad air, was historically ascribed to miasma from swamp air. The disease produces a cycle of fevers and chills every third (tertian) or every fourth (quartan) day. Our current knowledge makes it clear that one particular species, *Plasmodium malariae*, has a life cycle 1 day longer (quartan) than the other *Plasmodium* species, *Plasmodium vivax* and *Plasmodium ovale*, that regularly infect humans and have an established fever cycle (tertian). The most serious form of malaria is from the species *Plasmodium falciparum*, which does not normally produce a fever cycle. But until the discovery of the parasites in the late 19th century, doctors had to diagnose the disease based upon the clinical symptoms of fever and chills, occurring in tertian or quartan cycles.

Therapy for malaria came long before discovery of the parasite. The most common story of the origin of malaria treatment is almost certainly an exaggerated saga. The story involves the Countess of Chinchón, wife of the Viceroy of Peru. She was reportedly cured of her cyclical febrile episodes, termed *agues*, by ingesting

Germ Theory: Medical Pioneers in Infectious Diseases, Second Edition. Robert P. Gaynes.
© 2023 American Society for Microbiology.

bark of the Cinchona tree around the late 1620s (1). Her husband may have been the one who suffered from malaria. Jesuit priests of Peru documented the Amerindian practice of making a powder from the bark of the Cinchona tree and drinking it in a beverage to treat a variety of agues. In 1633, an Augustinian monk, Fray Antonio de la Calancha, made the first written record in Latin of the successful use of the bark. He described its widespread use for ague in Lima, Peru (2). How the bark got from Peru to Europe is a matter of great debate. Either the Countess and her husband or the Jesuit priests may have introduced it to Spain and Italy sometime in the 1630s. An ecclesiastic, Cardinal Juan de Lugo, is credited with purchasing large amounts of the bark and distributing it with great success in Rome, where malaria was endemic at the time. From there the costly bark was taken via church couriers to other Italian cities, England, Germany, and Belgium. Spain had received the bark directly from Peru. The chronicle of the bark of the Cinchona tree was filled with intrigue after demand in Europe outpaced supply. Cheating became common. Other pulverized bark was substituted, creating some doubt about the effectiveness of treatment. In the 17th century there were no assays, biological or chemical, to ascertain whether the material was actually going to be effective. Indeed, no one had any idea what the active ingredient was. As European powers attempted to colonize malarial regions of the globe, several countries attempted to retrieve the Cinchona tree to cultivate the so-called fever tree back home. They generally failed in their attempts to bring live trees back to Europe. The French attempted to cultivate it in Algeria, but the climate was too dry. Even the name of the tree was poorly handled in Europe. In 1742, the Swedish naturalist Linnaeus included the tree in his famous taxonomy of the natural world, dubbing it "Cinchona." It should have been spelled Chinchona (from the region where it was first found), but the error was never corrected. When political instability occurred in South America in the late 18th century, the supply of the Cinchona bark in Europe was seriously threatened.

Scientists attempted to discover the active ingredient of the Cinchona bark for years. Finally, in 1820, two French chemists isolated two active compounds from the bark: quinine and cinchonine. Physicians quickly discovered that it was quinine that was the effective treatment of tertian and quartan fevers. Extraction of quinine from the bark proceeded. Attempts to synthesize quinine chemically failed for over 100 years until 1944, when two Harvard scientists, Robert Woodward and William Doering, finally succeeded. But the failed attempts to synthesize quinine had an unexpected windfall. In 1847, William Henry Perkin, an 18-year-old chemistry student, tried to synthesize quinine but by accident produced the first aniline dye, mauveine (3).

The link between the development of dyes and therapy for infectious diseases is a curious one. Paul Ehrlich is found right in the middle of it. To understand this connection, we must go back to the origin of the European dye industry and then to the life of Paul Ehrlich and his early theories. In the early part of the 19th century gas, mostly ethylene gas, was used in the lamps that illuminated streets and buildings. The ethylene came from coal distillation, a process, you will recall, first developed in the medieval Islamic culture. Around 1812, the demand for ethylene gas accelerated the distillation of coal. Coal tar was the by-product of this process, accumulating in large amounts until a new use was discovered. Boiling and distilling tar produced two oils: creosote and pitch. Creosote was used to preserve wood. Pitch was sold to manufacture asphalt. After the discovery of the first aniline dye in 1847, a German chemist, August Hoffmann, determined in 1850 that creosote contained both benzene and aniline. Hoffmann was able to use creosote and other waste products to manufacture a variety of aniline dyes. Germany quickly became the center of the European dye industry, producing aniline dyes in an extensive variety of colors.

Among other uses, the new dyes produced in Germany had medical applications in staining tissues. Rudolf Virchow's scientific expertise placed Germany as the leading cellular pathology center in the world in some small part because of the availability of various aniline dyes to stain human tissues. At first, Paul Ehrlich also used these dyes to study human tissues, but later he began to speculate on their therapeutic uses. His efforts set in motion the discovery of new therapeutics and paved the way for an entire pharmaceutical industry for the next century. Curiously, aniline dyes have links to another of Ehrlich's contributions that sparked a new field—immunology. The achievements of Paul Ehrlich are sometimes overshadowed by those of some of his contemporaries, like Robert Koch and Joseph Lister. But Ehrlich provided a vital impetus for a revolutionary change in treatment and in the understanding of the body's defenses against infections.

EARLY INFLUENCES

Paul Ehrlich was born on 14 March 1854 in the Prussian town of Strehlen (now Sztrelin, in Poland). His father, Ismar Ehrlich, was a prosperous Jewish innkeeper, described as a man of cheerful manner but detached and odd, sometimes sitting at a window for hours, talking hurriedly to himself. His mother, Rosa Weigert, seemed to compensate for this idiosyncratic behavior, personally tending to customers of the inn and taking wonderful care of the household, which included three daughters in addition to their son. Ehrlich's grandfather imparted to the boy an interest in natural science and chemistry. Paul Ehrlich seemed to inherit his father's speedy pattern of speech that seemed to show impatience and nervousness, but he exhibited his mother's warmth as well. His secretary and biographer, Martha Marquardt, described Ehrlich speaking with vivid forms of expression, often with visual references (4). His first cousin, Karl Weigert, was 9 years Ehrlich's senior. Weigert exhibited his own scientific prowess as a professor of pathological anatomy and saw the scientific potential in his younger relative. Weigert exerted a profound influence on Ehrlich and remained a close friend throughout his life.

EHRLICH'S DISCOVERY OF THE MAST CELL

Ehrlich attended the University of Breslau, near his hometown, for a term. Ehrlich was disappointed in his studies, showing interest only in biology and histology, though he showed an exceptional grasp of organic chemistry. Ehrlich transferred to the University of Strasbourg in the middle of the 1870s. The prominence of the German chemical industry in aniline dyes frequently found biological applications at the university. Ehrlich became fascinated, almost obsessed, with the subject. On his own initiative, he began to experiment with histologic staining of tissues, almost to the exclusion of his other university courses. Fortunately for Ehrlich, one of his professors, Heinrich Waldeyer, recognized Ehrlich's interest and understood his peculiarities. Waldeyer gave Ehrlich relatively free rein in his laboratory. Ehrlich quickly rewarded his professor with a discovery. Using various dyes, Ehrlich identified a cell with prominent granules that could be selectively stained with certain dyes, which he called mast cells. We know these cells to have granules filled with a chemical called histamine. Mast cells, with their histamine granules, are prominent in tissues of persons with allergic reactions. Histamine is the chemical that produces the flushing and watery eyes in someone having an allergic reaction. The chemical also changes the permeability of capillaries, allowing white blood cells and other proteins to engage foreign substances. The change in permeability of capillaries and the resulting flood of cells and proteins in tissues caused Ehrlich to believe that the cells had a nutrient role. He named them *Mastzellen*, from the German *Mast*, meaning feed, and *zellen*, meaning cells. In his third term, Ehrlich passed his Physikum, the German equivalent of an examination over the material from the first and second years of medical school.

FIRST MEETING WITH ROBERT KOCH

Ehrlich returned to Breslau on the advice of his cousin, Karl Weigert, who encouraged him to pursue an opportunity to work with Julius Cohnheim, another prominent pathologist. During his time at Breslau, Ehrlich was working in Cohnheim's laboratory when he was introduced to a visiting physician from nearby Wöllstein who had come to the university to demonstrate his work on the anthrax bacillus. The physician, Robert Koch, was introduced to Ehrlich and later told, "That is 'little Ehrlich.' He is very good at staining, but he will never pass his examinations" (5).

Such was the humble first meeting of two future Nobel Prize-winning physicians who would work together years later.

DOCTORAL DISSERTATION: THEORY AND PRACTICE OF HISTOLOGIC STAINING

Ehrlich finally passed his Breslau examinations with a mark of "excellent." Ehrlich then completed his medical studies at the University Leipzig, graduating with a Doctor of Medicine degree in 1878. He wrote his doctoral dissertation on staining methods with aniline dyes and their significance in medicine. It was titled "Contributions to the Theory and Practice of Histologic Staining. Part I. The Chemical Conception of Staining. Part II. The Aniline Dyes from Chemical, Technological and Histologic Aspects." His thesis, revolutionary for its time, emphasized the *theory* of staining tissues: i.e., why do some tissues take up certain stains? Ehrlich reasoned that there was a chemical bonding of different substances to protoplasm, a process not often considered by pathologists and biologists who were busy simply using the stains. Ehrlich argued that (i) staining reactions are purely chemical in nature; (ii) staining reactions show a marked degree of specificity; and (iii) to a large extent, structure defines function—i.e., the chemical side chains that are attached to the aniline core of a dye define the strength of the attachment to the cell. These theoretical concepts, the importance of chemical reactions in cells and the specificity where structure defines function, would influence Ehrlich's lifework. Review of the original manuscript of the thesis also showed Ehrlich's peculiar style of handwriting in the margins, where he shunned capitalization and most accepted forms of punctuation (5). In 1878, Ehrlich passed his state examination to become a physician and began his professional career.

THE CHARITÉ HOSPITAL IN BERLIN

At the age of 24, Paul Ehrlich secured his first position at the Charité Hospital in Berlin. Ehrlich was fortunate to have Friedrich von Frerichs as his chief. From very early in his career, von Frerichs recognized the unusual and distinctive capabilities of his young assistant. Ehrlich was skilled in diagnosis, but von Frerichs did not burden him with excessive clinical duties. He gave Ehrlich a free hand to work in his laboratory and carry out experiments. Ehrlich quickly earned a good reputation for staining preparations of blood from hospitalized patients. He was an expert stainer with exacting techniques. Ehrlich soon garnered such a reputation that crowds of students, followed by university professors, came to him to learn his staining techniques. Ehrlich's staining of blood smears helped to distinguish among the various kinds of white blood cells, notably basophils, eosinophils, and neutrophils. His work set the foundation for the developing branch of medicine called hematology. Ehrlich was not content to merely stain blood or tissues. He strove to understand how the stain got into the cell, speculating that its entry was dependent on the size of the molecule. Ehrlich tried to determine to what component of the cell the stain was binding.

IMPROVING THE IDENTIFICATION OF THE TUBERCLE BACILLUS

After several years in his position in Berlin, Ehrlich attended a meeting of the Physiological Society of Berlin in 1882. It was at this sensational event that Robert Koch announced his discovery of the tubercle bacillus. Ehrlich later described the evening as the greatest scientific experience of his life. On the night of the lecture, an inspired Ehrlich experimented with dyes. He improved upon Koch's staining technique the very next day. The enhancement came from a happy accident. Ehrlich had left some slides with stained tissues containing the tubercle bacteria on top of a small stove to dry overnight. Without seeing the slides, the cleaning lady had lit the fire in the stove in the morning. When Ehrlich came into the laboratory, he rushed to save his specimens, fearing the worst. But the heat had improved the staining, allowing easier identification of the bacilli under the microscope. The mishap led to one of Ehrlich's "Big G's." In this case, he used the German word *Glück*, or luck, which Ehrlich said was important for successful work. He also included *Geduld, Geschick, und Geld*—patience, ability, and money—in this adage. Ehrlich summoned Koch immediately to tell him of his discovery. Koch acknowledged the improvement in a publication. Thus began a long period of collaboration between the two men. Ehrlich participated in later trials of tuberculin as a remedy for tuberculosis. Even as it became clear that tuberculin was a treatment failure, Ehrlich stood by Koch.

A year later, Ehrlich married Hedwig Pinkus, the daughter of a Jewish industrialist, who was to remain his understanding, faithful, and lifelong companion. The couple had two daughters. They, like all visitors to Ehrlich's office, had to endure his practice of smoking strong, black cigars—a lifetime habit that occasionally presented a quandary when Ehrlich repeatedly and insistently offered the cigars to guests.

In 1885, Ehrlich published a basic research monograph titled "The Oxygen Requirement of the Organism" that held a key position in his developing theories. Ehrlich extended the view that was evident in his doctoral dissertation concerning the study of dyes and intracellular physiology by introducing the idea that color changes and staining properties of dyes are related to the oxidation/reduction capacity of the cytoplasm of a particular cell (6). While the connection from this monograph to infectious diseases, immunity, and chemotherapy may not be readily apparent, it formed the basis of Ehrlich's side chain theory (see below).

Additionally, Ehrlich speculated that since most bacteria had a high oxygen requirement, cells that provided a reducing environment, i.e., cells that bound oxygen strongly themselves, would provide a hostile setting for such microorganisms and thus would be involved in immunity.

DEATH OF VON FRERICHS

Paul Ehrlich's career changed with the sudden death of his first chief, von Frerichs, in 1885. His successor, Carl Gerhardt, did not allow Ehrlich the freedom to which he had become accustomed. Gerhardt forced Ehrlich to increase his clinical duties, which Ehrlich disliked. To make matters worse, Ehrlich acquired tuberculosis, which he diagnosed himself from staining his own sputum. He used the disease as a reason to quit his job at Charité Hospital and move his family to the dry climate of Egypt for the next 2 years. His cousin, Felix Pinkus, wrote much later that pulmonary tuberculosis was not the primary explanation as to why Ehrlich left Berlin:

> Ehrlich's mind could not bear any bonds. Just as a highly-strung racehorse would end by quivering helplessly in the yoke, breaking down as a result of nervous excitement without advancing or making any effective effort, so Ehrlich's body would pine away when his spirit was fettered . . . After the sudden death of his patron and friend, von Frerichs, the greatest disaster which had befallen Ehrlich, he was forced into drudgery, was compelled to march the old road of clinical routine. He could not endure this, and visibly began to fade away until he had to free himself in order to save his life. His illness was called tuberculosis of the lung, and he had the clinical symptoms of that disease. But what he was suffering from mostly was constraint (7).

BACK TO BERLIN

In 1889, the Ehrlichs returned to Berlin. Strong and healthy again, Paul Ehrlich was content to set up his own laboratory on a small scale, financed by his father-in-law, and work independently. Ehrlich worked on plant toxins, ricin and abrin, in his meager laboratory. In an 1891 publication, Ehrlich wrote,

> In the course of my investigations on the relationship which exists for a vast number of substances between chemical constitution, the distribution within individual organs, and physiological activity, as was quite obvious from systematic experiments, I was also necessarily led to the meaningful study of poisonous proteins . . . the same relationship ought to be of interest for an understanding of infectious diseases (6).

Ehrlich connected chemical science to immunity with his work on studies with ricin that he carried out in his laboratory. He showed that using small, increasing doses of ricin could produce a tolerance or immunity to their effects in animals. He showed that there was a substance present in blood that counteracted the effects of ricin. Most importantly and for the first time, Ehrlich distinguished between active immunity (giving ever-increasing doses of ricin to an animal to actively make it immune) and passive immunity (giving an experimental animal the substance from the blood of another animal which had been made immune). Ehrlich noted that active immunity has a long duration; passive immunity is appreciably shorter. These concepts were groundbreaking and years ahead of their time, as we shall see.

Robert Koch, who had been appointed the director of the newly founded Institute for Infectious Diseases, offered Ehrlich a position at the Institute in 1891. Ehrlich leapt at the offer. Ehrlich was now positioned to help and eventually lead in what would become one of the most extraordinary periods in the history of infectious diseases.

DISCOVERY OF THE DIPHTHERIA ANTITOXIN AND SERUM THERAPY

Modern therapy for infectious diseases did not begin with penicillin, sulfa drugs, or even chemical/drug therapies. Modern therapy for infectious diseases started with a serum treatment for diphtheria. The story behind the "miracle treatment" began a few years before the time Ehrlich joined Koch's staff. In 1884, Friedrich Loeffler, working in Berlin under Koch, had characterized the causative agent of diphtheria, *Corynebacterium diphtheriae*. The disease was an upper respiratory illness characterized by sore throat, low-grade fever, and a characteristic adherent membrane, called a pseudo membrane since it was mostly comprised of bacteria and acellular material, on the tonsils. Swelling of the tonsils or associated lymph nodes in the neck could cause breathing difficulties, sometimes completely obstructing the upper airway. But the virulence of the disease was largely related, as we know today, to a toxin produced by the causative bacterium. As diphtheria progressed, the toxin would be carried throughout the body. The toxin was quite potent; very small amounts could cause serious complications, mostly to the heart and the nervous system. We now know that toxins would interrupt protein synthesis of cells, leading to heart failure or an interruption of the heart's rhythm; either could lead to death. Alternatively, nerve function could be affected, leading to paralysis, notably of the respiratory muscles. Unless placed on a ventilator, which was unavailable at the turn of the 20th century, the paralyzed person would die of respiratory failure. Tragically, diphtheria was generally a disease of children. In 1892, an estimated 50,000 children had diphtheria in Germany alone; over 50% died from the disease (8).

In 1888, Émile Roux and Andre Yersin from Institut Pasteur in Paris isolated the diphtheria toxin. They showed that it was the toxin that could produce the disease known as diphtheria. Efforts at the Institute for Infectious Diseases in

Berlin were aimed at finding a way to abate the intoxication induced by the bacterial poison. Emil von Behring took a position at the Institute in mid-1889. He had been fascinated with bacteriological research for several years and eagerly accepted a position with Koch. Behring undertook an experiment where he infected guinea pigs with *Corynebacterium diphtheriae* and then treated them with iodotrichloride. A few of the infected animals survived. Behring then reinfected the surviving animals with what should have been lethal amounts of the toxin-producing diphtheria bacterium. This time, none developed disease. Behring then performed a crucial experiment that led to an entirely new therapy. He injected a new group of guinea pigs with the diphtheria toxin but also injected them with sera from animals that had survived his initial experiments. The serum-treated guinea pigs survived! The principle of an antitoxic effect of serum therapy had been established. So had a new term originated by von Behring—antitoxin. Simultaneously, working at the Institute for Infectious Diseases as well, Japanese-born Shibasaburo Kitasato found analogous results working with rabbits and the tetanus toxin. The possibility of treatment, even cure, of these deadly diseases using serum antitoxins could not be denied.

CELLULAR IMMUNITY AND HUMORAL IMMUNITY

The experiments by Kitasato and von Behring helped to develop a new theory of immunity. In 1884, a Russian scientist working at Institut Pasteur named Ilya Mechnikov had discovered the principle of phagocytosis, the cellular process where a particle is engulfed by the cell membrane. This finding led to the concept of cellular immunity. But the investigators at Koch's Institute, among others at the time, had formed a thesis that something other than a cell might be responsible for immunity—the notion of humoral immunity. Humoral immunity, as we know it today, refers to the formation of antibody and accessory acellular processes to deal with foreign substances. But it was only a theory. To develop practical production of serum to treat humans was a formidable obstacle. This was the situation when Paul Ehrlich began his work at the Institute for Infectious Diseases in Berlin. Ehrlich became acquainted with Emil von Behring and developed a close personal relationship with the man, who, coincidentally, was the same age as Ehrlich, born only 1 day later.

STANDARDIZATION OF DIPHTHERIA ANTITOXIN

The two friends began to tackle the practical problem of serum production for treatment of diphtheria. Ehrlich had experimental experience giving ever-increasing doses of ricin to animals so that they could tolerate up to a 1,000-fold-increased dose of toxin that would have killed an untreated animal.

He worked out the precise method of measuring the curative value of the antitoxin using units defined in relation to a fixed standard. Soon, potent antitoxin to the diphtheria toxin was obtained from horses. British physiologist Sir Henry Dale wrote in his 1954 reminiscence of Paul Ehrlich,

> Ehrlich knew from the first . . . that a biological measurement, like any other measurement, has no real meaning unless it is made in relation to a fixed, invariable standard. And it may be noted yet further that von Behring's contribution to the practice of active immunization against diphtheria did not extend much beyond the initial discovery and recommendation; so that in any account of its practical and systematic application, now so widely successful, it would be necessary to mention the names of a number of others [in addition to Ehrlich] who were responsible for its first trials (9).

In collaboration with the company Meister Lucious and Brüning in Höchst, a district of Frankfurt, large-scale production of serum containing diphtheria antitoxin began. Diphtheria patients on the pediatric wards of several Berlin hospitals were given the standardized serum, with great success. Ehrlich carefully assessed the results of the experimental therapy, noting three tenets:

Treatment needed to be initiated at the onset of disease.

The later in the course of the disease, the higher the serum quantities needed for cure.

Minimal dosing could be recommended, depending upon the severity of the disease.

The success of antitoxin for diphtheria made von Behring a worldwide celebrity; he became known as the "Children's Savior." von Behring loved the limelight but did not always credit Ehrlich and others who had helped him. Soon a rift developed between the two friends. In 1893, von Behring wrote out a contract with Ehrlich, splitting the acquisition costs and describing the distribution of any revenues from the serum production. But von Behring and a factory representative called Ehrlich urgently to a meeting. von Behring spoke of a state institute for research where Ehrlich could be the director. von Behring said he would use all his influence to get Ehrlich appointed to this position. von Behring pointed out that in the event of this appointment, Ehrlich would have to resign from his contract due to conflict-of-interest issues with someone who was a government employee. Ehrlich was delighted with the prospect of becoming a director of a research institute, so much so that he paid inadequate attention to the specifics of the contract. When von Behring's influence was not sufficient to create the research institute and keep his promise, Ehrlich was crushed. Meanwhile, von Behring established the "Behring-Works" and kept much of the income from the success of the serum treatment, making him a wealthy man. By 1894, von Behring had left Berlin. The two men never saw each other again. In 1899, Ehrlich wrote a friend who inquired about the experience,

> I am not surprised at what you tell me about Behring. He has reaped as he sowed . . . He owes his success with the diphtheria serum, especially his big material success, to *me*. When we started to work together his serum contained only 1/4 to 1/2 a unit of antitoxin per cc, while mine had thirty. I had worked with ruminants, goats, and cows, whereas he had used horses, which give much stronger antitoxins. When, with great difficulty and after nearly nine months of hard work, he had, by using my methods, reached 100-150 units, he tried to bring me into discredit, pretending that I knew nothing about immunization and that he obtained a much higher unitage than I did, etc. . . . I always get wild whenever I think of that dark period and the way in which Behring tried to hide our scientific partnership. But the revenge has come. He can see how far he has got without me since our separation (10).

Behring actually received the first Nobel Prize in Medicine for "his" work on serum therapy in 1901. In his Nobel address, he made only one passing mention of Ehrlich's name and never mentioned Kitasato's name at all. But Ehrlich turned out to be correct. von Behring never did achieve much significant scientific success beyond his antitoxin discovery. In the end, Ehrlich's quantitative approach to standardizing the diphtheria antitoxin had a lasting effect. Thorvald Madsen, who became one of the world's authorities in immunology in the early days of the field, stated,

> Ehrlich's immunity unit plays the same role for antitoxin measurement as does the Standard Meter for the measurement of length (6).

However, Paul Ehrlich was just getting warmed up.

THE STEGLITZ INSTITUTE AND THE ROYAL PRUSSIAN INSTITUTE FOR EXPERIMENTAL THERAPY IN FRANKFURT

In 1896, Paul Ehrlich did achieve his dream of becoming the director of an institute when he was invited to become Director of the State Institute for the Investigation and Control of Sera in Steglitz, a borough of Berlin. The small German government institute was hampered by meager budgets and limited resources. Ehrlich did not mind. He was finally a director. Ehrlich's enthusiasm and energy in the laboratory were obvious to any visitor. Ehrlich was quoted as saying, "As long as I have a water-tap, a flame, and some blotting paper, I can work just as well in a barn."

Ehrlich remained productive during his time in Steglitz, publishing scientific papers and standardizing antisera including diphtheria and tetanus. But for much of his time in the tiny institute, Ehrlich was aware of a coming change in his career. In the late summer of 1896, the mayor of Frankfurt, Franz Adickes, announced the founding of an institute for experimental therapy in his city. Three years later, in 1899, the Royal Institute for Experimental Therapy, under the direction of Paul Ehrlich, opened in Frankfurt. There, Ehrlich would spend the rest of his life and make some of the most important discoveries of his career.

THE SIDE CHAIN THEORY—THE FIRST THEORY OF ANTIBODY PRODUCTION

In Frankfurt, Ehrlich and his colleagues continued to regulate sera as they had done in Steglitz, so the Royal Institute for Experimental Therapy was often referred to as the Serum Institute. But Ehrlich had time available to perform research in other areas of interest. From his time in Steglitz he had been working on a theory that developed from his visual mind, from his staining theory in his doctoral dissertation, and from the work on the diphtheria antitoxin. It became known as the side chain theory. When a government official came to visit Ehrlich in 1897, he queried Ehrlich about a recent publication on the subject, asking him to explain, in person, what this theory was. His personal secretary, Martha Marquardt, later recorded Ehrlich explaining his own theory.

> Cells have the ability to attract foreign chemical substances which have specific chemical relationship to the substances of the cell itself. Whenever such substances come in contact with the cell, a chemical binding takes place. This is as close, and as well adjusted, as a key is to its lock. . . . Still another, better picture . . . In order to incorporate the chemical stuff, the cell will, so to speak, stretch out arms, "receptors", to catch and get hold of the substances. The group giving the foreign substance its affinity for the cell receptor, the haptophore group, will be caught by the receptor, the catching arm, and thus anchored to the cell protoplasm. If the foreign substance has also a toxophore group, giving it poisonous properties, the cell may be killed. If it survives, the receptors will be regenerated in excess, some will float off into the serum, and their function as a specific antibody for foreign substance, having an affinity for it (11).

While the side chain theory got many of the mechanisms of antibody formation wrong, it was the first model of antibody formation in immunity. The side chain theory anticipated theories of later immunologists, including Niels Jerne's network theory of antibodies and Frank Macfarlane Burnet's theory of clonal selection, by over 50 years (6, 12, 13)! Ehrlich had trouble convincing his colleagues, who thought the idea was madness. There were problems with this theory that began to cause difficulties for Ehrlich. For example, critics asked how the body could possibly have prepared receptors for such a bewildering variety of potential antigens, biological and chemical. The side chain theory fell out of favor, awaiting immunologists and geneticists who determined the mechanisms by which immunologic diversity could occur. Ehrlich became impatient with the criticism. He countered some, but not all, of the criticisms with further experiments, but he began to turn his attention to developing a receptor theory using chemicals, not the body's immune defenses. His side chain theory suggested to Ehrlich that there was a specific interaction between the antitoxin and toxin, as specific as a key fitting a lock. He began to take his concept of chemical specificity with dyes and cells from his doctoral thesis and put it together with the key-and-lock receptor theory. He theorized that it should be possible to isolate or synthesize small chemicals that would act specifically on an infecting microorganism, leaving the host unaffected. He would spend most of the remainder of his life looking for the "magic bullet."

THE MAGIC BULLET: THE DAWN OF CHEMOTHERAPY FOR INFECTIOUS DISEASES

Today, we take for granted the concept of treating infectious diseases using drugs. But aside from quinine, chemicals or drugs that could effectively treat infections were rare before the beginning of the 20th century. There was no standardized method to search for them. Quinine was discovered by accident. It took over a century to even figure out what the active ingredient in Cinchona bark was. Many of the major causative bacteria were identified in the late 19th century. But diagnosis did not immediately lead to treatment. How would one even start a search for such medicines? Paul Ehrlich developed the theory and the method.

Ehrlich's search for therapeutic agents started with his work with dyes, especially methylene blue. Ehrlich found that this dye seemed to have a special affinity for nerve fibers of certain animals. He injected it into living animals and found that it stained a tiny parasitic worm in the bladder of a living frog. He removed the worm and could see that the methylene blue had stained the nerve and muscle fibers of the worm as it crawled along. He reasoned that, like the lock-and-key type of specificity he had theorized about during his work on immunity, some dyes might have specificity for parasites but could cause the parasite to die and leave the host relatively untouched. In describing his concept, Ehrlich wrote,

> Curative substances—a priori—must directly destroy the microbes provoking the disease; not by an action from a distance but only when the chemical compound is fixed by the parasites. The parasites can only be killed if the chemical has a specific

affinity for them and binds to them. This is a very difficult task because it is necessary to find chemical compounds, which have a strong destructive effect upon the parasites, but which do not at all, or only to a minimum extent, attack or damage the organs of the body. There must be a planned chemical synthesis: proceeding from a chemical substance with a recognizable activity, making derivatives from it, and then trying each one of these to discover the degree of its activity and effectiveness. This we call *chemotherapy* (14).

Paul Ehrlich coined the term chemotherapy and was eager to put it to use. In 1904, Ehrlich began working with trypanosomes. These protozoa produced a disease in mice similar to African sleeping sickness in humans. Ehrlich had noted that one particular dye, benzopurpurine, could remain inside a living organism for a long period. He tested it in animals with trypanosomiasis. It seemed to have a small but definite effect on survival in the treated compared to the untreated mice with the disease. He believed that the reason for its suboptimal effect was related to its poor solubility. A colleague at a German pharmaceutical company made the compound more soluble by synthesizing a derivative with an additional sulfonic acid group. Ehrlich tried it in the infected mice again, with great success. The dye, now called trypan red, after the action on trypanosomes, was the world's first chemotherapeutic agent. Unfortunately, the dye did not work well as a treatment for African sleeping sickness in humans, but the theory now had proven merit.

Investigators in Liverpool, England, had reported in 1906 some experimental success in treating trypanosomal infections with an arsenical compound called Atoxyl. Ehrlich had tried it but did not find success with the chemical. Moreover, use of the compound in humans with sleeping sickness showed a serious toxic effect on the optic nerve, running the risk of blindness. Atoxyl was virtually abandoned. But Ehrlich was not ready to give up. He investigated the chemical structure of Atoxyl and found differences from the accepted chemical structure. He began to experiment with derivatives of the compound.

In September of 1906, Ehrlich's laboratory became part of the new Georg Speyer Haus for Chemotherapeutical Research. The institute was endowed by the wealthy widow of Georg Speyer with the expressed purpose of allowing Ehrlich to continue his work on chemotherapy and was built next to Ehrlich's existing lab. Ehrlich and his assistants began to see some progress with various derivatives of Atoxyl. They tested hundreds. Considerable activity was noted against trypanosomes with arsenophenylglycin, compound 418. The testing continued until Ehrlich came to compound 606, chemically known as dioxy-diamino-arseno-benzene dihydrochloride. The records show that this compound showed no activity against trypanosomes. In retrospect, the reasons to explain compound 606's lack of activity were unclear. The activity of compound 418 was so active in animals that it may have overshadowed compound 606's activity. Alternatively, some of the testing may not have been properly performed. Either way, further experiments with compound 606 were curtailed when Ehrlich's assistant for these experiments left his lab.

THE NOBEL PRIZE

In 1908, Ehrlich received news that he and Ilya Mechnikov of Institut Pasteur had been awarded the Nobel Prize in Medicine for their groundbreaking studies in immunology. In the midst of his work, Ehrlich headed to Stockholm, Sweden, for the award, clutching two boxes of cigars under his arm. Fearing he would be unable to secure his favorite brand in Sweden, he brought his own supply for the trip. In his Nobel address, Ehrlich discussed his work on humoral immunity that had brought him to that day, specifically the side chain theory. He also included a discussion of his ongoing work on chemotherapy, explaining how it was an outgrowth of his side chain theory and his work with dyes (15).

Early on, the Nobel Prize committee seemed eager to award prizes in medicine to those working on infectious diseases and immunology. Five of the first eight Nobel Prizes in Medicine were awarded to men who worked on infectious diseases and/or the immunologic response to infection (Table 12.1). Since 1909, there have been 93 years in which a Nobel Prize in Medicine has been awarded. In only 20 of those years was the Nobel Prize awarded for an infectious disease or immunologic discovery. The predominance of awards in the first decade of the 20th century probably represents both the enormous activity in the field and the relative importance of infectious diseases in society at the turn of the 20th century.

When Ehrlich returned to Frankfurt from the ceremonies in Stockholm, his many friends organized a party to suit the man. Rather than a stuffy, formal dinner with speeches from various dignitaries, a beer party was held in a large hall. Ladies were not admitted. Even Ehrlich's wife and two daughters could only watch from the galleries above.

TABLE 12.1 Some early recipients of the Nobel Prize

Year	Recipient and reason for award
1901	Emil Adolf von Behring, for his work on serum therapy, especially its application against diphtheria
1902	Ronald Ross, for his work on malaria
1905	Robert Koch, for his investigations and discoveries in relation to tuberculosis
1907	Charles Louis Alphonse Laveran, in recognition of his work on the role played by protozoa in causing disease
1908	Paul Ehrlich and Ilya Ilyich Mechnikov, in recognition of their work on immunity

COMPOUND 606—SALVARSAN

In the spring of 1909, Sahachiro Hata, a pupil of Ehrlich's old friend, Kitasato, came to Frankfurt to work with Ehrlich (Fig. 12.1). Hata came to work on syphilis, not trypanosomiasis. In 1905, Fritz Richard Schaudinn and Erich Hoffmann had determined the causative bacterium for syphilis, *Treponema pallidum*. No one had yet ascertained methods to artificially cultivate the microorganism, but

FIGURE 12.1 *Paul Ehrlich (left) and Sahachiro Hata (right). Courtesy of the Hata Memorial Museum, Shimane.*

Hata had developed an animal model of the disease in rabbits. Ehrlich suggested trying the arsenicals in the syphilis rabbit model, simply because he reasoned that both parasites had high metabolic rates. Martha Marquardt describes the effort that went into the testing.

> No outsider can ever realize the amount of work involved in these long series of animal experiments, with treatments that had to be repeated and repeated for months on end. No one can grasp what meticulous care, what expenditure and amount of time were involved. To get some idea of it, we must bear in mind that Arsenophenylglycin had the number 418, Salvarsan the number 606. This meant that these two substances were the 418th and 606th of the preparations which Ehrlich worked out. People often, when writing or speaking about Ehrlich's work, refer to 606 as the 606th *experiment* that Ehrlich made. This is not correct, for 606 is the number of the *substance* with which, as with all the previous ones, very numerous animal experiments were made. The amount of detailed work, which all this involved, is beyond imagination (16).

Hata reported his first trials to Ehrlich, saying that he believed that compound 606 was very efficacious. Ehrlich's previous experience with compound 606 against trypanosomes had shown nothing. Ehrlich made Hata repeat the experiments over and over again. The treatments with compound 606 were amazingly effective in rabbits infected with syphilis. Hata was a tireless worker, but his patience began to wear thin. Ehrlich was finally convinced that compound 606 was active in syphilis. The 606 compound remained the most effective of any substance Hata tested. With definitive proof of its activity against syphilis in rabbits, Ehrlich's excitement could barely be contained. Other investigators had tried compound 606 on dogs, with good success. So, Ehrlich quietly gave some of his physician colleagues a little of the compound to test on humans. His friend Julius Iversen, at the Obuchow Hospital in St. Petersburg, used the experimental 606 compound to treat a number of hospitalized patients with relapsing fever and primary syphilis, reporting that they were completely cured. The compound was less effective against later stages of syphilis after the disease had caused paralysis, but some improvement was seen. Ehrlich and Hata prepared manuscripts for publication, but in April 1910, it was time to announce the effectiveness of compound 606 to the world.

THE PUBLIC ANNOUNCEMENT OF SALVARSAN AT THE CONGRESS OF INTERNAL MEDICINE IN APRIL 1910

Ehrlich chose the Congress of Internal Medicine in Wiesbaden, Germany, to make a public announcement of his discovery. Ehrlich, Hata, Iversen, and others described their work. The medical community greeted the news with great enthusiasm.

The newspapers got wind of the presentations and generated spectacular headlines but used the misleading name "Ehrlich-Hata 606" for the compound. Finding a new substance to treat syphilis created the same sort of commotion that had occurred in Berlin when Robert Koch announced his "cure" for tuberculosis. There were important differences. Ehrlich did not intend to relive the mistakes made by his mentor, Robert Koch. He had both laboratory and clinical evidence to back up his claims of compound 606's efficacy. Because he had been working with compound 606 since 1907, Ehrlich had patented the drug. He had control over its production and distribution. Ehrlich was not going down the same road with compound 606 that he had been down with serum therapy of diphtheria. The response to the announcement, especially the public response, was immense. This new discovery was the first scientifically grounded treatment for a disease that affected millions of people. After the Congress, Ehrlich was swamped with requests. Visitors appeared at his laboratory from across the globe. Ehrlich had planned to perform further clinical testing of compound 606 before the Congress began. Ehrlich graciously sent each visitor to the nearest clinic since he had no clinic facilities. But Ehrlich was exacting in his instructions to each clinic that did receive a supply of the drug. The clinics had to promise to carefully select patients only with the primary stage of syphilis, where compound 606 seemed most effective. They had to keep careful, detailed records of their care and send a copy to Ehrlich.

The preparation and handling of large quantities of compound 606 became an important part of Ehrlich's duties. A contract with a chemical factory in Höchst required that the factory suspend other production to work solely on compound 606. A name, Salvarsan, was patented, since compound 606's chemical name, dioxy-diamino-arseno-benzene dihydrochloride, was monstrous.

Ehrlich distributed compound 606 to any clinic or doctor that asked for it at the beginning. But the demand soon outpaced supply. Ehrlich wanted to keep informed of the amount of Salvarsan that was distributed, the doses, and the names of the recipients. His usual compulsive industrious work ethic was pushed to the limit. Surviving seemingly on mineral water and cigars, Ehrlich managed to maintain an enormous workload, keeping careful records of drug, clinics, and patients, yet still finding kind words for everyone who appeared with requests. Ehrlich became even more of a celebrity. Successful reports poured in; so did money. But Ehrlich was tremendously generous. He helped collaborators and coworkers and their families financially, often without ever mentioning these great kindnesses to the people he saw every day.

TROUBLES INTRODUCING SALVARSAN TO HUMAN MEDICINE

One facet of Ehrlich's efforts to introduce Salvarsan to human medicine that has gone generally unnoticed is that he had the first chemotherapeutic agent in history but no guidance as to *how* to introduce the drug. Every problem imaginable occurred. Ehrlich had to deal with and solve each problem. Notably, Ehrlich set principles for the introduction of new agents that have largely remained unchanged. His notes provide direction to us all:

1) Unlike experiments with animals, it is not possible in the treatment of man to fix or use the "*dosis maxima bene tolerata*" or the maximal well-tolerated dose. It is therefore necessary to take great pains over carefully conducted experiments, beginning with very small doses, and gradually increasing them in order to find out the most efficacious dose.

2) In human beings primary sensitivities and acquired hypersensitivities very often exist. More than half of all the medical substances used produce undesired reactions, even in small doses. With the new specific chemical preparations, we must always be prepared for this possibility, and realize that because of their powerful action grave results may sometimes occur.

3) These primary sensitivities and acquired hypersensitivities are a very great hindrance. For this reason, the new treatment ought only to be tried out under conditions where the most careful and continual attention to the patient is possible.

4) Before treating a patient with the new preparations it is well to find out whether in his case there has been damage to any particular organ or part of the body so that a "*locus minoris resistentiae*," or site of lowered resistance, is to be feared, which might be particularly threatened by the new preparation. . . .Before the new preparations are released for use in general medical practice very extensive tests must be made, and thorough experience of indications, contraindications, undesirable reactions, etc., must be obtained in order to avoid unfortunate results which would bring this new field of investigation into discredit (17).

These principles bear a striking resemblance to phases I, II, and III for clinical trials, required by the U.S. Food and Drug Administration before approval of a pharmaceutical. Ehrlich followed his own principles, giving great attention to reports of adverse reactions. The early problems were less about Salvarsan than the circumstances surrounding its preparation, handling, and administration. Salvarsan had to be produced without exposing the drug to air or oxidation of the chemical would occur. The oxidative form could be toxic, causing blindness or even death. Once sent to the clinic, the solution containing Salvarsan was also in danger of oxidation, causing the color of the solution to change from light yellow to dark brown. Eventually, Ehrlich ordered that the drug should be delivered in single-dose vials. Unfortunately, doctors made solutions of Salvarsan using tap water, leading to abscesses and necrosis at the sites of intramuscular injection. When Ehrlich learned of this practice, he examined leftover preparations used in clinics that were experiencing problems with the injections. Under the microscope he saw solutions teeming with bacteria. Ehrlich ordered the manufacturer to send freshly distilled (and eventually sterile) water with the Salvarsan with explicit instructions.

Despite successful reports and managing distribution and use problems, Ehrlich was aware that he had only built a foundation for a grand house. More reports surfaced demonstrating further problems with injections of Salavarsan. Some of the mishandling of Salvarsan led to doubts about its effectiveness. Ehrlich wrote,

> It is almost unbelievable. Always this carelessness with the water, lack of skill in giving the injections, puncture of the muscle, necrosis . . . It is nothing but incompetence which should have been overcome long ago. And the unfortunate patient must pay for it (18).

As experience with the drug increased and problems settled down, Ehrlich took a much-needed rest. But while on vacation, he received a telegram that caused terrific alarm. A patient had lost hearing after a Salvarsan injection. Trembling with excitement, Ehrlich fired back a telegram to the sender in Vienna demanding to know more of the circumstances. He headed straight back to Frankfurt. Ehrlich ultimately determined that the deafness was caused by a relapse of the disease, not the Salvarsan itself. The treating physician had used too low a dose, causing the relapse due to insufficient treatment. With larger doses of Salvarsan, the deafness completely disappeared. To avoid many of the problems with intramuscular injection with Salvarsan, Ehrlich advocated intravenous administration. Ehrlich worked tirelessly on the drug for which he became best known. His cousin, Felix Pinkus, later wrote,

> In spite of all the new and enlightening progress which Paul Ehrlich's work revealed, he always had the feeling that he had only just begun. What the world considered to be the final result of Ehrlich's intensive research was to Ehrlich himself only the beginning of still greater explorations into the unknown. He never enjoyed a success in tranquility and peace, but soon left behind the result just attained, and rushed forward in a profusion of deductions which carried him on with equal rapidity to a still higher peak of perfection; then this point, once reached, was in its turn again surpassed (18).

Ehrlich and his assistants continued to test hundreds of preparations. Compound 914 showed its usefulness in animals. Ehrlich determined the maximal tolerated dose and its activity against syphilis in animals and then humans. While its activity was not as great as that of Salvarsan, compound 914, or Neosalvarsan, as it became known, was more soluble and much easier to produce and handle. To be sure, Neosalvarsan was an effective treatment for human syphilis (Fig. 12.2).

The introductions of Salvarsan and Neosalvarsan ushered in a complete transformation not just in syphilis but also in the concept of chemotherapy. Ehrlich had developed a means to search for and test new drugs. He pioneered principles to

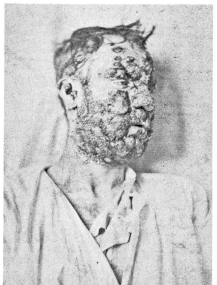

FIGURE 12.2 *(Left) Before treatment of syphilis. (Right) After successful treatment of syphilis with Neosalvarsan, 1915. Used with permission of the Mütter Museum of the College of Physicians of Philadelphia, Philadelphia, PA.*

determine tolerability, efficacy, and the necessity to monitor for adverse effects. He demanded intravenous injection, which required careful technique and sterile materials. It is now used as a mode of administration throughout the world. But the introduction of treatment for syphilis is the discovery for which Ehrlich is best remembered. For more than 40 years, Ehrlich's discovery remained the mainstay of syphilis treatment. As he said in his Nobel Address, "My dear colleagues, for seven years of misfortune, I had one moment of good luck."

AWARDS AND HONORS

Towards the end of his life, many honors were bestowed upon Paul Ehrlich. In 1913 the city of Frankfurt renamed the street in front of the Georg Speyer Haus "Paul Ehrlichstrasse." Many universities, both in Germany and abroad, awarded honorary degrees to Ehrlich. He received decorations from the German government, including the esteemed title *Wirklicher Geheimrat,* or privy councilor; a person of this rank in the German empire of the early 20th century would have been addressed as "Your High Excellency." According to his secretary, while Ehrlich took pleasure in these awards, some of his most satisfying joys came from simple postcards from cured patients, one of which he always kept in his wallet.

LAST YEARS

When the First World War broke out in 1914, Ehrlich expressed great concern: "But this war is pure madness. No good can possibly come of it." The war brought changes to the Georg Speyer Haus and the Serum Institute. Many men were called away to military service. As his institute deteriorated, so did Ehrlich's health after a small stroke in 1914. He purchased a car to help him get to his laboratory from his home and back again, but as the war began, the military requisitioned the car. Ehrlich went back to a horse-drawn carriage. For the first time in his life, Ehrlich no longer displayed vigor in his work. His declining health and the war greatly depressed him. In August of 1915, Ehrlich, his wife, daughter, and grandchildren went on vacation. Ehrlich suffered a second stroke, dying on 20 August 1915. He was buried in the Jewish Cemetery in Frankfurt. An obituary notice in *The Times*, in London, even in midst of the war, read,

> The vast number of problems he set himself bear witness to the strength of his imagination. He opened, "new doors to the unknown" and the whole world at this hour is his debtor.

Sadly, after Adolf Hitler and the Nazi Party came to power in Germany in 1933, the street sign honoring Paul Ehrlichstrasse was removed and many of Ehrlich's papers were burned because he was Jewish. His widow and children were persecuted as Jews until they left Germany in 1939. However, after World War II ended, the city of Frankfurt honored its famous resident posthumously, renaming the Institute for Experimental Therapy the Paul Ehrlich Institute and establishing the Paul Ehrlich Prize, which is a biennial award given on 14 March, Ehrlich's birthday, in one of Ehrlich's fields of research: immunology, cancer research, hematology, microbiology, and chemotherapy. Ehrlich's grandson collected his grandfather's remaining papers, donating them to Rockefeller University in the United States. Through the singular efforts of Paul Ehrlich, the world had its first chemotherapeutic agent for an infectious disease and a model on which to search for more such agents. But scientists soon discovered that finding a magic bullet is exceedingly difficult.

REFERENCES

1. **Jaramillo-Arango J.** 1949. A critical review of the basic facts in the history of Cinchona. *J Linnaean Soc* **53**:1272–1309. http://dx.doi.org/10.1111/j.1095-8339.1949.tb00419.x.
2. **Lee MR.** 2002. Plants against malaria. Part 1: cinchona or the Peruvian bark. *J R Coll Physicians Edinb* **32**:189–196.
3. **Garfield S.** 2000. p 30–34. *In Mauve: How One Man Invented a Colour That Changed the World.* Faber and Faber, London, United Kingdom.
4. **Marquardt M.** 1951. p 1–12. *In Paul Ehrlich.* Henry Schuman, New York, NY.

5. **Marquardt M.** 1951. p 13–18. *In Paul Ehrlich*. Henry Schuman, New York, NY.
6. **Silverstein AM.** 1999. Paul Ehrlich's passion: the origins of his receptor immunology. *Cell Immunol* **194:**213–221. http://dx.doi.org/10.1006/cimm.1999.1505.
7. **Marquardt M.** 1951. p 19–28. *In Paul Ehrlich*. Henry Schuman, New York, NY.
8. **Winau F, Winau R.** 2002. Emil von Behring and serum therapy. *Microbes Infect* **4:**185–188. http://dx.doi.org/10.1016/S1286-4579(01)01526-X.
9. **Dale H.** 1954. Paul Ehrlich. *BMJ* **1:**659–663. http://dx.doi.org/10.1136/bmj.1.4863.659.
10. **Marquardt M.** 1951. p 29–40. *In Paul Ehrlich*. Henry Schuman, New York, NY.
11. **Marquardt M.** 1951. p 65–76. *In Paul Ehrlich*. Henry Schuman, New York, NY.
12. **Burnet FM.** 1957. A modification of Jerne's theory of antibody production using the concept of clonal selection. *Aust J Sci* **20:**67.
13. **Jerne NK.** 1955. The natural-selection theory of antibody formation. *Proc Natl Acad Sci USA* **41:**849–857. http://dx.doi.org/10.1073/pnas.41.11.849.
14. **Sherman IW.** 2007. p 83–103. *In Twelve Diseases That Changed Our World*. ASM Press, Washington, DC. http://dx.doi.org/10.1128/9781555816346.ch6.
15. **Nobelprize.org.** 31 October 2010. Paul Ehrlich—Nobel Lecture. Nobelprize.org. http://nobelprize.org/nobel_prizes/medicine/laureates/1908/ehrlich-lecture.html.
16. **Marquardt M.** 1951. p 163–175. *In Paul Ehrlich*. Henry Schuman, New York, NY.
17. **Marquardt M.** 1951. p 188–194. *In Paul Ehrlich*. Henry Schuman, New York, NY.
18. **Marquardt M.** 1951. p 195–206. *In Paul Ehrlich*. Henry Schuman, New York, NY.

13   Lillian Wald and the Foundations of Modern Public Health

Infectious diseases are not chance events but events that occur as a result of the infected host's contact with their environment. Nowhere in medicine does the environment play such a crucial role in disease than in the field of infectious diseases. These connections have only been accurately assessed in the last 120 years. A disquieting number of infectious diseases have been associated with low social and economic status, including human immunodeficiency virus (HIV)/AIDS, infectious diarrhea, meningitis, hepatitis, and sexually transmitted diseases (1). Tuberculosis (TB) may be the single best example of the interplay between an infectious disease and the host's environment. TB rates are heavily influenced by social, environmental, and epidemiological circumstances and have been inversely correlated with the income of various populations. Treatment of TB alone is not enough to control the disease. Socioeconomic factors may be more important determinants of epidemiological trends than treatment programs (2).

As the germ theory of disease was taking shape in the latter part of the 19th century, deplorable living conditions in major European and American cities made them rife for TB, typhoid fever, and even cholera. Governmental public health departments either did not exist or were fledgling offices. A system for dealing with the health needs of a population of people and improving their surroundings came from a nurse, Lillian Wald (Fig. 13.1). Understanding the essential role played by social and economic factors in disease, Wald sought to correct them in one of the most challenging areas in the world, the Lower East Side of New York City in the 1890s.

Germ Theory: Medical Pioneers in Infectious Diseases, Second Edition. Robert P. Gaynes.
© 2023 American Society for Microbiology.

FIGURE 13.1 *Lillian Wald as a young nurse in uniform.*

Lillian Wald is the first woman and first nurse profiled in this book. Wald's achievements are generally not considered by those cataloguing events on the timeline of advances in infectious diseases. During a seminar on the individuals profiled in this book that I gave at the Centers for Disease Control and Prevention,

I asked the audience if anyone had heard of Lillian Wald. Within the large auditorium, no hand was raised. Yet her accomplishments and her influence in public health are so pervasive, especially in the United States, that they serve as the underlying foundation of the public health infrastructure. Wald's story, with her impact on so many aspects of American life, is a tale worth knowing.

EARLY INFLUENCES

Lillian Wald was born on 10 March 1867 in Cincinnati, OH. She was the third of four children. Her father, Max Wald, was a successful merchant who moved from Cincinnati to Dayton, OH, and then in 1878 to Rochester, NY. Wald considered Rochester her hometown. Lillian's childhood was happy. Her grandfather, a successful merchant himself, indulged her. Lillian never had to personally deal with the poverty that would challenge her life's work. There were two distinguishing features of Lillian's background that would affect her in subtle but profound ways. First, she was a first-generation American. Both her mother's and her father's families were European immigrants who fled the 1848 revolutions in Poland and Germany. She grew up with an understanding of the American immigrant point of view. Second, Lillian Wald had a liberal Jewish upbringing that would later motivate her life's work with the largely Eastern European Jewish immigrants of the Lower East Side of Manhattan.

During Lillian Wald's education in Rochester, she demonstrated great skills in languages, along with proficiency in math, science, and art. In 1889, she enrolled in the nursing program at New York Hospital. Upon graduation in 1891, she worked for a year at the New York Juvenile Asylum but left to become a doctor. Wald began taking courses at New York's Women's Medical College.

"BAPTISM OF FIRE"

During her first few months at the Women's Medical College, Wald accepted an invitation to organize some nursing classes for immigrants on the Lower East Side of Manhattan. It was an experience that altered everything for Lillian Wald. Few people can point to a single circumstance that changes their life's calling. The plea of a small child brought that circumstance to Lillian Wald. At the conclusion of one of the classes, a small child desperately grabbed Wald to take her to her ill mother. Wald described the experience:

> Over broken asphalt, over dirty mattresses and heaps of refuse we went . . . There were two rooms and a family of seven not only lived here but shared their quarters with boarders . . . [I felt] ashamed of being a part of society that permitted such conditions to exist . . . What I had seen had shown me where my path lay (3).

Wald called the experience her "baptism of fire." All her vague notions of her life's ambitions disappeared. She *knew* what she should do with her life. Lillian confided in the woman who had sponsored the class that she was teaching, Mrs. Fanny Loeb. Mrs. Loeb was the wife of a wealthy New York financier, Solomon Loeb. She told Mrs. Loeb,

> We wouldn't let animals live in a place like that. And these are human beings—brave people, good people. They must be, because they struggled to come to this country in order to give their children a better chance than they had . . . And they found filth and degradation (4).

WALD'S PROPOSAL: A NURSING SERVICE

With an understanding and compassion for the immigrant's plight, Wald proposed a bold plan to Mrs. Loeb. Wald wanted to live in that section of the city, perhaps with another nurse. The people of the neighborhood would get to know the nurses as friends so that they could bring nursing help directly into the homes of the residents. Loeb was impressed with Wald's sense of purpose and honesty. The proposal was wildly innovative and strangely practical. Loeb was so persuaded by Wald's passion that she brought the idea back to her small group of wealthy, earnest people in New York who were devoted to financially helping the city's poor. One of those in the group was her son-in-law, Jacob Schiff. Schiff was a German immigrant who had lived in New York since his arrival in 1865. He had married into the Loeb family but had his own energy and intelligence to have climbed near the top of the financial world. Schiff had a great sense of responsibility, always giving one-tenth of whatever he made to charity. He believed that helping people help themselves was the best way to spend his money. He helped fund Wald's initial venture into the Lower East Side of New York, which would become the beginning of a long financial and social relationship.

Lillian Wald spoke of her proposal to Mary Brewster, a fellow graduate of New York Hospital's nursing program. Mary was slender, even fragile, and a quiet woman. She was nothing like the enthusiastic and energetic Wald, except that both were passionate about being of use to the world. They both saw the needs of the people of the Lower East Side and knew that they needed help.

THE LOWER EAST SIDE OF NEW YORK IN THE 1890S

In 1893, the population of New York City was a million and a half. Nearly three-quarters were foreign-born or first-generation Americans. Most worked in factories or in their homes, making only enough money to barely survive. Children were sent to work to augment the family income as soon as they could be hired.

The Lower East Side of the city became a mass of befuddled immigrants who shared tenements of unspeakable misery and filth. The immigrants, many of whom had moved from a rural existence, could not lean on their old ways. Many spoke little English. There was no one to instruct them on the ways of their new country.

The confusion of immigrants was not lost on the industrialists of the time, who grew rich on this cheap source of labor. For these newcomers, there were no unions or governmental organizations regulating the working or living conditions. Landlords grew fat on the rent from their tenements. The Lower East Side quickly became one of the worst slums in the world. About 190,000 people lived on 500 acres; in one section nearly 1,000 people lived on only 1 acre of land. Lillian Wald and Mary Brewster set out to find a home amid this horrid squalor.

The two women had one requirement: a bathroom. Finding suitable accommodations for two young women in the Lower East Side would be no easy feat without the requirement. But finding a place to stay with the miracle of modern plumbing became an immense challenge. They searched over cobblestone streets where residents dumped their garbage since there were no sanitation regulations on the Lower East Side. Eventually, they found a two-room apartment on the upper floor of a tenement on Riverton Street and moved in.

THE PUBLIC HEALTH NURSE

Lillian Wald coined the term "public health nurse" in 1893 for nurses who worked outside hospitals in poor and middle-class communities.

> Our basic idea was that the nurse's peculiar introduction to the patient and her organic relationship with the neighborhood should constitute the starting point for a universal service to the region . . . We considered ourselves best described by the term "public health nurses." (5)

Specializing in both preventative care and the preservation of health, these nurses responded to referrals from physicians and patients, giving free treatment or charging according to the resources of the patient.

Wald's first order of business was the procurement of authority. She rather presumptuously approached the president of the Board of Health. She tried to convince him that the poor were the city's responsibility. The president remained unimpressed and concerned about giving two inexperienced nurses the right to use "Board of Health" on any badge. Wald reminded the president that she was not asking for money. Then, Wald promised to report on the sanitary conditions of the neighborhood. The president thought that ongoing reports on the sanitary conditions would be useful to him. Knowing that Wald and Brewster had the backing of two prominent New York City citizens, Jacob Schiff and Mrs. Loeb, he finally agreed.

In July 1893, Lillian Wald and Mary Brewster donned blue uniforms with their badges, "Under the Auspices of the Board of Health," and began their work. They intended to nurse the sick but quickly found "terrible filth everywhere." The only water for many buildings came from sinks in the hallway that were often clogged with smelly refuse. The floors reeked with garbage odor. Among the first families that she was called to see, there was a small child who had been badly bitten by rats. The exhausted mother had been trying to care for her child, her husband (who had severe rheumatoid arthritis), and another child whose mother had died. Wald helped the rat-bitten child and enlisted the aid of all the tenants of the building in a general housekeeping effort. She found the janitor, threatening some vague action if he did not improve the conditions of the building. He did, but only after Wald told him she would return to monitor the situation.

Each day Wald and Brewster saw dozens of patients, some in their residence on Riverdale but most in their own homes. Illness was invariably accompanied by unemployment, hunger, neglect, or ignorance. The nurses tended to the sick but quickly began to teach the importance of health maintenance. Immigrants had an innate fear of institutions, especially hospitals. Wald recognized the fear. Once, the nurses found a household with several ill children. Their mother had gone without sleep for 4 days trying to nurse them rather than take them to a hospital.

> [The nurses took over for the mother] ... without mentioning the hospital at all. But when the worried woman awoke and saw the cleanliness Mary Brewster had created in the little room, the instruments, and materials she had brought with her, and the skill with which the nurse fed an unconscious child, she realized suddenly how inadequate her own best efforts had been. The woman asked if that was what a hospital would do. [Mary replied that] a hospital would do much more (6).

Wald slowly convinced the residents of the neighborhood that in some circumstances, a hospital could give better care than the nurses could. After 2 months, another nurse joined Lillian Wald and Mary Brewster. Wald kept Jacob Schiff informed of their efforts and needs, which were ever increasing. Slowly, the residents began to trust the nurses. The nurses' reputation began to extend outside the neighborhood to hospitals, clinic staffs, doctors, Board of Health officials, policemen, and leaders of charitable organizations.

The nursing service quickly expanded with financial support from Jacob Schiff as they developed credibility with all these organizations, leading to a more efficient and coordinated response whenever a request came from the Nursing Service. Wald described her approach, its evolution, and ultimate impact:

> The service, through covering so wide a territory, is capable of control and supervision. The division into districts, with separate staffs for contagious and obstetrical cases, may be compared to the hospital division into wards. Like the hospital, it has a system of bedside notes, case records, and an established etiquette between physicians, nurses, and patients. Those that can best be cared for in hospitals are sent there, the sifting process being accomplished by the doctors and nurses working together. Approximately ten percent of our patients are sent to the hospitals . . .
>
> The work begun from the top floor of the tenement comprised in simple forms those varied lines of activity which have since been developed into the many highly specialized branches of public health nursing now covering the United States and engaging thousands of nurses . . .
>
> The [newly formed] New York Commission drafted the new health law in New York State (1913). "The advent of trained nursing marks not only a new era in the treatment of the sick, but a new era in public health administration." This Commission also created the position of Director of the Division of Public Health Nursing in the state department of health (5).

TB AND THE NURSING SERVICE

Like many before her, Wald appreciated the relationship between poverty and TB. She spent some of her early monies on disinfectants and sputum cups for TB patients. Although the prevailing textbook view was that Jews were nearly immune to the disease, the density of people on the Lower East Side, especially in the textile industry, led to TB's nickname, the "tailors' disease." Wald and her nurses led a campaign of education and home visits, leaving written instructions to families with known TB cases. Wald described the effect.

> . . . although it is pre-eminently a disease of poverty and can never be successfully combated without dealing with its underlying economic causes—bad housing, bad workshops, undernourishment, and so on—the most immediate attack lies in education for personal hygiene. For this, the approach to the families through the nurse and her ability to apply scientific truth to the problems of human living have been found to be invaluable.
>
> The National Association for the Study and Prevention of Tuberculosis in its report for 1915 states that the tuberculosis death rate in the registration area of the United States has declined from 167.7 in 1905 to 127.7 in 1913 per 100,000 population: a net saving to this country of over 200,000 lives from this one disease. [Note: this decline occurred more than 25 years before effective antituberculous antibiotics.] (7)

SCHOOL NURSES INTRODUCED IN NEW YORK CITY SCHOOLS

In 1902, Wald became involved with the school system when she came upon a 12-year-old boy who was denied entrance to school because of a chronic skin condition, eczema. He had been given medicine, but no one ever instructed the boy in its use. He was unable to read the directions. Embarrassed, he hid the medicine and never used it. After questioning the boy about his condition and discovering the existence of his medicine, Wald instructed the boy in the medicine's use, with good effect. However, she needed to intervene with the school to ensure his reentrance. Wald's interaction with the school system prompted her to maintain a list of students who had been wrongly prevented from attending school. She quickly realized that she needed to keep another list, those students who were attending school but should have been kept home. One such child had scarlet fever. Wald rapidly separated this child from the other children. She took the child directly to the Board of Health president, demonstrating to him the calamity that could have occurred had this child been allowed to remain in the classroom. Wald told the president that she recognized that it was not the teachers' fault or their responsibility to be knowledgeable about contagious illnesses among the children. She suggested that one of her nurses be placed within certain schools to help. By the end of 1 month the nurse had given 893 treatments, visited 137 homes, and returned to their classrooms 25 children who had not been receiving any care until the nurse took over. The school system immediately recognized the value of the nurse's work. The School Board voted money for the employment of a dozen school nurses—the first school system in the world to do so!

Hundreds of communities throughout the United States and the world followed New York's example. The ubiquitous school nurses in today's schools owe their existence to Lillian Wald's common-sense approach to meeting the needs of the children in her community. A school nurse's impact on infectious diseases should not be devalued. This nurse can prevent outbreaks in highly susceptible populations of children by removing children with infectious diseases from the school. More importantly, the school nurse is the one who ultimately certifies that children have had the required immunizations before school entrance.

THE HOUSE ON HENRY STREET

Jacob Schiff recognized the immense value of Wald and her nurses to the neighborhood. He never hesitated in his financial support for their work. In 1895, Schiff found a larger and more suitable place than their previous residences in the Lower East Side of the city, a home on Henry Street. Ironically, the Henry Street Nurse's Settlement House was located next door to the house where Lillian Wald had taught her first class. Wald soon became known as the "Angel of Henry Street."

Shortly after the move to Henry Street, Mary Brewster, whose health had never been strong, had to leave the nursing service, dying shortly thereafter. The loss of the person who started the work together with Wald was difficult for her. But Wald always seemed to find new nurses to replace or augment any losses. At a meeting of the National Conference of Charities and Corrections, Wald solicited women of talent and spirit to join her public health nurses. Expansion occurred incrementally, initially with the help of Jacob Schiff, but after a few years, the City of New York began to pitch in financially. By 1903, 18 New York City district nursing service centers treated 4,500 patients a year all around the city. By 1910, 54 nurses ran a convalescent center, three country homes, and several first aid stations.

During the period between 1903 and 1910, a gradual shift occurred in focus of the settlement nursing service. Numerous boards of health began working with the Henry Street Settlement and other visiting nursing services to develop preventative programs for school children, infants and mothers, and patients with TB. Inspired by Wald's vision, these nurses became the foot soldiers of a modern campaign for public health. By 1915, 100 nurses cared for more than 26,575 patients and made more than 227,000 home visits in New York City alone. Around 1920, public health nurses began to specialize in venereal diseases, TB, maternal and child welfare, and mental illnesses. Wald was somewhat opposed to the public health nurses' specialization. She believed in a broadly focused role for nursing for social betterment (8).

COLUMBIA UNIVERSITY AND THE DEPARTMENT OF NURSING AND HEALTH

In 1910, Columbia University in New York created the Department of Nursing and Health, establishing the first nursing department in any American institution of higher learning. At Wald's suggestion, the department affiliated with the Henry Street Settlement House. Nursing students would receive preclinical training and then gain field experience with Wald and her nurses. This affiliation was the earliest example of university-based field training of nurses outside hospitals.

NATIONWIDE INSURANCE COVERAGE FOR HOME-BASED CARE

For the first 15 years or so, the Henry Street Settlement House Nurses received most of their financial support from private sources, notably Jacob Schiff. Wald recognized the precarious nature of the charitable financing, which could disappear at any time. More importantly, the financial model that relied on charity could not be expanded to the nation in its current state. In 1909, Wald proposed an innovative experiment. Convinced that she could prove that home-based care could be cost-effective for the growing insurance industry, Wald proposed to one insurance company, Metropolitan Life Insurance Company, that the company hire visiting nurses to care for their policyholders during their illnesses. The funding for the nurses would occur through a modest fee per policy. The company cautiously agreed to test the idea using nurses from the Henry Street Settlement House to visit their policyholders. After only 3 months, the cost-effectiveness was impressive enough to the company's directors that they extended the plan across all of New York City. The company saw such improvement that the number of death claims that it had to pay out reduced substantially; by 1914 the nursing services had contributed to a 12.8% decline in the mortality rate of policyholders (8). By 1916 the company made a visiting nurse available to 90% of its 10.5 million policyholders across the United States and Canada. At an initial cost of 5 cents per policy and 50 cents per nursing visit, Metropolitan Life Insurance Company provided a reliable source of funds that was not subject to the fluctuations of charitable support. By 1925, the Company claimed that over 240,000 lives had been saved, with a net savings of 43 million dollars. In the 1930s several events contributed to shrinking the viability of this business model: the increasing cost of visiting nurses, the rising number of cancelled policies during the Great Depression, and the overall improvement in public health for contagious diseases. The benefits of a visiting nurse were less evident to the insurance industry for patients with chronic diseases.

ESTABLISHMENT OF A NATIONAL PUBLIC HEALTH NURSING SERVICE

In 1909 there were about 1,400 public health nurses in the United States. In 1912, Lillian Wald was elected as the first president of the National Organization for Public Health Nursing. As part of her vision for public health nursing, Wald helped to create a National Public Health Nursing Service. Wald's concept began with the American Red Cross, which, up to that time, had chiefly been identified with wartime activities. Wald proposed using the agency to standardize public health nursing throughout the United States. Initially, the Red Cross showed little interest until Jacob Schiff and another benefactor offered to provide the needed financial support to begin the plan. The Red Cross developed national policies and procedures, but the local chapters did the hiring and supervision of qualified nurses. In less than 20 years there were nearly 10,000 public health nurses in the United States. But Wald anticipated that the Red Cross and private philanthropists would have difficulty continuing to support these nurses. From the beginning, Wald envisioned that the support would eventually need to come from municipal, county, state, and federal agencies. In 1944 the Visiting Nurse Service of the Settlement separated from the Henry Street Settlement to become the Visiting Nurse Service of New York. By 1947 less than a quarter of all public health nurses were working under the auspices of the Red Cross or charitable organizations. Local, state, and federal governments employed most of the public health nurses, just as Wald had foreseen. However,

Wald's vision for public health nursing did not materialize in its entirety. Most health departments abandoned any venture into treatment or curative activities for public health nurses. The pursuits became largely preventative in nature. Analysis of vital health statistics and monitoring or surveillance of diseases were also important functions of the public health nurse. As a consequence of the limited government-funded public health nursing activities, private visiting nurse associations sprang up to treat the sick in their homes. While treatment and prevention were activities that Wald combined in the duties of her nurses, these functions became separate and distinct. However, the foundation of the public health departments' essential functions was derived directly from the vision of Lillian Wald, who wrote,

> The famous Dr. William H. Welch [first Dean] of Johns Hopkins University declared that America has made three original contributions to public health: the sanitation of the Canal Zone, the State Tuberculosis Laboratories instituted by Dr. Hermann Biggs, and the public health nurse (9).

JOINT BOARD OF SANITARY CONTROL

Lillian Wald's intimate relationship with the residents of the Lower East Side thrust her into problems that she never considered when starting her work. In 1910, a strike among cloak makers led Wald into fundraising, picketing, and raising public awareness of their unhealthy workplace conditions. When the strike was settled, the terms of the settlement included the formation of an agency called the Joint Board of Sanitary Control, a new experiment in the sanitary control of an industry by organized employers, organized workers, and representatives of the public. Wald served on the board as a public representative, working to eliminate abusive work programs and establishing a minimum wage for women workers. This agency monitored standards of ventilation, fire protection, pollution, and other sanitation issues in the factories.

Wald also worked tirelessly to abolish child labor. After years of efforts with private funds, Wald campaigned for federal legislation and a federal agency to help deal with children's social and health problems. Finally, in 1912, President William Taft established the Federal Children's Bureau. Wald turned down Taft's offer to head the agency, believing she would be more useful at Henry Street. Taft appointed Julia Lathrop as the first Bureau Chief, the first woman ever to head a government agency in the United States. The Bureau's mission was to investigate and report on infant mortality, birth rates, orphanages, juvenile courts, and other social issues of that time. Today, the Children's Bureau is one of two bureaus within the Administration on Children, Youth and Families, Administration for Children and Families, of the Department of Health and Human Services, but their establishment can be attributed to Wald's efforts.

WALD'S OTHER ACHIEVEMENTS AND ACTIVITIES

Wald used the Henry Street Settlement for a variety of purposes that were unrelated to public health nursing. It was a recreation center for the neighborhood, including one of New York City's first playgrounds to provide a safe environment for children. Wald was also instrumental in the improvement and upkeep of Seward Park, which became the first municipal playground in New York City.

The Settlement had special classes for "backward" children and study rooms where Wald helped children with their homework. The Henry Street Settlement served primarily Jewish residents of the Lower East Side but was nondenominational and integrated. Wald was a founding member of the National Association for the Advancement of Colored People, allowing use of the Henry Street Settlement for its early meetings.

Serving as an advocate for residents of the Lower East Side, Wald was constantly confronted with social injustices. With her energy and passion, it is not surprising that Wald fell into politics, which she approached with the same enthusiasm as her work with the Nursing Service. Many of Wald's political efforts centered on children and working women. She campaigned for women's suffrage in 1915. Wald was an ardent pacifist and opposed U.S. entry into World War I. As President of the American Union Against Militarism, she lobbied Woodrow Wilson's administration for mediation to the conflict and away from active military involvement. Once the United States entered the war, she continued to serve as an advocate for the immigrants to forestall any abridgments of their civil liberties. During the 1918 influenza pandemic, Wald headed the Red Cross Nurses Emergency Council. Under her direction, Henry Street cleared all cases of influenza and organized the efforts of thousands of volunteers. After the war, Wald continued to head the Henry Street Settlement and remained politically active for a variety of causes. She became friendly with Eleanor Roosevelt and supported Franklin Delano Roosevelt's (FDR's) policies. Many of Roosevelt's administrative appointees were former Lower East Side residents and clients of the Henry Street Settlement, including Adolph A. Berle, Jr., an original member of FDR's "Brain Trust"; Frances Perkins, Roosevelt's Secretary of Labor; Henry W. Morgenthau, Jr., FDR's Secretary of the Treasury; and Sidney Hillman, a key figure in the founding of the Congress of Industrial Organizations and major supporter of the Democratic Party during FDR's presidency. Wald was delighted with FDR's New Deal and believed that the ideas formulated experientially at Henry Street were being put to use in government during FDR's presidency.

AWARDS AND HONORS

In 1930, Wald's health forced her to retire to a country house in Westport, CT. After 40 years as headworker of the Henry Street Settlement, she had collected numerous honors, including the Gold Medal of the National Institute of Social Sciences, honorary degrees from Mount Holyoke College and Smith College, the Lincoln Medallion for her work as Outstanding Citizen of New York, and the Rotary Club Medal. In 1922, the *New York Times* named Lillian Wald as one of the 12 greatest living American women. In 1993, Wald was inducted into the National Women's Hall of Fame.

LATER YEARS

After her retirement from the Henry Street Settlement in 1930, Wald lived in Connecticut. She never married. Her efforts and travels in the 1920s took their toll on her health. She used her retirement to write the second of two semiautobiographical books, *The Windows on Henry Street*. In 1940, Wald died from a cerebral hemorrhage at age 73. Thousands mourned at her funeral. A few months later, over 2,500 people crammed into Carnegie Hall in New York for a service memorializing Lillian Wald. Perhaps the greatest testament to Lillian Wald is that the Henry Street Settlement and the Visiting Nurse Service of New York continue more than 100 years after Wald founded them. Over 100 years ago, Lillian Wald developed an alternative solution to the health care problems of her time, providing a unifying structure for health care and preventative care to an underserved, uninsured population. In health care and public health, times have changed, but today's underserved, uninsured populations who have needs like those of those residents of the Lower East Side of New York would welcome another Angel of Henry Street.

REFERENCES

1. **Semenza JC.** 2010. Strategies to intervene on social determinants of infectious diseases. *Euro Surveill* **15:**32–39. http://dx.doi.org/10.2807/ese.15.27.19611-en.
2. **Frieden TR.** 2009. Lessons from tuberculosis control for public health. *Int J Tuberc Lung Dis* **13:**421–428.
3. **Wald L.** 1915. p 1–25. *In The House on Henry Street.* Henry Hold and Co, New York, NY.
4. **Williams B.** 1948. p 75–85. *In Lillian Wald, Angel of Henry Street.* Julian Messner, Inc, New York, NY.
5. **Wald L.** 1915. p 26–43. *In The House on Henry Street.* Henry Hold and Co, New York, NY.
6. **Williams B.** 1948. p 96–105. *In Lillian Wald, Angel of Henry Street.* Julian Messner, Inc, New York, NY.
7. **Wald L.** 1915. p 44–65. *In The House on Henry Street.* Henry Hold and Co, New York, NY.
8. **Buhler-Wilkerson K.** 1993. Bringing care to the people: Lillian Wald's legacy to public health nursing. *Am J Public Health* **83:**1778–1786. http://dx.doi.org/10.2105/AJPH.83.12.1778.
9. **Wald L.** 1934. p 70–110. *In Windows on Henry Street.* Little, Brown, and Company, Boston, MA.

14 Alexander Fleming and the Discovery of Penicillin

During the first 2 decades of the 20th century, therapy for infectious diseases remained rather meager. Progress in serum therapy produced antitoxins from horse sera against diphtheria, streptococci, Shiga toxin-associated dysentery, tetanus, and gas gangrene. There was also progress in passive immunization with antibodies from horse sera directed against some bacteria, including pneumococci, meningococci, and leptospira, and a few viral infections, including polio and distemper. But problems with serum therapy became evident quickly. Anaphylaxis often occurred when horse serum was used. The human body reacts to foreign proteins present in horse serum with its own antibodies. The most serious difficulty, immediate hypersensitivity reaction, is triggered by an antibody called immunoglobulin E (IgE) and can result in a severe, systemic response that can lead to shock or respiratory compromise. A more delayed reaction called serum sickness can also occur. In serum sickness, a different antibody is formed—IgG. This antibody, combined with a foreign protein, can deposit throughout the body. This IgG-mediated response was more common when large amounts of serum were injected, such as for pneumococcal pneumonia. Rashes, joint pains, swollen lymph nodes, fever, enlarged spleen, and kidney damage were common problems associated with serum sickness. As a result, the use of horse serum was largely abandoned by the 1930s.

Germ Theory: Medical Pioneers in Infectious Diseases, Second Edition. Robert P. Gaynes.
© 2023 American Society for Microbiology.

PROGRESS IN CHEMOTHERAPY OF INFECTIOUS DISEASES

If Paul Ehrlich had lived considerably longer, further chemotherapeutic discovery may have occurred at his Institute. Unfortunately, Salvarsan and Neosalvarsan remained the only chemotherapy for bacterial diseases for over 2 decades.

The connection between chemotherapy of infectious diseases and dyes did not end with the death of Paul Ehrlich. In the German dye industry, during investigations of azo dyes, a graduate student made *para*-aminobenzene sulfonamide in 1908 (1). Dyes were made from this compound that chemically bound with proteins in wool. Heinrich Hoerlein, an executive at the German chemical trust IG Farben, headed a team of scientists at their subsidiary, Bayer Laboratory, that believed, like Ehrlich, that coal tar dye could also bind to bacteria and parasites as a "magic bullet." After years of fruitless effort with hundreds of dyes, Bayer researchers found one that actually worked in protecting mice from bacteria—Prontosil. The story of Prontosil and its related compounds intertwined during the first few decades of the 20th century with the story of penicillin, although the contrasts between the developments of these two antibacterial drugs are striking.

THE BEGINNING OF SULFONAMIDES

Gerhard Domagk led the investigations into the bacteriological activity of Prontosil, the forerunner of sulfonamides. Domagk was a German veteran of World War I. He had been wounded and spent most of the war witnessing the horror of battlefield wound infections from his work as a medic. After the war, he completed his medical studies, receiving his Doctor of Medicine degree in 1921. In 1927, he began working at Bayer Laboratories, a research division of the IG Farben conglomerate in Germany. With the help of chemists, Domagk and his team were testing some 30 chemicals a week for antibacterial activity.

THE BEGINNINGS OF THE PENICILLINS

Penicillin was discovered entirely by accident rather than by the methodical trial-and-error process that led to Prontosil. At nearly the same time that Prontosil was found, Alexander Fleming stumbled over the antibacterial compound that was produced by another microorganism, a mold. But his 1929 discovery would not find its way into clinical use for a decade. The reason for this delay had much to do with Fleming himself.

ALEXANDER FLEMING: EARLY INFLUENCES

Alexander Fleming was born on 6 August 1881 in southwest Scotland (2). He was the second youngest of eight children. His father had two daughters and two sons by his first wife, who passed away. He remarried at the age of 60. His second marriage resulted in a son and a daughter before Alec (as he was known) was born. A younger brother was born 2 years later, in 1883. His father had a stroke and died when Alec was only 7 years old. Fleming's extended family was largely responsible for raising the boy. He attended school on the moors until 1893, when he went to Kilmarnock Academy, the alma mater of Robert Louis Stevenson and Robert Burns. At the age of 13, Alec moved to London. He was enrolled in Regent Street Polytechnic, where he did quite well in school. When Alec was 16, he took an apprenticeship with a shipping firm. Following the tradition of the males in his family, when Alec turned 18, he joined the London Scottish Rifle Volunteers. His short stature (he was 5 ft 6 in. tall) led to much teasing. Fleming did not have a prodigious career with the Volunteers, although he was an expert shot with a rifle. After 4 years out of school, Alec passed an examination that qualified him for entrance into medical school.

Fleming did well enough on his examination that he had his choice of 12 medical schools in London. He had no specific knowledge of any of the schools, but he had played water polo against a team from St. Mary's, the school he chose. Fleming began at St. Mary's Hospital school in 1901; there he would spend the next 50 years. He did well in his preclinical work, earning numerous awards. He also did nicely in the clinical years, performing well enough for a fellowship in the Royal College of Surgeons. He concentrated on obstetrics at first. Then Alec decided to become a surgeon. He amazed his classmates by passing all four parts of the board examinations for the Royal College of Surgeons and Physicians with seemingly little studying and effort. Fleming's career took a rapid turn when he was offered a position not in surgery but in pathology, under the direction of one of the most brilliant but controversial faculty members, Sir Almroth Wright. But Fleming was unconvinced that the opportunity was right for him. His friend J. Freeman tried to persuade Fleming that the position would be a good one:

> I plugged the fact that Almroth Wright's laboratory would make a good observation post from which he could keep an eye open for a chance to get into surgery. I told him, too, that he would find work in the lab interesting, and that company there was congenial. The lab, at the time, consisted of only one room where the staff lived a sort of communal life. (3)

ALMROTH WRIGHT AND THE INOCULATION DEPARTMENT OF ST. MARY'S HOSPITAL

Almroth Wright was the director of the Inoculation Department at St. Mary's and a force with which to be reckoned. A giant of a man, Wright had an opinion about everything. Wright knew George Bernard Shaw, who used Wright as the model for the physician character of his play *The Doctor's Dilemma*. Supremely confident and authoritarian, Wright could speak 7 languages and read 11. Fleming, always chided for his small size, was taciturn but had a dry wit. Surprisingly, these two opposite personalities meshed well together. Although he had no intention of becoming a bacteriologist, Fleming just went with the flow, publishing his first paper with Wright in 1908. Wright's primary interest, indeed, passion, was vaccination. His efforts in vaccines brought revenues to St. Mary's, making Wright's scientific "republic," as he called it, nearly financially independent. Wright's research interests attempted to reconcile the prevailing but seemingly opposing views of immunity: cellular immunity and humoral immunity. Wright would say that the components of humoral immunity were what the body "butters" the disease germs with to make the white cells (phagocytes) eat them.

Wright was also a friend of Paul Ehrlich. When Ehrlich developed Salvarsan, he gave some to Wright for clinical testing. It was Fleming who ran the clinical trial in London. Fleming was the first to use the chemotherapeutic agent in England and developed a reputation for treating syphilis. Wright was relatively uninterested in research related to chemotherapeutic agents. Fleming, on the other hand, seemed intrigued by the research but still felt the calling of surgery. After less than a year, Fleming took and passed all the clinical requirements for full accreditation as a surgeon, but he continued in his role in Wright's department.

FLEMING AND THE FIRST WORLD WAR

With the outbreak of World War I, Fleming's life took another dramatic shift. Wright went to the front lines, taking Fleming and another colleague, Leonard Coleman, along with him. From a military hospital, they witnessed the appalling slaughter in the war. Fleming saw firsthand the toll that battlefield infections took on the troops. Wright assigned Fleming the task of culturing and identifying the bacteria causing the wound infections. Fleming's research showed that 90% of the infections were caused by *Clostridium welchii* (now known as *Clostridium perfringens*); staphylococci and streptococci were also prominent. Lister's methods of antisepsis were the standard of care, but Fleming noted that the antiseptics were not penetrating all the areas of the wounds. He showed that it was not possible to sterilize a wound with the then-known antiseptics. The white blood cells could be shown to help destroy the bacteria gathered in the wound. But Fleming devised experiments to show that the antiseptics were killing off the white blood cells, or phagocytes, and doing more harm than good (4). Fleming's comments were not well received by the medical community, which viewed his research as an attack on Lister himself. Leading the assault was Sir William Cheyne, who was President of the Royal College of Surgeons and a friend of Lister. However, Fleming had the good fortune of having Wright covering his back. In a masterfully worded response to Cheyne's attacks, Wright turned the tables, using Cheyne's own words to support Fleming's and his own position (5). Fleming realized that antiseptics were not the answer for these infected wounds. In Fleming's own prophetic words,

> What we are looking for is some chemical substance which can be injected without danger into the blood stream for the purpose of destroying the bacilli of infection, as Salvarsan destroys the spirochetes (6).

He could not have dreamed that he was going to be the one to discover just such a substance.

During the chaos and carnage of World War I, Fleming found the time to get married in 1915 while on leave in England. When Fleming returned to France to announce his marriage, his colleagues believed it was just a practical joke. Well known for such gags, Fleming had difficulty convincing everyone that he had actually married. His bride, Sarah McElroy of Ireland, had a twin sister who married Alec's brother. As the war ended, Fleming remained on the continent in an unsuccessful attempt to determine the cause of the 1918 influenza epidemic. He eventually returned to St. Mary's, where he and his wife, known as Sareen, began to build a life together.

THE DISCOVERY OF LYSOZYME

Fleming's discovery of penicillin was preceded by another discovery, that of lysozyme. He was not tidy in his laboratory. A colleague who had come to work in Fleming's lab wrote about the experience there:

> I went to St. Mary's to work in the Inoculation Department with Fleming. From the very first he started to pull my leg about excessive tidiness. Each evening I put my bench in order and threw away anything I had no further use for. Fleming told me that I was a great deal too careful. He, for his part, kept his cultures sometimes for two or three weeks to see whether by chance any unexpected or interesting phenomenon had appeared. The sequel was to prove how right he was and that, if he had been as neat as I am, he would probably have found out nothing new. (7)

The importance of Fleming's disorderly nature and its role in the discovery of penicillin is well known, but it proved important in an earlier development, the discovery of lysozyme. Fleming had cultured some mucus from his own nose 2 weeks earlier. There were golden bacteria covering the petri dish, but Fleming noticed something else. Surrounding the blob of mucus there was a zone where no bacteria grew at all. Fleming began to culture other secretions, looking for the substance that seemed to dissolve bacteria. He realized that freshly cultured mucus was even more potent. However, the dissolving, or lytic, power of lysozyme could be shown in tears, saliva, sputum, and other body fluids. Fleming excitedly described his observations:

> It was possessed of extraordinary power. Up till then I used to wonder at the much slower action of the antiserum which, when added to an infected broth warmed in an incubator or in the water bath, takes some considerable time to dissolve the microbes, and then only incompletely. But when I studied this new substance, I put into a test-tube a thick, milky suspension of bacteria, added a drop of tear, and held the tube for a few seconds in the palm of my hand. The contents became perfectly clear. I had never seen anything like it. (8)

Fleming investigated this substance, which seemed to be the body's natural defense against invading microbes. It appeared to have the qualities of an enzyme and was termed lysozyme, since it dissolved the microbes. Luck had been Fleming's ally in lysozyme's discovery. First, Fleming stumbled into finding the zone of inhibition around the mucus that he had cultured. Second, Fleming was fortunate that the microbe on the original plate was sensitive to the action of lysozyme. Fleming began to assess if other bacteria were susceptible to its dissolving effect, hoping that lysozyme was going to be the substance he had been searching for since the war. Unfortunately, Fleming's luck ran out. He quickly determined that lysozyme's effects were limited. The nonpathogenic bacteria were affected, but more danger-ous pathogens such as the streptococcus were not destroyed. Fleming wrote that he was not surprised by this observation. Fleming reasoned that it was only natural that pathogenic bacteria would be resistant to the action of lysozyme. If they were not resistant to its action, they would not be pathogenic! Lysozyme would have destroyed them. Fleming presented his findings on lysozyme to the Medical Research Club, a scientific body founded in 1891. The physicians present did not ask a single question following Fleming's presentation, an ominous sign that the club found the paper to be essentially worthless. Fleming's quiet, disorganized presentation style did not help matters. Even the dynamic Almroth Wright could not inject interest in lysozyme when he presented the work to the Royal Society in London in 1922. In the period between 1922 and 1927, Fleming published six papers on lysozyme. He tried to isolate the substance to study it chemically but had little success since he lacked sufficient training in chemistry. He was disappointed that lysozyme was not the magic bullet for treating bacteria. Despite the frigid reception his discovery had in the medical community, Fleming thought that lysozyme might play an important role in the future. He was right. Eventually, other scientists appreciated Fleming's work on lysozyme. Lysozyme is still used today by scientists in the laboratory to gently dissolve the capsules surrounding bacteria. Lysozyme also protects foodstuffs such as Russian caviar and has been added to wine as a preservative.

WORK ON ANTISEPTICS

During the 1920s, Fleming continued to work on lysozyme, but he also worked on antiseptics, inspired by his observations during the war. Specifically, he devised experiments to study antiseptics' action on bacteria and on white blood cells. Fleming observed that antiseptics kill white blood cells at concentrations much lower than those required to kill bacteria. Fleming concluded,

> These experiments show that there is little hope that any of the antiseptics in common use could be successfully introduced in the bloodstream to destroy the circulating bacteria in cases of septicemia. (6)

Fleming's safe chemical substance that destroyed pathogenic bacteria remained elusive.

THE DISCOVERY OF PENICILLIN: "THAT'S FUNNY"

Due to its importance in medicine, the story of penicillin's discovery has become shrouded in legend and distorted truths. It started as a simple laboratory observation. Fleming had been invited to contribute an article on staphylococci for the Medical Research Council's *A System of Bacteriology*. He was working with a colleague, Merlin Pryce, and studying the various forms and mutants of the bacteria. Fleming had dozens of cultures of staphylococci in his laboratory on any given day. In August of 1928, Fleming had returned from a vacation when Pryce went to see him in his lab. While the two coworkers were speaking, Fleming picked up a 2-week-old culture dish that had been contaminated with a mold—not an unusual event. Fleming told Pryce,

> As soon as you uncover a culture dish something tiresome is sure to happen. Things fall out of the air. (8)

Then Fleming stopped talking. He kept looking at the dish. Finally, Fleming said, "That's funny." The dish had several large colonies of staphylococci and one colony of mold in pure culture, approximately 20 mm in diameter. What was "funny" was the zone around the mold where hardly any bacteria were present. The finding was similar to his lysozyme discovery but with an important difference. Whatever was causing the inhibition of pathogenic bacteria was coming from another living organism. Fleming hardly looked up as he scraped a piece of the mold with a scalpel and placed it in a tube of broth. Pryce watched this process and later wrote,

What struck me was that he didn't confine himself to observing but took action at once. Lots of people observe a phenomenon, feeling that it may be important, but they don't get beyond being surprised—after which, they forget. That was never the case with Fleming. I remember another incident, also from the time when I was working with him. One of my cultures had not been successful, and he told me to be sure of getting everything possible out of my mistakes. That was characteristic of his whole attitude to life. (8)

Fleming kept the petri dish. It was to be his most prized possession for the rest of his life. The dish eventually found its way to the British Museum. I recall seeing it there in a glass case. Although completely dried, an outline of the mold and zone of inhibition were still evident. Sitting next to the petri dish in the case was the original manuscript from Paul McCartney's song "Yesterday," although I never understood the Museum's reasoning for placing those two items next to each other.

Fleming thought he might have found the substance that he had been looking for since the war. He found the evidence on the plate compelling, even though his colleagues seemed to only feign interest when Fleming showed them the plate. Fleming pressed on, giving this accidental finding all his attention. The mold contamination that had frequently occurred on his older culture plates had never looked like this one. Fleming grew the mold on an agar plate so he could test other bacteria in its presence. He found that certain microbes—streptococci, staphylococci, and diphtheria organisms—were affected by the mold's substance, whereas others, such as typhoid bacilli, were not. Fleming then grew the mold in broth so he could extract the material that seemed to be responsible for the effect on bacteria. After several days of growth of the mold in broth culture, the liquid turned yellow. Fleming took a filtrate of the yellow liquid and found that it was as active on bacteria as the mold on the agar plate. The same types of bacteria were affected. Fleming diluted the yellow liquid, only to find that a surprisingly weak solution—a 500th the strength of the original—still affected staphylococci. Fleming had to know just what this mold was. He showed it to a colleague, C. J. La Touche, St. Mary's mycologist, who decided it was *Penicillium rubrum*. Later, Fleming found that La Touche had misidentified the mold. It was actually *Penicillium notatum*.

Fleming found the substance produced by this mold to have unique properties. He showed that it did not simply inhibit certain bacteria (bacteriostatic); it killed the bacteria (bactericidal) at concentrations that were several hundred times more potent than any known antiseptic. Even more important, Fleming thought, the substance had no toxic effects on white blood cells. Fleming wrote,

It was the first substance I had ever tested which was more antibacterial than it was antileukocytic and it was this especially which convinced me that some day when it could be concentrated and rendered more stable it would be used for treatment of infections. (3)

A DECIDEDLY UNSTABLE SUBSTANCE

The "some day" would be further away from Fleming's initial observation than he would have liked. Fleming tested other molds to see if they also produced this kind of substance. None of them did. He began to realize how unusual his find was. Fleming injected the unpurified substance into animals in extraordinarily concentrated solutions, with no ill effects on them. But he found that this substance was decidedly unstable. During his work with lysozyme, Fleming, who readily admitted that he was not a chemist, had perfected some rudimentary methods to analyze and isolate compounds. Fleming and his assistant, Stuart Craddock, were optimistic but had the difficult job of isolating and characterizing this substance. When the substance was left in broth at room temperature, its antibacterial activity rapidly diminished. The substance was also affected if it was in acid solutions, with diminished activity. Alkaline solutions made it more stable. As Craddock wrote,

> We were full of hope when we started but as we went on, week after week after week, we could get nothing but this glutinous mass which, quite apart from anything else, would not keep. The concentrated product retained its power for about a week, but after a fortnight it became inert. (9)

Eventually, Fleming and Craddock gave up on their attempts to extract the substance. Still, Fleming was full of hope that he had something of scientific importance.

FIRST PRESENTATION OF PENICILLIN'S DISCOVERY

Fleming prepared a paper for presentation to the same group that had greeted his discovery of lysozyme with a cold shoulder, the Medical Research Club in London. On 13 February 1929, Fleming read his paper to the group. Sir Henry Dale, chairman of the Club, recalled his presentation:

> He [Fleming] was very shy, and excessively modest, in his presentation, he gave it in a half-hearted sort of way, shrugging his shoulders as though he were deprecating the importance of what he said. . .. All the same the elegance and beauty of his observations made a great impression. (10)

However great the impression may have been, the silence following Fleming's presentation was deafening. It was another signal from the Medical Research Club that they thought that the paper was utterly worthless. The physicians present had heard the first presentation of one of the greatest advances in the history of medical science, and not a single question was asked! Fleming was appalled, though he gave no outward sign of it. He had enough confidence in his discovery to prepare a manuscript for *British Journal of Experimental Pathology* (11). Fleming gave the substance in his "mould broth filtrate" the name *"penicillin"* in this paper, deriving it from the name of the mold. Little notice of his paper occurred at the time. But Fleming continued his interest in penicillin.

FIRST ATTEMPT TO PURIFY PENICILLIN

In 1932, Fleming gave a chemist, Harold Raistrick, at the London School of Tropical Medicine and Hygiene strains of his *Penicillium* mold. Raistrick assembled a team to attempt to isolate penicillin. The team met with scientific and personal difficulties, including the untimely death of a young chemist on the team. Raistrick and his team were able to grow the *Penicillium* mold on synthetic media and extract penicillin in ether. But when they tried to evaporate the ether to obtain pure penicillin, the powder that remained had no antibacterial activity at all. With the inopportune breakup of his team and the instability of the substance that they were attempting to isolate, Raistrick stopped his work, saying that unstable material "would never be of practical use in clinical medicine" (12). Fleming was disappointed when a renowned chemist could not isolate penicillin. He ceased work on the substance but never gave up hope that someday someone would solve the problem of penicillin isolation. He saved the mold and gave samples to any person who requested it, a policy that became unexpectedly fortuitous but would later cause concern when the drums of the Second World War began to beat. For nearly 10 years following its discovery, penicillin would lay dormant on the scientific shelves. It would take progress in another drug, Prontosil, to help pull penicillin back down.

PRONTOSIL: EARLY WORK LEADS TO SUCCESS

In the years from 1927 to 1932, Gerhard Domagk was busy testing potential antimicrobial compounds against a virulent streptococcus in a mouse model of infection and *in vitro*. Progress for Domagk and his team was dreadfully slow but for completely different reasons than those that slowed Fleming. His team's results were erratic, especially in the animal testing. Many scientists believed that chemotherapy for infectious diseases, as Ehrlich had envisioned, would never come to pass. Finally, in 1932, progress improved when Domagk worked out the testing of an azo dye derivative, sulfonamidochrysoidine (also known as KI-730). The KI-730 substance, dubbed Prontosil, worked in his animal model, protecting mice from injected streptococcal bacteria, but did not work against the same streptococcus *in vitro*. At the time, the investigators had no explanation for the activity of the Prontosil in the mouse model but not in the test tube. The company, IG Farben, was issued a patent for Prontosil in 1932. Domagk published the results in

1935 (13). The reason for the 3-year publication delay was never explained, but clearly, in 1932, people at IG Farben thought they were onto something. In 1936, Domagk and his colleagues were preparing to begin human trials for Prontosil when fate intervened. Domagk's own daughter became seriously ill with a strepto-coccal infection. He used the drug on his own daughter, with dramatic success. Spurred on by this triumph, Domagk made Prontosil available for wider human testing. Fleming's colleague Leonard Colebrook was among those in England test-ing the drug and the more soluble Prontosil S. Colebrook successfully treated some human puerperal infections, although there were conflicting results from other investigators (14). However, Gerhard Domagk had made a chemotherapeu-tic agent for the pathogenic streptococcal and other bacterial diseases a reality. For this contribution, he was awarded the Nobel Prize in Medicine in 1939.

THE DISCOVERY OF SULFANILAMIDE

In France, scientists at Institut Pasteur, including Ernest Fourneau, made an important discovery in 1935. A part of the Prontosil molecule, sulfanilamide, was an effective antimicrobial agent itself. Fourneau's research seemed to explain the unusual finding with Prontosil—its activity *in vivo* but not *in vitro*. Prontosil broke down in an animal (or human) body to derivatives; one was sulfanilamide, which was toxic to bacteria. The activity of the derivative made the German patent on Prontosil of little consequence to the French. Sulfanilamide began to be tested and marketed in France. In the course of 2 short years, sulfanilamide was on the market in Great Britain, France, and the United States. The success of sulfanilamide changed everyone's thinking about chemotherapy of bacteria. In his thorough review of the origins of sulfa drugs, John Lesch pointed out,

> One of the remarkable things about this story is that the skepticism about bacterial chemotherapy, so pervasive in early 1935, was everywhere dissipated by the Spring of 1937. (15)

With the discovery that a widely available dye, sulfanilamide, had antibacterial activity, organic chemists went to work modifying the chemical, which turned out to be relatively easy. The first derivative, sulfapyridine, was available by 1938. From 1935 to 1945, more than 5,000 new sulfa drugs were prepared (15).

RENEWAL OF INTEREST IN PENICILLIN

The developments in sulfa drugs did not escape the notice of Alexander Fleming. But he was cautious in his optimism. Before the modern antimicrobial era had even begun, Fleming predicted its Achilles' heel. Fleming foresaw the development of resistance to the sulfa drugs:

> This may be due to one of two causes. Either the more sensitive organisms have been eliminated by the drug, while the naturally less sensitive have survived. And, in reproducing themselves, have engendered whole resistant generations: or as a result of insufficient treatment, a microbe, once vulnerable, has acquired the power to resist. (10)

Amid the commotion over Prontosil, Fleming knew he had something even better. He confided to a colleague in 1935,

> I've got something much better than Prontosil, but no one'll listen to me, I can't get anyone to be interested in it, nor a chemist who will extract it for me.... I don't know. It's too unstable. It will have to be purified, and I can't do it myself. (10)

Fleming had no way of knowing when he made those statements in 1935 that a team was assembling in nearby Oxford, England, that would do what he could not.

THE OXFORD TEAM

Howard Walter Florey headed the team at Oxford that was to isolate penicillin. But he did not start out at Oxford intending to study penicillin. Florey was interested in immunity and how substances like lysozyme might be involved in protecting the body against bacteria. He was born in Adelaide, Australia, in 1898. He completed his medical education in Oxford, England, on a Rhodes scholarship in 1924. From Oxford, Florey went to Cambridge and then spent 1 year in the United States on a Rockefeller Fellowship, a year where he was to make important contacts for his later work on penicillin. He returned to Cambridge and received his Ph.D. in 1927. After spending 4 years at both Cambridge and then the University of Sheffield, Florey came back to Oxford as a professor of pathology in 1935 and as the head of a laboratory studying immunity to bacteria. Florey's laboratory continually suffered from paltry budgets, forcing Florey to fight for funds, which was not a chore he relished. But Florey was up to the task. His biographer, Gwyn McFarlance, met and described the man:

> A rough, tough Australian, completely uncompromising, rather prickly, very energetic and tense as a coiled spring. And he brought to his work this extraordinary dedication, which was very infectious, in such a way that he really could collect a team of people who became almost as dedicated and enthusiastic as himself. (16)

Florey wanted to study the effects of lysozyme on bacteria. He put together a team of scientists at Oxford that had expertise in areas where he did not. Florey's complex and driven personality created distance and sometimes problems interacting with other members of his group. One member that Florey managed to attract to his group was Ernst Chain. Chain was born in Berlin in 1906. By 1933, Chain had received his doctorate in a relatively new field, biochemistry. Chain, a Jewish scientist, left Berlin the day the Nazis came to power and fled to England. He worked first in London and then Cambridge before Florey, who was looking for a biochemist, asked him to join his team in Oxford in 1935. Florey recognized the importance of biochemistry in his work and gave Chain a relatively free hand. Chain believed that most biological phenomena such as the action of toxins and bacterial lysis could be explained in chemical terms. Chain began working on lysozyme, which he believed, correctly, to be an enzyme. Chain's laboratory experiments isolated the enzymatic protein lysozyme and determined the chemical from the bacterial cell that was the substrate for lysozyme—a polysaccharide. In 1938, Chain looked in the scientific literature for other substances that might lyse bacteria or inhibit their growth. He found 200 papers on bacterial growth inhibition from substances that were produced by other microorganisms. But little was known about the chemical properties of these substances. The success of Prontosil and sulfa drugs invigorated hope that chemotherapy of microbes could be a reality. After discussing the matter with Florey, the two men agreed that this was a fruitful area for research and developed a plan. Florey would examine the biological nature of a substance, including testing it in animal models of infection; Chain would examine the chemical properties of any found substances. Chain described his next step:

> One of the most impressive and best described phenomena of bacterial antagonism which I found during the literature search was described in 1929 by the same bacteriologist who had discovered lysozyme some seven years earlier, Alexander Fleming. He had shown that a mould, a penicillium species which had settled on one of his Petri dishes, later identified as *Penicillium notatum*, had growth inhibiting properties against a number of pathogenic bacteria. I had come across this paper early in 1938 and on reading it I immediately became interested. The reason was that according to Fleming's description the mould had strong bacteriolytic properties against the staphylococcus. (17)

At first, Chain believed that penicillin was an enzyme like lysozyme. This belief was strengthened by his study of Raistrick's difficulties in isolating it because of the substance's instability (18), a frequent characteristic of enzymes. But Chain was wrong about penicillin being an enzyme, as he soon found out.

THE ISOLATION OF PARTIALLY PURIFIED PENICILLIN

Chain's first step was to get a sample of *Penicillium* mold. By coincidence, a strain of Fleming's mold was stored just down the hall. One of Oxford's scientists and Florey's predecessor, George Dreyer, was interested in bacteriophages, virus-like particles that invade bacteria, causing lysis of the cell. Dreyer had asked Fleming for a subculture of his mold, mistakenly thinking that a bacteriophage was responsible for Fleming's observation of the inhibition of staphylococci on his famous plate. Dreyer soon realized that the effect that Fleming observed was not due to a bacteriophage and stopped his work on it. But he saved the mold. Chain obtained a sample of it. Neither Florey nor Chain had any experience growing molds. Florey recruited help from a third member of the team, Norman Heatley, who became responsible for the microbiological aspects for the team. Florey tasked Heatley with growing large quantities of the mold to get sufficient amounts of penicillin. Personality clashes among the three men were evident from the beginning. Heatley refused to report to Chain, demanding that he would continue working only if he reported directly to Florey. However, the team's sense of purpose won out. Chain began his work on penicillin in 1939, just as World War II broke out. Chain was familiar with techniques to extract enzymes from his work on lysozyme. He used a freeze-drying technique that had been only recently developed and isolated partially purified penicillin. Chain's chemical studies of penicillin confirmed its instability, especially in acid solutions or at high temperatures, but the work also showed that it was not an enzyme. Penicillin was a relatively small molecule. The investigators tested the substance on a mouse, injecting it with a whopping 20 to 30 mg of partially purified penicillin, with no adverse effects on the animal at all. They repeated Fleming's experiments on white blood cells, finding that penicillin did not affect them. They also tested penicillin *in vitro* against various bacteria, confirming Fleming's findings on the spectrum of antibacterial activity.

The process of purifying penicillin was exceedingly difficult at first, but slowly Chain improved the process, which was markedly different from developing derivatives of sulfanilamide for sulfa drugs. Sulfa drugs were made from a widely available dye that organic chemists could easily modify. Penicillin had to be isolated from a living organism that produced a highly active, unstable, antibacterial substance but in minute quantities.

PENICILLIN TESTING IN ANIMALS—MIRACULOUS RESULTS

From all their work, the Oxford team believed that they had a chemotherapeutic agent that needed to be tested in an animal model of infection. In May 1940, they tested penicillin on a group of mice that had been infected with staphylococci, streptococci, and *Clostridium septicum*. Heatley stayed in the laboratory all night to find that all the animals treated with a small quantity of penicillin survived, while all the control animals died—a miraculous result! More animal experiments followed, with the same results, leading to publication in *Lancet* (19). The authors commented in the article that penicillin was also active against anaerobic bacteria, including those that caused gangrene. Florey and his team realized that penicillin had great medical potential.

Fleming knew nothing of the Oxford team until he read their *Lancet* paper in 1940. Surprised and delighted, Fleming went to Oxford to see Florey and Chain. "Shocked" was a better way to describe Chain's reaction when Fleming showed up in his lab. Chain thought Fleming was dead! Chain, then, described his reaction to Fleming:

> He struck me as a man who had difficulty expressing himself, though he gave the impression of being somebody with a very warm heart doing all he could to appear cold and distant. (20)

In actuality, Fleming could not have been more pleased to see what the Oxford team was doing. On his return to St. Mary's, Fleming told his assistant, Craddock,

> They have turned out to be the successful chemists I should have liked to have with me in 1929. (19)

THE FIRST HUMAN TRIALS OF PENICILLIN

The time had come to begin human trials of penicillin. The Oxford team was uneasy since they still had little purified penicillin on hand. Treating patients would be risky without large-scale production. Florey went to industry leaders, but all flatly refused. It may be difficult to imagine that companies would refuse to put resources into isolating one of the greatest medical treasures, penicillin. But the timing of world events intervened. In June 1940, after the German offensive in which France fell, Great Britain was in serious danger of an invasion. Factories were fully occupied with producing war materials, with no resources to spare. So, Florey told his team to produce a hundred liters of mold culture per week and extract penicillin themselves.

THE FIRST PENICILLIN PATIENT: THE OXFORD POLICEMAN

By February 1941, they had a small supply of penicillin for a daring human experiment. An Oxford policeman who had a small scratch at the corner of his mouth became the first test patient. The small wound had become infected with staphylococci that spread throughout his body, causing serious illness with abscesses everywhere. He was dying. The doctors tried sulfa drugs, without success, and gave the man no chance of survival.

On 12 February 1941, the first patient ever to receive penicillin received an intravenous injection of 200 mg. In 24 hours, he showed significant improvement. The penicillin was continued, and he received a blood transfusion. But concerns were raised that there may not be enough penicillin to completely treat the patient. Arrangements were made to collect the man's urine and extract whatever amounts of penicillin that they could. The Oxford team knew from their animal experiments that penicillin was excreted in the urine, largely unchanged. By the end of 4 days of penicillin, the policeman was able to eat. He no longer had fevers. However, the 4 days of penicillin treatment were not enough, and the supply of the drug was exhausted. The man managed to live a few more days, but the infection recurred. On 15 March, he died.

Florey knew that penicillin had proven itself, but the man's death presented problems proving it to the world. Critics would say that the blood transfusion could have been responsible for his temporary improvement. The Oxford team became more determined to obtain an adequate supply of penicillin. Three more patients were treated, two with spectacular success. The third was a child with coma who was improving until a blood vessel ruptured, resulting in death. With these and a few more cases, penicillin's triumph could no longer be questioned. The team published their entire research, including the first nine humans treated with penicillin, in *Lancet* (21). In the publication they also described in detail their methods of growing the mold and purifying penicillin. But incessant bombing made large-scale production of penicillin in England impossible. Florey had no choice but to turn to the United States for help.

PENICILLIN PRODUCTION IN THE UNITED STATES

In June 1941, Florey made arrangements to head to America. He chose to take Heatley with him, not Chain. Chain never forgave Florey for leaving him behind, leading to a feud lasting years. Florey and Heatley had to travel through Lisbon, Portugal, in the heat of June. Concerned about the security of taking a culture of the precious *Penicillium* mold in a vial that could be stolen, Heatley suggested that they smear their coats with the *Penicillium* strain for safety on their journey. When they arrived in the United States, Florey turned to his American contacts for help. Eventually, the two men were led to Charles Thom, the Principal Mycologist of the U.S. Department of Agriculture, and Orville May, then Director of the U.S. Department of Agriculture facility in Peoria, IL. It was Thom who corrected the earlier noted error in identification of *Penicillium notatum*, which had been initially identified as *Penicillium rubrum*. Thom also recognized the rarity of this *P. notatum* strain because only 1 other strain in his collection of 1,000 *Penicillium* strains produced penicillin. The strain that was eventually used in mass production was a third strain, *P. chrysogenum*, found in a moldy cantaloupe in a Peoria market, which produced 6 times more penicillin than Fleming's strain.

The efforts of Florey and Heatley now diverged. Heatley stayed in Peoria to help with the large-scale culturing of the mold. Florey headed back east to interest the U.S. government and pharmaceutical companies in penicillin.

LARGE-SCALE CULTIVATION OF *PENICILLIUM*

Heatley spent 6 months in Peoria and shared all his knowledge of cultivating Fleming's mold with two members of the U.S. Department of Agriculture facility there, R. D. Coghill and A. J. Moyer. However, a component of the media that Heatley used in England was unavailable in Peoria. Moyer suggested an alternative: corn steep liquor. Peoria was in the middle of the wet corn milling industry. Corn steep liquor, a waste product of manufacture of cornstarch, was available in large quantities. Corn steep liquor was used with a deep fermentation technique for *Penicillium*. This process produced exponentially greater amounts of penicillin in the filtrate of the mold than the Oxford team had ever been able to produce. Coghill noted,

> One of the least understood miracles connected with [penicillin] is that Florey and Heatley were directed to our laboratory in Peoria—the only laboratory where the corn-steep liquor magic would have been discovered. (16)

Another wonder of the penicillin story is the astonishing rarity of the *Penicillium* strain that produced the substance. During a 5-year period, Peoria researchers examined 1,000 other molds to see if they produced penicillin. Only three did: Fleming's strain, another *Penicillium* strain in the facility's collection, and a mold found on a rotting cantaloupe in a Peoria market that produced 6 times more penicillin than Fleming's mold. Consider the odds for Fleming to have made his discovery!

Meanwhile, Florey tried to interest U.S. pharmaceutical firms in the commercial production of penicillin. Amazingly, he was initially spurned at every turn. Finally, with some U.S. government funding, he coordinated penicillin production with several companies, among which were Merck & Company, E. R. Squibb and Sons, Charles Pfizer and Company, Bristol Laboratories, Abbott Laboratories, Winthrop Chemical Company, Cutter Laboratories, and Parke, Davis, and Company. Researchers at these drug companies developed a new technique for producing enormous quantities of penicillin-producing *Penicillium* spp.: deep-tank fermentation. This process adapted a fermentation process performed in shallow dishes to deep tanks by bubbling air through the tank while agitating it with an electric stirrer to aerate and stimulate the growth of tremendous quantities of the mold (22).

WARTIME PENICILLIN PRODUCTION IN THE UNITED STATES

In December 1941, world events once again influenced penicillin's history. With the entry of the United States into World War II, the U.S. government took over penicillin production, which rapidly escalated. At the beginning of 1942, there was barely enough penicillin to treat a single patient. By the end of 1942, penicillin supplies allowed the first large-scale use of the drug to treat patients burned in Boston's Cocoanut Grove tragedy. The Cocoanut Grove was one of Boston's most popular nightspots. On 28 November 1942, 800 people were jammed into the place when a fire quickly took over. A better firetrap could not have been deliberately designed. Side exits had been bolted to keep patrons from skipping out on their bills. Windows were boarded up due to the war's blackout. The lights went out, sending people to the only exit they knew—the front revolving door. Firemen had to break down the door, finding bodies of the dead six deep. In all, over 490 people lost their lives. A great many survivors were burned or suffered smoke inhalation injuries. Hospitals all over the city were overwhelmed, especially Boston City Hospital and Massachusetts General Hospital (MGH). MGH received many of the burn patients. A tragedy of this magnitude (it was the second most disastrous U.S. fire, after Chicago's Iroquois Theater fire in 1903, in which 571 people died) made national headlines. On 12 December 1942, two chemists from Merck raced, with police escorts, to Boston's MGH with penicillin, to help treat dozens of burn victims in the first extensive clinical use of the drug in the United States (23).

After this clinical trial, penicillin production became incredibly successful, largely because of the unparalleled cooperation between Great Britain and the United States, and an unprecedented U.S. government/private industry partnership. By September 1943, the stock was sufficient to satisfy all the demands of the Allied Armed Forces.

PENICILLIN AND PATENTS

Issuing a patent for penicillin was a controversial topic from the beginning. Chain raised the issue early on in his work at Oxford. He believed that obtaining a patent was essential. Florey and others in England held a much different view: patents were unethical for such a lifesaving drug and contrary to the research traditions in Great Britain. Indeed, penicillin was challenging the basic notions of a patent, considering that it was a natural product, produced by another living microorganism, raising a legal question of whether it could even be the subject of a patent. The prevailing view in the United Kingdom at the time was that a process could be patented, but a chemical could not. The publication in *Lancet* added to the discord among the Oxford team, since the article openly disclosed extensive details of all their methods for penicillin production and isolation (19). The American view on patents was more aligned with Chain's view. Merck and A. J. Moyer each filed patents on the process of penicillin production, with no opposition. Eventually, at the war's end, British scientists were faced with paying royalties for a discovery made in England. Resentment toward the United States remained for a considerable period over penicillin patents. Chain publicly rebuked the British community for not acting as he had suggested. This disagreement, together with his personality conflicts with members of the Oxford team, prompted Chain's move to Italy in 1948.

PENICILLIN USE IN ENGLAND

After Florey returned to England, he and his team were impatiently awaiting supplies of penicillin from the United States at the beginning of 1942. The first supply from Merck did not arrive until April 1942 and was not sufficient to meet the many needs. So, the Oxford team had to continue its own meager efforts to produce penicillin. Florey's wife, Ethel Florey, headed a group of individuals, dubbed the P-Patrol, whose work was to secure the urine of any patient treated with penicillin in order to extract the precious drug.

PUBLIC AWARENESS OF PENICILLIN: THE FLEMING MYTH

In the summer of 1942, Fleming received a call from his brother, Robert, about a colleague who was dying of streptococcal meningitis, a life-threatening infection of the membranes surrounding the brain and spinal cord. Fleming immediately contacted Florey requesting penicillin. Florey gave Fleming nearly his entire supply. Fleming took a courageous step of injecting the penicillin into the fluid around the spinal cord, known as an intrathecal injection. No one had ever injected the drug that way, but the man was dying. The treatment worked: the man recovered. The *Times* of London carried a story about the incredible treatment, but no names of any doctors or researchers were mentioned. Almroth Wright saw the article and corrected the oversight in his usual flamboyant style:

> Sir: In the leading article on penicillin in your issue yesterday you refrained from putting the laurel wreath for this discovery round anyone's brow. I would, with your permission, supplement your article by pointing out that, on the principle of *palmam qui meruit ferat*, it should be decreed to Professor Alexander Fleming of this laboratory. For he is the discoverer of penicillin and was the author of the original suggestion that this substance might prove to have important applications in medicine. (16)

Up to that point the media had paid little attention to penicillin. But Wright's comments in the *Times* started a frenzy of media attention. Almost immediately, Sir Robert Robinson of Oxford sent a letter to the *Times* to acknowledge the role of Florey, stating that if Fleming deserved a laurel wreath, Florey deserved at least a handsome bouquet. The press descended on St. Mary's. Fleming was happy to comply with requests for interviews, finding the whole affair somewhat amusing. He quickly became something of a celebrity. Florey, on the other hand, was in no mood to court the media. In fact, he prohibited any of the Oxford team from giving media interviews. So, the press focused on Fleming, generally ignoring the Oxford team's contributions. The press portrayed Fleming as the sole inventor of the miracle drug. The adoration in the media and the distorted truths about penicillin's discovery did not endear Fleming to the members of the Oxford team.

In 1943, Fleming received one honor after another, including Fellowship in the Royal Society. The "Fleming Myth" was born (24). But Fleming always acknowledged the role that chance played in his discovery. He once gave this piece of advice:

> Never neglect an extraordinary appearance or happening. It may be—usually is, in fact—a false alarm which leads to nothing, but it may on the other hand be the clue provided by fate to lead you to some important advance. But I warn you of the danger of first sitting and waiting till chance offers something. We must work, and work hard. Pasteur's often quoted dictum that Fortune favors the prepared mind is undoubtedly true, for the unprepared mind cannot see the outstretched hand of opportunity. (25)

SECRECY IN WARTIME ENGLAND

The British government went to great lengths to prevent the means for producing penicillin from falling into enemy hands. However, news about penicillin leaked out. A Swiss company (CIBA, Basel, Switzerland) wrote to Florey requesting *P. notatum*. Concerned about responding, Florey contacted the British government. Agents attempted to track down where Fleming's *Penicillium* cultures had been distributed. Fleming wrote, "During the past 10 years I have sent out a very large number of cultures of *Penicillium* to all sorts of places, but as far as I can remember NONE have gone to Germany" (26). Florey believed that, without the mold, no one in Germany could produce penicillin, even though his publication had provided a "blueprint" for its small-scale manufacture. Florey was wrong, and so was Fleming.

Fleming had sent a culture of *Penicillium* strains to "Dr. H. Schmidt" in Germany in the 1930s. Schmidt was unable to get the strains to grow, but even though the Germans did not have a viable strain, other Europeans did.

PENICILLIN PRODUCTION DURING WORLD WAR II ON THE CONTINENT OF EUROPE

France

Someone at Institut Pasteur in France had Fleming's strain. In 1942, efforts began at Institut Pasteur and Rhône-Poulenc to produce penicillin. Eventually, German officials found out, and in early 1944, the Germans asked the French for their *P. notatum*. They were given a false strain that did not produce penicillin. With limited supplies, the French produced only enough penicillin to treat approximately 30 patients before the war's end.

The Netherlands

The situation in the Netherlands was different. The Centraalbureau voor Schimmelcultures (CBS) near Utrecht had the largest fungal collection in the world. A published list of their strains in 1937 included *P. notatum*. A letter found at CBS shows that in February 1942 the Nazis asked CBS to send their strain of *P. notatum* to Dr. Schmidt in Germany, mentioning penicillin in the letter. CBS told the Germans they did not have Fleming's strain of *P. notatum*. In fact, they did. In the 1930s, Fleming had sent his strain to Johanna Westerdijk, the CBS director. Westerdijk could not refuse the German request for their strain of *P. notatum* but sent them the one that did not produce penicillin.

Efforts to produce penicillin in the Netherlands went underground at a company in Delft, the Nederlandsche Gist-en Spiritusfabriek (the Netherlands Yeast and Spirit Factory; NG&SF). After the German occupation in 1940, NG&SF was still allowed to function. Because Delft was not bombed in the war, NG&SF's efforts were unaffected. In early 1943, NG&SF's executive officer, F. G. Waller, secretly wrote to Westerdijk at CBS, asking for any *Penicillium* strains that produced penicillin. In January 1944, Westerdijk sent all of CBS's *Penicillium* strains to NG&SF.

Four reports in NG&SF records detailed their efforts (27). In the first report, NG&SF scientists tested 18 *Penicillium* strains from CBS; they found 1 strain with the greatest antibacterial activity, which was coded P-6 and was identified as *P. baculatum*. The second report discussed how NG&SF scientists then isolated an extract from P-6. They gave the substance in the extract the code name Bacinol after the species from which it was derived and to keep the Germans unaware of what they were doing. As Waller wrote, "When we first started looking, in 1943, only one publication was available, that of Fleming in 1929. It was on that basis we started our research." NG&SF researchers then had help from an unanticipated source. In 1939, Andries Querido was employed by NG&SF as a part-time advisor. By January 1943, however, his Jewish background limited his visits. On his last visit in the summer of 1944, Querido met someone in Amsterdam's Central Train Station who gave him a copy of the latest *Schweizerische Medizinische Wochenschrift* (*Swiss Medical Journal*), which he passed on to the NG&SF scientists. The June 1944 issue contained an article entirely devoted to penicillin, showing the results that the Allies had achieved, including details of penicillin growth in corn steep extract, the scaling up of penicillin production, the measurement of strength by the Oxford unit, results of animal and human studies, and identification of the bacteria known to be susceptible to penicillin. The third report described how NG&SF scientists isolated Bacinol from the extract using the information supplied secretly by Querido.

Large-scale production would be difficult to perform and to keep secret from the Germans, especially with a German guard on site. However, NG&SF scientists used an obvious ploy to keep the German guard, who knew nothing about microbiology, at bay: they kept him drunk. "We did have a German guard whose job it was to keep us under surveillance, but he liked gin, so we made sure he got a lot. He slept most afternoons." NG&SF scientists used milk bottles for growing large quantities of *Penicillium* mold. From July 1944 until March 1945, production of Bacinol continued, as detailed in the fourth report. At the end of the war, the NG&SF team still did not know if Bacinol was actually penicillin until they tested it against some penicillin from England, proving it to be the same compound. NG&SF began marketing the penicillin they produced in January 1946. Although the original building where Bacinol was produced was demolished, NG&SF named a new building in honor of their WWII efforts (Fig. 14.1).

Denmark

Like the Netherlands, Demark succeeded in making penicillin in secret from the Nazis, but Danish efforts are less well known outside of Denmark because much of the documentation remains only in Danish. Peder Worning, a microbiologist at Hvidovre Hospital outside of Copenhagen, wrote about the Danish penicillin production during the war in his 2014 book *Der er flere bakterier I et gram lort end der er mennesker i hele verden* (FADL's Forlag, Denmark) and was kind enough to translate his writing into English, which may mark the first time this story has appeared in English. The account begins at Løvens Kemiske Fabrik, now LEO Pharma. The president, Knud Abildgaard, had foresight right from the start of the war in September 1939. He had purchased large quantities of the most important chemicals so the factory could continue production for several years. When Denmark was occupied on 9 April 1940, Abildgaard called the staff together and said that if they were required to work for the occupying forces, he would close the factory. The Germans heard about this statement and immediately stopped all import permits for the factory, but the stockpile of chemicals allowed work to continue.

FIGURE 14.1 *Bacinol 2, a building named in honor of the site of efforts in Delft, The Netherlands to produce penicillin during World War II and the drug produced by The Netherlands Yeast and Spirit Factory.*

In 1942, Abildgaard heard of Florey and Chain's success with penicillin and obtained a copy of their 1941 *Lancet* article through some connections he had in Sweden. He showed Kai Adolf (K. A.) Jensen, a professor of pathology, the article and asked if he would head a group to try to start production of penicillin at the factory. Jensen agreed, but several difficulties had to be overcome before penicillin could be produced, chief among them obtaining a strain of *Penicillium* that produced penicillin. In Alexander Fleming's original 1929 paper, he described the fungus falling from the air onto a culture plate. K. A. Jensen wondered if he could isolate a penicillin-producing mold in the same manner. He set up a number of petri dishes around his laboratory, at his home, and at the homes of friends and acquaintances. As with many aspects of the penicillin story, good fortune intervened. On a back staircase in Nørrebro, Copenhagen, a *Penicillium* spore that produced significantly more penicillin than Fleming's original mold fell into one of his petri dishes. Using this strain as a starting point and the 1941 *Lancet* paper as a guide, Jensen's group developed a method of extracting the penicillin and producing it in pure form. The following year, 1943, they performed the first experiments on mice. Just as in the work of Florey, Heatley, and Chain, penicillin-treated mice survived infection with bacteria; the untreated ones died. Further findings showing the effectiveness of their production were achieved with some cows from Abildgaard's farm. Several cows acquired mastitis. Through the washing of the udders with cell-free extract from the *Penicillium* mold, the cows were cured. With the successful treatment of both mice and cows, Abildgaard and Jensen thought the time had come to produce penicillin on a larger scale to start treating sick people.

By September 1944, Abildgaard and Jensen produced enough penicillin to successfully treat people. But penicillin's manufacture had to be kept secret from the Germans or they would seize the production and use it for their soldiers. Rumors about the drug's production began to fly. One rumor held that they were not making real penicillin, another was that the penicillin was so impure that it gave patients a fever, and a third was that the factory was collaborating with the Nazi forces and supplying penicillin to the German army. If people believed that the factory was collaborating with the Nazis, members of the resistance might sabotage the factory, halting all production. One of the first people to be treated with the limited supply of penicillin was a doctor from Aarhus, who was cured of pneumonia. He was so grateful that he gave an interview to a newspaper. The secret was out! Abildgaard held a press conference where he said that production was still in the experimental stage, that they had only tested it on very few people, and that nothing was being shipped out of the country. But Abildgaard's situation with the Nazis became too dangerous and he had to flee to Sweden. Jensen became responsible for penicillin production. The Germans came and wanted the Danish penicillin, but Jensen managed to convince them that their penicillin was so impure that the patients using it often developed high fever. Using one of the many false rumors about their production to his advantage, Jensen convinced the Germans that the Danish penicillin was not worth seizing, and production quietly continued.

The Danes secretly treated wounded resistance fighters who needed penicillin. Jensen oversaw the distribution of the penicillin from the factory to the country's hospitals. When the hospitals called him, he would distribute what they had produced by ambulance or by a porter. For hospitals in the western part of Denmark, Klaus Jensen, K. A. Jensen's son, would take the penicillin ampules to the Central Train Station and give them to a train driver so that the hospital staff could collect the medicine when the train arrived, keeping the distribution secret from the Nazis.

The Nazis eventually succeeded in making penicillin by October 1944. However, Allied air raids crippled mass production of the drug (28).

POSTWAR PENICILLIN PRODUCTION

The penicillin production in the Netherlands and Denmark turned out to be of more than historical interest. Because these countries had researched and developed their own penicillin using their own mold cultures and used their own production methods, they were not embroiled in any patent clashes. The marketing of their penicillin eventually increased penicillin supply and decreased prices, although supply of penicillin remained a problem in other countries for several years following the war.

THE NOBEL PRIZE

In 1944, the press contained rumors that Fleming was going to win the Nobel Prize in Medicine. The rumors were not true, but the Queen of England knighted both Fleming and Florey. The next year, 1945, Fleming, Florey, and Chain shared the Nobel Prize in Medicine. Fleming remained gracious and humble. At a banquet during the Nobel award ceremonies, Fleming spoke about the lessons from the story of penicillin:

> The first is that teamwork may inhibit the primary initiation of something quite new but once a clue has been obtained teamwork may be absolutely necessary to bring the discovery to full advantage.
>
> The second is that destiny may play a large part in discovery. It was destiny which contaminated my culture plate in 1928—it was destiny which led Chain and Florey in 1938 to investigate penicillin instead of the many other antibiotics which had then been described and it was destiny that timed their work to come to fruition in wartime when penicillin was most needed. (29)

Fleming went on to a period of continuous honors after the Nobel Prize. After the war was over, he and his wife traveled extensively as Fleming was showered with the honors, including his likeness on a stamp (Fig. 14.2). He had an audience with the Pope on three separate occasions. Despite his popularity and celebration, Fleming never did become an accomplished conversationalist. As one colleague described it,

> [Conversing with Fleming] was like playing tennis with a man who, when he received a service, put the ball in his pocket. (16)

In 1946, Fleming assumed the title of Principal of St. Mary's Institute of Pathology, as it was now called, when Wright retired. A year later, Wright died. Wright's passing was a source of deep sorrow for Fleming. Philip Wilcox of the Institute described the two men, who were so dissimilar but such friends:

> Fleming was an easy man to get on with, and to me he always seemed to be unruffled and utterly lacking in fussiness or strained nerves. He was calm, easy-going, docile, never detached from the world around him or over-engrossed in his work. In this

FIGURE 14.2 *Alexander Fleming's image on a European postage stamp.*

respect he was more "worldly" than Sir Almroth Wright, who gave one the feeling that he was a man with a gigantic brain, concentrated on the world of bacteria, and caring little for sport or gaiety. (30)

During the next 3 years, Fleming's wife, Sareen, grew progressively weaker from an unknown illness. She died in October 1949.

After Sareen's death, Fleming continued to work in his laboratory. There, he met Amalia Koutsouri-Vourekas, who remained with Fleming as a research assistant for several years. Amalia was fluent in three languages. She served as a translator for foreign visitors and accompanied Fleming on some trips. She returned to Athens in 1951. From 1951 to 1954, Fleming was Rector of Edinburgh University. But Fleming was emotionally shut down during this time, confiding in no one. He claimed that people put too much stock in what he said, so he felt it best to be cautious. In 1953, at a meeting of the World Health Organization in Athens, his former research assistant, Amalia, showed Fleming the city and served as his guide at the conference. He was delighted. After only 3 days in Greece, Fleming proposed to Amalia. They were married shortly afterwards in London. Fleming was 71; she was 39. Fleming enjoyed introducing his new wife to friends. His colleague and friend Professor Roger Lee of Harvard wrote,

Alec would sit down and sigh at times and explain that he was not a desk man, nor a travelling man; he was a laboratory man who would like to get back to the bench. I never knew how he *did* the travelling, the speeches, etc. He was always accommodating, and everyone loved him. Over the years I have had many communications from Alec, practically all of them short and brief . . . his letters were longer when he was discussing Amalia. (31)

In March 1955, after only 2 years of marriage, Alexander Fleming died suddenly of a heart attack at his home. He was 73 years old. Fleming was buried in the crypt of St. Paul's Cathedral, a high honor in England. His longtime friend at St. Mary's, C. A. Pannett, delivered the funeral oration:

> Looking back on his career, we find woven into the web of his life a number of apparently irrelevant chance events without one of which it would probably not have reached its climax. . . . His choice of a profession, his selection of a medical school, his deviation into bacteriology, his meeting with Almroth Wright, the nature of the work he did with him, the chance drop of a tear, the chance fall of a mould, all these events were surely not due to mere chance. We can almost see the finger of God pointing to the direction his career should take at every turn. (32)

THE CHEMICAL STRUCTURE OF PENICILLIN

Attempts to determine the exact chemical structure of penicillin began in Robert Robinson's laboratory at Oxford in the early 1940s but were hampered by the instability of the compound. A huge collaborative that included dozens of laboratories in both the United Kingdom and the United States was established in 1943 to determine the chemical structure and if penicillin could be synthetically produced. By the end of 1945, researchers showed that there was not a single penicillin molecule but a family of structures sharing an unusual ring structure that contained the usual carbon, hydrogen, and oxygen atoms of organic compounds but also nitrogen and sulfur atoms in an arrangement called a β-lactam ring, raising hopes that chemical synthesis would be possible. R. B. Woodward, who won the Nobel Prize in Chemistry in 1965 and was involved in the collaborative, described the difficulties of synthetic penicillin:

> Despite the best efforts of probably the largest number of chemists ever concentrated upon a single objective the synthetic problem had not been solved when the program was brought to a close at the end of the War. (33)

In 1957, researchers at the Massachusetts Institute of Technology managed to synthesize penicillin in small quantities, but the processes were not economically viable for mass production. The pharmaceutical industry had to continue purifying the substance from *Penicillium*.

THE IMPACT OF PENICILLIN ON CHEMOTHERAPY OF BACTERIAL DISEASES

The introduction of penicillin into clinical practice led to the modern antibiotic era, with the discovery in the 1940s and 1950s of many drugs that were isolated from other microorganisms. The substances were no longer simply chemicals or chemotherapy. They were isolated from other microorganisms, leading to a new term: antibiotics. Among the many antibiotics that appeared during these decades were streptomycin (Selman Waksman won the Nobel Prize in Medicine in 1962 for its discovery), chloramphenicol, chlortetracycline, erythromycin, vancomycin, and kanamycin. Their discoveries occurred through modification of a method that Fleming had stumbled upon: methodically determining if a microorganism produced a substance that inhibited bacteria. Isolation and purification of any of these substances required rapid expansion of the fermentation process first developed for penicillin (34). The fermentation process took the pharmaceutical companies in entirely new directions since the industry had emerged from organic chemistry beginnings with sulfa drugs. Borrowing from the development of sulfa drugs, the industry began chemically modifying the penicillin molecule. Beecham Research Laboratories introduced semisynthetic penicillins in the early 1960s, including methicillin (1960), ampicillin (1961), and cloxacillin (1962). Other penicillin derivatives and a related class of antibiotics, the cephalosporins, were introduced shortly afterwards. Nearly all of the developments of the last 75 years in antibiotic discovery have their roots in the efforts of early investigators such as Domagk, Fleming, Florey, and Chain.

REFERENCES

1. **Feldman HA.** 1972. The beginning of antimicrobial therapy: introduction of the sulfonamides and penicillins. *J Infect Dis* **125**(Suppl):22–46. http://dx.doi.org/10.1093/infdis/125. Supplement_1.S22.
2. **Ligon BL.** 2004. Sir Alexander Fleming: Scottish researcher who discovered penicillin. *Semin Pediatr Infect Dis* **15**:58–64. http://dx.doi.org/10.1053/j.spid.2004.02.002.
3. **Maurois A.** 1959. *The Life of Sir Alexander Fleming: Discoverer of Penicillin*, p 30–38. E P Dutton and Co, New York, NY.
4. **Fleming A.** 1915. On the bacteriology of septic wounds. *Lancet* **ii:**638–643. http://dx.doi.org/10.1016/S0140-6736(00)54169-5.
5. **Wright A.** 1916. As to how septic war wounds should be treated. *Lancet* **188:**503–514. http://dx.doi.org/10.1016/S0140-6736(00)99139-6.
6. **Maurois A.** 1959. *The Life of Sir Alexander Fleming: Discoverer of Penicillin*, p 83–97. E P Dutton and Co, New York, NY.
7. **Maurois A.** 1959. *The Life of Sir Alexander Fleming: Discoverer of Penicillin*, p 109–122. E P Dutton and Co, New York, NY.
8. **Maurois A.** 1959. *The Life of Sir Alexander Fleming: Discoverer of Penicillin*, p 123–130. E P Dutton and Co, New York, NY.
9. **Maurois A.** 1959. *The Life of Sir Alexander Fleming: Discoverer of Penicillin*, p 131–142. E P Dutton and Co, New York, NY.

10. **Maurois A.** 1959. *The Life of Sir Alexander Fleming: Discoverer of Penicillin*, p 143–158. E P Dutton and Co, New York, NY.

11. **Fleming A.** 1929. On the antibacterial action of cultures of a *Penicillium*, with special reference to their use in the isolation of *B. influenzae*. *Br J Exp Pathol* **X:**3–13.

12. **Birkinshaw JH.** 1972. Harold Raistrick, 1890–1971. *Biogr Mem Fellows R Soc* **18:**489–509. http://dx.doi.org/10.1098/rsbm.1972.0017.

13. **Domagk G.** 1986. A contribution to the chemotherapy of bacterial infections. *Rev Infect Dis* **8:** 163–166. (Translated by T Brock from Domagk G. 1935. Ein Beitrag zur Chemotherapie der bakterillen Infektionen. *Dtsch Med Wochenschr* **61:**250–253.) http://dx.doi.org/10.1055/s-0028-1129486.

14. **Colebrook L, Kenny M.** 1936. Treatment of human puerperal infections, and of experimental infections in mice with Prontosil. *Lancet* **227:**1279–1286. http://dx.doi.org/10.1016/S0140-6736(01)20734-X.

15. **Lesch J.** 2007. *The First Miracle Drugs. How the Sulfa Drugs Transformed Medicine*. Oxford University Press, New York, NY.

16. **Ligon BL.** 2004. Sir Howard Walter Florey—the force behind the development of penicillin. *Semin Pediatr Infect Dis* **15:**109–114. http://dx.doi.org/10.1053/j.spid.2004.04.001.

17. **Chain E.** 1971. Thirty years of penicillin therapy. *Proc R Soc Lond B Biol Sci* **179:**293–319. http://dx.doi.org/10.1098/rspb.1971.0098.

18. **Clutterbuck PW, Lovell R, Raistrick H.** 1932. The formation from glucose by members of the *Penicillium chrysogenum* series of a pigment, an alkali-soluble protein and penicillin—the antibacterial substance of Fleming. *Biochem J* **26:**1907–1918. http://dx.doi.org/10.1042/bj0261907.

19. **Chain E, Florey HW, Gardner NG, Heatley NG, Jennings MA, Orr-Ewing J, Sanders AG.** 1940. Penicillin as a chemotherapeutic agent. *Lancet* **236:**226–228. http://dx.doi.org/10.1016/S0140-6736(01)08728-1.

20. **Maurois A.** 1959. *The Life of Sir Alexander Fleming: Discoverer of Penicillin*, p 159–175. E P Dutton and Co, New York, NY.

21. **Abraham EP, Gardner AD, Chain E, Heatley NG, Fletcher CM, Jennings MA, Florey HW.** 1941. Further observations on penicillin. *Lancet* **238:**177–188. http://dx.doi.org/10.1016/S0140-6736(00)72122-2.

22. **American Chemical Society.** 19 November 1999. Discovery and development of penicillin: international historic chemical landmark. https://www.acs.org/content/acs/en/education/whatischemistry/landmarks/flemingpenicillin.html.

23. **Levy SB.** 1992. *The Antibiotic Paradox*, p 1–12. Plenum Press, New York, NY. http://dx.doi.org/10.1007/978-1-4899-6042-9.

24. **Spink WW.** 1959. Penicillin and the "Fleming myth" on reading "The Life of Sir Alexander Fleming" by Andre Maurois. *Minn Med* **42:**1447–1450.

25. **Maurois A.** 1959. *The Life of Sir Alexander Fleming: Discoverer of Penicillin*, p 193–205. E P Dutton and Co, New York, NY.

26. **Shama G.** 2009. Zones of inhibition? The transfer of information relating to penicillin in Europe during World War II, p 133–158. *In* Laskin AI (ed), *Advances in Applied Microbiology*, vol 69. Academic Press, New York, NY.

27. **Burns M, van Dijck PWM.** 2002. The development of the penicillin production process in Delft, The Netherlands, during World War II under Nazi occupation. *Adv Appl Microbiol* **51:**185–200. http://dx.doi.org/10.1016/S0065-2164(02)51006-6.

28. **Shama G, Reinarz J.** 2002. Allied intelligence reports on wartime German penicillin research and production. *Hist Stud Phys Biol Sci* **32:**347–367. http://dx.doi.org/10.1525/hsps.2002.32.2.347.

29. **Nobelprize.org.** 27 March 2001. Sir Alexander Fleming: banquet speech. http://nobelprize.org/nobel_prizes/medicine/laureates/1945/fleming-speech.html.

30. **Maurois A.** 1959. *The Life of Sir Alexander Fleming: Discoverer of Penicillin*, p 221–240. E P Dutton and Co, New York, NY.

31. **Maurois A.** 1959. *The Life of Sir Alexander Fleming: Discoverer of Penicillin*, p 260–273. E P Dutton and Co, New York, NY.

32. **Maurois A.** 1959. *The Life of Sir Alexander Fleming: Discoverer of Penicillin*, p 274–280. E P Dutton and Co, New York, NY.

33. **Bentley R.** 2009. Different roads to discovery; Prontosil (hence sulfa drugs) and penicillin (hence β-lactams). *J Ind Microbiol Biotechnol* **36:**775–786. http://dx.doi.org/10.1007/s10295-009-0553-8.

34. **Rolinson GN.** 1998. Forty years of β-lactam research. *J Antimicrob Chemother* **41:**589–603. http://dx.doi.org/10.1093/jac/41.6.589.

15 Françoise Barré-Sinoussi and the Discovery of the Human Immunodeficiency Virus

On 5 June 1981, the scientific community at large became aware of a cluster of unusual infections among five homosexual men in Los Angeles, California (1). This report was the first account of a syndrome that would ultimately become known as acquired immune deficiency syndrome (AIDS), a syndrome that would perplex and devastate the world. The World Health Organization estimates that, as of 2020, 39 million people have died from AIDS since 1981; another 37.7 million in 2020 live with human immunodeficiency virus (HIV)/AIDS (2). The search for the cause of this syndrome would bring about the discovery of the most important microorganism of the 20th century. However, the road to the discovery was convoluted and brought about one of the most rancorous disputes in the history of science.

Initially, the scientific world paid little attention to this cluster whose cause, while unclear, "suggested the possibility of a cellular-immune dysfunction related to a common exposure that predisposes individuals to opportunistic infections" (1). A few months later, an account appeared of a group of men in New York and California with an unusually aggressive cancer called Kaposi's sarcoma (3). By the end of 1981, 270 cases of this new syndrome had been reported in gay men; 121 had died. The condition was initially called gay-related immune deficiency (or GRID) by the media. Because the initial reports were among gay men, early theories of the cause centered around lifestyle, including "immune overload" from multiple sexually transmitted infections, use of amyl or butyl nitrate

Germ Theory: Medical Pioneers in Infectious Diseases, Second Edition. Robert P. Gaynes.
© 2023 American Society for Microbiology.

"poppers" to heighten sexual response, or even an immune reaction to exposure to multiple sexual partners. But when similar cases were reported among hemophiliacs and Haitians (leading some to erroneously believe it had originated in Haiti), the scientific and lay community began to take notice and express concern that would lead to dreadful displays of intolerance and discrimination (4). In September 1981, the Centers for Disease Control (CDC) coined the term "acquired immune deficiency syndrome (AIDS)." By 1982, epidemiologic evidence had accumulated to strongly suggest that AIDS was an infectious disease whose etiologic agent could be transmitted sexually and by exposure to contaminated blood or blood products (5). The implications of these findings meant the blood supply could not be safeguarded until the agent was found and the blood supply could be tested for the agent. Screening potential donors for their sexual proclivity or using surrogate markers such as markers for hepatitis B was controversial, costly, and imperfect. Indeed, the blood supply remained unsafe in the early 1980s, when many people acquired AIDS via blood transfusions. An entire generation of hemophiliacs, approximately 15,000 in the United States, became infected because of transfusion with contaminated blood products. Vertical transmission of the agent, i.e., transmission from mother to child, was reported. Sharing of contaminated needles among drug users furthered spread of AIDS. But no one knew what this agent was.

CLUES TO THE ETIOLOGIC AGENT OF AIDS

The discovery of the etiologic agent of AIDS was initially boosted not by virology but by immunology. One of the initial laboratory findings among AIDS patients was a decrease in a certain type of white cell called a CD4 cell. Among various types of white blood cells in the body, there is a category called lymphocytes that are broadly classified as B lymphocytes or T lymphocytes. B lymphocytes are involved primarily in formation of antibodies, which are specialized proteins that recognize and bind to foreign substances. The body's cell-mediated immune process involves T lymphocytes that attack and kill cells that the body identifies as foreign. Lymphocytes can be identified with laboratory markers; cells with one such marker, CD4 cells, are T lymphocytes that are central to the cell-mediated immune response, coordinating it much as a conductor might lead a symphony. Immunologists quickly determined that there was a substantial decrease in the numbers of CD4 cells in AIDS patients. When CD4 cells fell below 200 cells/mm^3 in the bloodstream, patients became vulnerable to infections to which others would not be susceptible, so-called opportunistic infections such as *Pneumocystis* pneumonia. AIDS patients with this low level of CD4 cells were also determined to be at risk for malignancies such as Kaposi's sarcoma. Whatever was causing

AIDS, the diminution of CD4 cells seemed to be central to the problem and suggested that the causative agent was responsible for this CD4 cell decline.

The finding that hemophiliacs had AIDS was also important in the search for an etiologic agent since the only class of microorganism small enough to evade the filters used to screen the clotting factor given to hemophiliacs was a virus. Therefore, most investigators believed the agent was either a new variant of a known human virus or a virus that had only recently been introduced into humans.

Early in 1983, an article in *Journal of the National Cancer Institute* suggested three groups of viruses as the most likely culprits (6). The first was hepatitis B virus (HBV). The epidemiology of patients with HBV infection was like AIDS patients, i.e., high prevalence in gay men and intravenous drug users. So a relative of HBV might be considered, although AIDS patients did not usually have the liver dysfunction that was characteristic of hepatitis virus infections. A second candidate was in the group known as herpesviruses, such as herpes simplex, herpes zoster, and especially cytomegalovirus (CMV) (7). However, studies did not identify a CMV strain that was specific to AIDS patients. The third group of viruses that merited special attention was the retroviruses.

REVERSE TRANSCRIPTASE AND RETROVIRUSES

The story of retroviruses goes back to the detection of an enzyme, reverse transcriptase, in 1970 (8). This discovery has been described as one of the most dramatic scientific moments of the second half of the 20th century because it overturned the central dogma of genetics, i.e., information encoded in genes flows from DNA through RNA to proteins, and never the reverse. Working in the 1950s, Howard Temin began working with the Rous sarcoma virus, known for its ability to induce sarcomas—malignant tumors of connective tissue cells—in chickens. This was an RNA virus. However, his experiments suggested that somehow the virus's RNA was being converted into DNA, which became a permanent part of the infected cell genome, in violation of the central dogma. Temin called this conversion of viral RNA into DNA a provirus. But he was unable to convince the scientific world of the existence of proviruses for over a decade. Like many innovators of paradigm shifts in thinking, he was derided over theorizing their existence. After a decade of work, Temin and his colleagues found unmistakable evidence of an enzyme termed reverse transcriptase that was responsible for making a provirus (6). Working with a different RNA virus, David Baltimore discovered the same enzyme, bolstering Temin's work and overturning conventional genetic wisdom. The impact of the discovery was immediate and validated the idea that these viruses, now called retroviruses, could cause cancer by incorporating into and altering the DNA of the host cells. Temin and Baltimore (along with

Renato Dulbecco) shared the Nobel Prize in Physiology or Medicine in 1975 for their discovery. In the 10 years that followed, their breakthrough initiated a largely futile search for human retroviruses that were associated with cancers. Only one human cancer is known to be directly caused by a retrovirus—human T-cell lymphotropic virus, or HTLV-1—which was discovered in Robert Gallo's laboratory at the U.S. National Institutes of Health (NIH).

DISCOVERY OF THE FIRST HUMAN RETROVIRUS—HTLV-1

The work in Robert Gallo's NIH laboratory was essential to pave the way to the discovery of HIV (9). In the 1970s, isolating a retrovirus in a laboratory meant that (i) one had to be able to choose the correct human (or animal) cell line, which could be cultured and maintained in the lab; (ii) the putative virus could infect this cell in the cell culture; (iii) there needed to be a detection method for the virus after the cell line was infected; and (iv) there needed to be a means to isolate the virus that had been propagated in the cells. The discovery of reverse transcriptase set off a flurry of attempts to find cancer-associated viruses using this general approach, employing the detection of reverse transcriptase as possible evidence for a virus. Unfortunately, most seemingly successful attempts to find a virus were frustrated by sample contamination from animal or human DNA enzymes such as DNA polymerase γ. In his quest to find a human retrovirus that was associated with a human cancer, leukemia, Gallo and his colleagues developed a technique to detect reverse transcriptase and distinguish it from cellular DNA polymerases (10). To determine that a potential human retrovirus could cause leukemia or lymphoma, one needs to maintain a culture of T lymphocytes in the laboratory. That task was made much easier with the discovery of T-cell growth factor, which was later named interleukin-2 (IL-2) by investigators in Gallo's lab (11). The stage was now set for the discovery of the first human retrovirus.

Gallo's group developed a T-cell line using newly discovered IL-2. However, the particular cell line used to discover the first human retrovirus came from a patient with a type of lymphoma termed mycosis fungoides, called such because of the nature of the skin lesions, which resembled (but were not) fungal infection. Their approach was to culture this T-cell line from this patient using IL-2, detect the presence of reverse transcriptase using their recently developed technique, and then attempt to isolate a virus. The investigators showed that the virus that they isolated was different from any known animal retroviruses. They demonstrated that this novel virus could infect and transform normal human T cells in culture. Transformation of a cell by a virus is the change in the biological properties of a cell that results from the regulation by viral genetic material on the functions of the infected cell and can confer certain properties including a change to malignancy.

Then they reisolated the virus. They further confirmed the presence of a human retrovirus by finding specific antibodies to this virus in the serum of the patient with mycosis fungoides. They demonstrated proviral DNA integrated in the DNA of the cells of the patient, from which the virus was isolated. Gallo and colleagues were able to isolate the first human retrovirus, first named human T-cell leukemia virus and now called human T-cell lymphotrophic virus, or HTLV-1 (12). Subsequent studies showed evidence of specific antibodies to this virus in five of six patients with adult T-cell leukemia in Japan (13). Proviral DNA was also detected in five of these Japanese patients but none of the healthy controls (14). While there was initial concern that two different viruses had been isolated, one in Japan and one at the NIH, efforts quickly showed that they were the same virus (15). A human retrovirus, HTLV-1, was isolated and associated with a human T-cell leukemia.

In 1982, Gallo and coworkers published on the discovery of a second human retrovirus, HTLV-2, in a patient with hairy cell leukemia. However, no malignancy or other disease has yet been clearly linked to infection with this virus (16). The laboratory techniques involved in the discovery of HTLV-1 and HTLV-2 would prove crucial in the search for the infectious agent that caused AIDS, which was spearheaded in the laboratory by a French woman, Françoise Barré-Sinoussi.

EARLY INFLUENCES

In June 2021, I had the opportunity to interview Françoise Barré-Sinoussi by telephone (you can listen to excerpts from the interview in this "Meet the Microbiologist" podcast: https://asm.org/Podcasts/MTM/Episodes/The-Discovery-of-HIV-with-Francoise-Barre-Sinoussi). Many of the comments about her and her quotations are from that interview, which I recorded with her permission. There was little in her upbringing to suggest a distinguished career in virology. She was an only child, born Françoise Sinoussi in Paris on 30 July 1947. Her father was a survey engineer who worked constructing buildings. Her mother did not work but had a strong interest in natural science. Her mother's interests may have influenced Françoise.

> My parents took me on vacation to Auvergne. I spent hours looking at the mountains, animals, insects . . . I liked to observe, ask myself questions, try to understand. I didn't know where it would take me. But it was a powerful attraction. (17)

A shy and reserved child, she learned quickly and excelled in the natural sciences. She expressed interest to her parents in attending university to study science. With no doctors or researchers in her family to guide her, she had no way of knowing the path that this schoolwork would take her on. After 2 years in classes at the University of Paris, she began to doubt the relevance of the classroom work.

She dismissed medicine from the outset, believing that the long and expensive studies would be too heavy a burden for her parents. Sinoussi looked to perform research in a laboratory simultaneously with her coursework at the university. She contacted many labs without success until a friend who worked in a laboratory offered to talk with her boss to see if he would be willing to take her on as a student. Her boss was Jean-Claude Chermann, who eventually would become head of the retrovirology unit at the Pasteur Institute. She began a part-time job that quickly became a full-time one. She immediately realized that this laboratory work was the direction that she wanted and essentially stopped going to classes. She would study at night, appearing at the university only for the examinations. While she continually cautions others against adopting this approach to classwork, she met with good success on her exams because she was putting into practice in the laboratory the methods that she would have been learning in class.

SINOUSSI AND THE PASTEUR INSTITUTE

Under the mentorship of Chermann, Françoise Sinoussi got her master's and Ph.D. degrees. Her work was on molecular control in the development of leukemia in mice infected with a retrovirus, murine leukemia virus. She became familiar with detecting reverse transcriptase in the lab, a methodology that would become essential for the detection of the virus that became known as HIV.

Being a woman at the Pasteur Institute was not without its challenges.

> At the end of my PhD, before going for my postdoc, I met the assistant to the director of the Pasteur Institute, to ask him if I could apply one day for a position at Pasteur. The guy looked at me and said, "As a researcher? No way. Women never have done anything in science. You had better think immediately to revise your career plan." That has been a driving force for me. To try to demonstrate to males what we can do. The guy at Pasteur, by the way, called me back many years later and said, "I would like to congratulate you, not only for what you did in the discovery of the AIDS virus but also because you have a lot of courage." (18)

WORK AT THE NATIONAL INSTITUTES OF HEALTH IN THE UNITED STATES

After her Ph.D., she moved to the United States in 1975 for her postdoctoral work on the genetics of the murine leukemia virus, specifically to identify the viral target of a specific gene product, the Fv1 gene product, and how it was implicated in the genetics of murine leukemia virus replication. That work took her to the NIH in Bethesda, Maryland, in the laboratory of Robert Bassin. During that period, she spent about a month learning techniques in Robert Gallo's laboratory, an experience that would later shape her role in a 1984 controversy. Working at the

NIH in the United States held several surprises for Sinoussi. In France, she was not accustomed to colleagues from other countries in the laboratory, which was commonplace in the NIH. The laboratory budget and, therefore, acquisition of reagents and materials was also considerably different—much harder in France, which required workarounds and innovations unnecessary in the U.S. But the greatest adjustment during this period was personal. About a year before she went to the U.S., she met Jean-Claude Barré in Paris. He had difficulty understanding the importance of her postdoctoral work and departure from France. He never visited her during the time she was in the U.S. When she returned to France in 1976, she was unsure if their relationship would continue. It did. They married in 1978.

RETURN TO FRANCE

Upon her return to France, she continued to work in the same laboratory on murine retroviruses. Her unit, now run by Luc Montagnier, was studying the link between retroviruses and cancers. She devoted herself to the research, not as an obligation but as a calling.

Even her wedding day in 1978 was affected by her work.

> I was working in the lab on that day, in the morning. At the end of the morning, around 11 or 11:30, I don't remember exactly, I received a phone call from my husband because I was still in the lab. All the family already had arrived at my parents' place. My future husband was there. I was the only one not there. He asked the very simple question, laughing at the same time because he knew me. He said, "Are you planning really to come?"

Now Françoise Barré-Sinoussi, she was one of the few women at the Pasteur Institute, indeed in science, at that time. She was faced with another choice, whether to have a family.

> I did not see myself with a child. I would have had a double guilt: not taking enough care of him and not giving enough to science. I didn't want to live torn like that, it would have been unbearable. My husband got it. He didn't have a schedule either, he was a sound engineer at Radio France and worked weekends or nights. We made this choice together. He has always supported me. Although it hasn't always been easy. (15)

Efforts in researching the relationship of retroviruses and cancer in animals went on in relative obscurity in Montagnier's unit at Pasteur Institute. But that dearth of attention was about to change.

ISOLATION OF A RETROVIRUS AT THE PASTEUR INSTITUTE

In December 1982, a French physician who had been caring for a patient with persistent, generalized lymphadenopathy called Luc Montagnier on the telephone. This syndrome, swollen lymph nodes throughout the body, often with no other symptoms, was noted to be a precursor to AIDS. The clinician wondered if a retrovirus, specifically HTLV-1, might be the causative agent. Montagnier called Barré-Sinoussi, who knew nothing about AIDS or the developing syndromes surrounding this new disease. Montagnier was working on the relationship between an oncogene and cancer. Barré-Sinoussi was more experienced in detecting reverse transcriptase from retroviruses. After discussion with Jean-Claude Chermann, a meeting was arranged with the group at Pasteur Institute and the clinicians. At this meeting, Barré-Sinoussi first learned about AIDS and its epidemiology, symptoms, and mortality. Shortly afterwards, the doctors gave Montagnier's lab a sample from a lymph node biopsy from the patient and the work began. Cells from the lymph node were cultured using techniques pioneered in Gallo's lab, namely cell culture with T-cell growth factor (IL-2), and stimulated with a compound called phytohemagglutinin (PHA) (9). A day or so later, the cell cultures were brought to Barré-Sinoussi's lab. After a few days, the PHA was removed from the cell culture and the cells were allowed to grow. Detection of reverse transcriptase would be the first indication that a retrovirus was present, again using previously described techniques (10). Every few days, Barré-Sinoussi would sample the cell cultures to detect the presence of reverse transcriptase, using several methods since this enzyme requires different conditions depending upon the retrovirus that produces it. Two to three days into the work, she began to find evidence of reverse transcriptase at low levels but did not mention it to anyone. Two days later, the level of reverse transcriptase was high enough that she was certain of its presence. She told both Chermann and Montagnier that there was evidence of a retrovirus in the cell cultures. Two days later, calamity struck the laboratory. The cells producing this reverse transcriptase activity began to die, raising the possibility that they could lose the would-be virus altogether. Thinking there was something toxic to the cells in the tissue cultures, the researchers scrambled to maintain viable lymphocytes to continue the research to find the retrovirus. Fortunately, Pasteur Institute had a blood bank. Barré-Sinoussi went across to the blood bank and got samples of blood containing normal lymphocytes. She added fresh, normal lymphocytes to the cell cultures with those dying cells containing the putative virus from the original patient. To her amazement, after 15 days she could detect significant levels of reverse transcriptase, which remained stable for an additional 15 to 20 days. Barré-Sinoussi showed that when a sample of a cell-free supernatant of these cultures was added to fresh lymphocytes from umbilical cord blood, reverse transcriptase activity could be detected, further confirmation that a retrovirus was involved. She sent samples to colleagues at Pasteur, who took pictures of infected lymphocytes using an electron microscope. The first pictures of the novel virus were seen (Fig. 15.1). The most distinguishing feature of this virus was that the infected T lymphocytes did not proliferate as in HTLV-1; rather, the researchers realized that the T lymphocytes from the original culture died due to this new virus, consistent with the clinical observations of diminishing CD4 cells in AIDS patients. The researchers performed experiments that successfully distinguished this virus from HTLV-1 in the lab. The group hurriedly prepared a paper, which was rejected by the journal *Nature* but later accepted by the journal *Science* (19). The authors wrote, "The role of this virus in the etiology of AIDS remains to be determined." The statement showed appropriate scientific caution since this virus had been isolated from only one patient who did not have AIDS but had a precursor to AIDS. However, Barré-Sinoussi and her colleagues were already confident.

> We were sure we had it. We could not really write that because if you want to make sure that a virus was the cause of a disease, you had to accumulate evidence that we did not have ourselves in hand at that time. So, we could not write that it was the cause of AIDS. But we were sure that was the cause of AIDS.

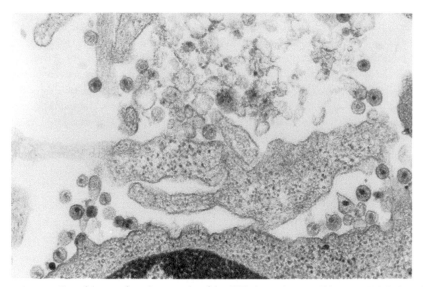

FIGURE 15.1 *One of the very first photographs of the AIDS virus, taken on 4 February 1983. A view of a section of a T lymphocyte infected with the virus isolated from an affected patient with generalized lymphadenopathy syndrome, which precedes AIDS. Photo by Charles Dauquet, courtesy of Institut Pasteur.*

The researchers isolated the same virus from other patients with AIDS. They also analyzed the serum of AIDS patients. They showed that there was an antibody in their serum that recognized this new virus, which was not in the serum of control patients. In early 1983, they put an immediate halt to all other research projects that were ongoing in the laboratory, including determining whether certain DNA sequences, known as MMTV sequences, could be associated with breast cancer, to work on this new virus exclusively.

LYMPHADENOPATHY-ASSOCIATED VIRUS (LAV)—THE CAUSE OF AIDS?

The task that lay before them was to convince the rest of the world that they had found the virus that caused AIDS. One place to start was at the prestigious Cold Spring Harbor Laboratory (CSHL) meetings in New York. CSHL meetings were intense and academic but with a summer camp character. Participants stayed in dormitory-style rooms on the nearly 100-acre campus, listening to presentations that often went on for 12 hours a day. In 1983, the yearly gathering of virologists at CSHL was the occasion for the first public presentation of the efforts of Barré-Sinoussi and her colleagues.

> I came to the U.S. in May 1983 for the Cold Spring Harbor meeting. That was the first time for myself that I presented at Cold Spring Harbor. That was exceptional that Cold Spring Harbor gave me the authorization to present. We were supposed to send abstracts for presentation at the end of January [1983]. At the end of January, we did not have enough data to send an abstract. . . . When I arrived at Cold Spring Harbor, I talked with some of my U.S. colleagues and said, "Look, we have this data and have this paper that will be published in *Science* very soon. Do you think the organizer will give me a few minutes to present the data?" . . . They said to me, "Yes, we give you five minutes." . . . Five minutes! . . . I was so stressed. There were no laser pointers but a stick [pointer]. I was so stressed that the stick fell on [Nobel laureate] David Baltimore who was sitting in the first row.

The reaction to her presentation was mixed; some were doubtful, but other people were enthused, particularly people from the CDC who came up to her afterwards to invite her to give a longer talk at the CDC. She recalled that her visit to the CDC initiated collaborations that would accelerate the confirmation that this virus, called LAV (lymphadenopathy-associated virus), was the cause of AIDS.

In February 1984, Dr. James Curran, then the head of the task force at the CDC looking into AIDS, attended a retrovirology meeting in Park City, Utah, where he heard a talk by Jean-Claude Chermann describing further progress in isolating the virus from other AIDS patients and the development of an LAV antibody test, which was positive in most of these patients. The day after the presentation, Dr. Curran presented the French findings at the weekly Tuesday morning seminar in Auditorium A at the CDC. I attended this seminar to hear from Dr. Curran what the French group had discovered. When Paul Ehrlich attended a meeting of the Physiological Society of Berlin in 1882 where Robert Koch announced his discovery of the tubercle bacillus, Ehrlich later described the evening as the greatest scientific experience of his life (chapter 12). Upon listening to Curran's presentation, I understood how Ehrlich may have felt that night. I can still vividly recall thinking to myself, "This is no longer just a seminar. This is history!" I knew I was one of a few hundred people who was aware that a new virus had been identified, which would become the most significant microorganism since the discovery of the germ theory of disease itself.

CONTROVERSY DEVELOPS: LAV OR HTLV-3?

Not everyone agreed that the French had found the cause of AIDS, however. Barré-Sinoussi's paper garnered little attention at the time or over the next 6 months, receiving only 27 citations in scientific journals (20). The 20 May 1983 issue of *Science* that included the Pasteur Institute's report contained other reports suggesting an association between AIDS and HTLV-1 infection. Max Essex and colleagues reported that at least 25% of AIDS patients had antibodies against HTLV-1 or a closely related virus (21). Robert Gallo's laboratory described detection of proviral HTLV-1 DNA in 2 of 33 AIDS patients and isolation of HTLV-1 from the T lymphocytes of another AIDS patient (22).

> We were aware about what was going on in Bob Gallo's lab because we were in contact with Bob Gallo since the identification of our virus. One of the first things that Jean-Claude Chermann did was to call Bob Gallo . . . in 1983. . . . We wanted to compare the isolate to HTLV-1. . . . I went to Bob Gallo's lab to make a presentation, by the way, in 1983. At the end of my talk, we had this discussion with Bob. Bob was telling me, "Françoise, I am sure it is an HTLV." I said, "No, Bob, it's not an HTLV!"

While Montagnier and colleagues were characterizing LAV and its association with AIDS, the Gallo laboratory began to change its focus from concentrating solely on HTLV-1 as the cause of AIDS to a broader search for a retrovirus distinct from HTLV-1 as the cause. The CDC and researchers in San Francisco had provided Gallo (along with the Pasteur Institute investigators) with clinical specimens in hopes of helping to determine the microbiologic cause of AIDS.

The world was unaware of what Gallo's lab had been doing until 23 April 1984, when Margaret Heckler, the Secretary of Health and Human Services; Ed Brandt, the Assistant Secretary for Health; and Robert Gallo held a press conference.

At the press conference, Heckler stated that "the probable cause of AIDS has been found—a variant of a known human cancer virus, called HTLV-3 . . . credit must go to our eminent Dr. Robert Gallo." Heckler also predicted that there would be a vaccine in a year or two. In the 4 May 1984 issue of *Science*, Gallo and colleagues reported the discovery of a new virus that he called HTLV-3. He showed that antibodies to this virus were detected in almost 90% of patients with AIDS, and that the virus could be isolated from a fairly large number of patients with AIDS and those patients with unexplained lymphadenopathy or "pre-AIDS." Gallo's laboratory was able to grow large amounts of virus in a specialized cell line of T lymphocytes. The authors suggested that HTLV-3 and LAV were different viruses but noted that "it is possible that this is due to insufficient characterization of LAV" (23).

An immediate question resulted: Were HTLV-3 and the French LAV the same virus? The answer had broad implications. First, who should get the credit for finding the cause of AIDS? Second, each laboratory had developed a blood test designed to detect antibodies to their newly discovered virus, a test that could be worth millions of dollars since it would be used to test millions of patients and blood donations to ensure the safety of the blood supply. One day after the press conference, Gallo applied for a patent on his antibody test. Pasteur Institute had applied for a U.S. patent on its antibody test a full 4 months before the April 1984 press conference.

CLAIMS AGAINST PATENTS FURTHER CONTROVERSY

In May 1985, the U.S. Patent and Trademark Office granted Gallo the patent on his blood test, which was to be used to detect antibodies to the virus. In December 1985, the Pasteur Institute filed claims against Gallo with the U.S. Claims Court. The dispute became very bitter and was even linked to national scientific prestige. The quarrel strained relations between the two labs and, particularly, between Gallo and Montagnier. For her part, Barré-Sinoussi tried to stay out of the clash with Gallo. She had worked in his lab; her lab tried to collaborate with members of the Gallo lab until the dispute prevented continuation of any partnership.

Eventually, the President of the United States and the Prime Minister of France tried to end the conflict in 1987 when they agreed, in an extraordinary settlement in science, to a 50-50 split of credit and patent royalties from work with HIV and from the antibody blood test (24). Jay Levy and colleagues from the University of California, San Francisco independently isolated the same virus, which they named AIDS-associated retrovirus (ARV) (25). In May 1986, the International Committee on Taxonomy of Viruses proposed calling the virus that caused AIDS human immunodeficiency virus (HIV) (26).

The French were not satisfied with the financial outcome of the agreement, however, when it became clear that the U.S. patented blood test was based on a laboratory contamination of a French virus. Not long after publication of Gallo's paper, in early 1985, a DNA analysis showed that the French and American viruses were virtually identical. Accusations were hurled at Gallo that he might have misappropriated the French virus. A convoluted, 4-year NIH/U.S. government investigation followed. Several years later, Gallo gave a scientific explanation of why his virus sample was so much like Montagnier's: a third strain of virus had contaminated and overgrown first the cultures in the French lab and then those in his own via a culture sent by Montagnier; i.e., contamination, not theft, was the explanation (27). Eventually, in 1993, the U.S. government dropped all accusations of misconduct against Gallo (28). The deal between the French and U.S. over patent royalties was renegotiated in 1994 (9).

The clash between Gallo and Montagnier, driven by the personalities of the two men, tended to overshadow the importance of the contribution—discovery of the virus. This breakthrough allowed researchers to make an astonishing number of advances to deal with AIDS in an extraordinarily short period, most within 4 years of the virus discovery. With a means to test the blood supply for exposure to the virus, transmission of AIDS by blood transfusion practically disappeared in countries where the detection of viral antibodies in blood donors was implemented (29). The virus's proven link to AIDS and its modes of transmission spurred the first programs for voluntary counseling and testing and subsequently prevention of sexual transmission. The knowledge that HIV could be cytopathogenic to CD4 lymphocytes became the basis for CD4 cell monitoring in AIDS/HIV-infected patients. The cloning and sequencing of HIV provided the necessary knowledge of the basis of the test to determine the viral load and to monitor resistance to therapy (30, 31). The characterization of viral reverse transcriptase activity provided the rationale for the first therapy for AIDS using azidothymidine (AZT) and the first therapeutic approach to prevent mother-to-child transmission (32, 33).

BARRÉ-SINOUSSI AND THE WORLD OF AIDS PATIENTS

The discovery of HIV had also plunged Barré-Sinoussi into the distressing world of AIDS and those who suffered from it. She recalled her first exposure in 1983 to an AIDS patient.

> I went with the clinicians to the intensive care service. I saw a guy who was really, really . . . dying. It was difficult for him to even speak and to move. . . . He took my hand, that I will remember forever. He took my hand, and I could see something on his lips. It seemed to me, I understood, "Thank you." But I was not sure because it

was so difficult for him to speak. So, I asked the clinician, "Did I understand well, he is saying 'thank you'?" He said, "Yes." I was surprised because that guy clearly was dying. I asked him, "Why are you saying thank you because, unfortunately, I cannot be of any help for you?" And that guy said to me, "Not for me . . . for the others." That is something that I have had in my mind . . . today . . . forever.

After this encounter, she understood how much AIDS patients were expecting from the scientists. She knew she had to honor those expectations. Some AIDS sufferers came to Pasteur to try to understand the nature of the damage caused by the virus or to ask for an experimental treatment on which her colleagues were working. Many AIDS patients arrived from abroad often in terrible conditions, spending all their money to come. The times after the discovery of HIV took a toll on Barré-Sinoussi.

> We bonded with patients we saw die. We were trying to give the best of ourselves to move forward as quickly as possible, while telling ourselves that it was not possible to come up with something overnight, we had to be realistic. It was extremely difficult psychologically. (16)

Several years after the discovery of HIV, her mentor, Jean-Claude Chermann, announced that he was leaving the Pasteur Institute to open a laboratory in Marseille, France. He invited Barré-Sinoussi to go with him, but for a variety of reasons, including her husband's desire to stay in Paris, she remained at Pasteur Institute.

BARRÉ-SINOUSSI AFTER DISCOVERY OF HIV AT PASTEUR INSTITUTE

After a conversation with the director of the Institute, Barré-Sinoussi submitted a proposal to create an administratively independent research unit. Following months of discussions, the request was denied. Barré-Sinoussi was now a researcher without a laboratory. She became unofficially associated with another laboratory at Pasteur in a complicated matrix, which left her in a peculiar position for funding, space, and scientific collaborators. Four years later, in 1992, the department of virology had a site visit by outside international experts, who were astonished that her proposal for a research laboratory had not been acceded to, which was finally done. When asked if she believed that her gender played a role in the initial rejection, Barré-Sinoussi said,

> Yes. But looking back, I don't regret what happened, it gave me a shell. I said to myself: nothing is going to happen to me anymore, I will show you what my team and I can do! In my life, this "you will see!" was a driving force. (16)

The creation of her own laboratory at Pasteur Institute did not solve all her problems. Her secretary and physician collaborators recognized profound changes in her behavior and mood and stepped in to help.

> After I got the lab, my own lab, I went through a depression myself. Probably, it was an accumulation of different things: the pressures we all had related to HIV/AIDS— working like crazy but also the fact that that most of the HIV community [of scientists/doctors] started to be close to HIV-positive patients and . . . most of them died. We were friends with them. It is very hard as a human being, not only as a scientist, to lose friends who have been fighting against this disease. Certainly, that was part of my depression. The difficulties that I had at Pasteur to obtain my own laboratory, and during that same period, my mother had a very bad disease. It was an accumulation [of things], lasting almost a year.

As Barré-Sinoussi improved, she renewed her efforts in her laboratory. She initiated research on viral and host determinants of HIV/AIDS pathogenesis. For example, she and her colleagues showed that HIV-1 can infect immature bone marrow T cells, but the massive viral replication occurs only when those cells differentiate to CD4 cells (34). Over the next decade, she also collaborated on HIV vaccine research, using primate models. But she was not limited to work in the laboratory at Pasteur Institute. Barré-Sinoussi promoted integration between HIV/AIDS research and actions in resource-limited countries through the Institute Pasteur International Network and the coordination of the AIDS research programs in Cambodia and Vietnam. She traveled frequently to build research capacity on-site in Africa and Asia.

In the mid-1990s, highly active anti-HIV therapy began to vastly improve the outlook of AIDS patients. This therapy was made possible by the basic understanding of the molecular biology of HIV, the virus she had discovered. During the 1990s and early 2000s, Barré-Sinoussi continued her work in the laboratory on molecular mechanisms of HIV, integrating basic science with clinical research (35).

In 2003, Gallo and Montagnier cowrote a Perspective article in the *New England Journal of Medicine* and seemed to have settled their disagreement (36). Perhaps recognizing the role their personalities played in the dispute, they wrote, "Many lessons can be drawn from this early intense period and most suggest that science requires greater modesty."

THE NOBEL PRIZE

The year 2008 was an eventful one for Françoise Barré-Sinoussi. Sadly, in February 2008, her husband of almost 30 years, Jean-Claude Barré, died. As the fall approached, recognizing the looming date of what would have been their 30th wedding anniversary, she deliberately planned travel to one of the international research sites in Cambodia so as not to be alone on that date in Paris where so many memories would remind her of her loss. She was working with Cambodia colleagues on a clinical trial in patients with HIV and tuberculosis coinfection. On 6 October 2008, the Nobel Committee awarded the Nobel Prize in Physiology or Medicine to Françoise Barré-Sinoussi and Luc Montagnier for their discovery of HIV, along with Harald zur Hausen for his discovery of human papillomavirus and its relationship to cervical cancer. In describing the reasons for the award to Barré-Sinoussi and Montagnier, the Nobel Committee wrote, "HIV has generated a novel pandemic. Never before has science and medicine been so quick to discover, identify the origin and provide treatment for a new disease entity. Successful anti-retroviral therapy results in life expectancies for persons with HIV infection now reaching levels similar to those of uninfected people." Upon hearing of the award in Paris, Montagnier issued a statement that included the comment that Robert Gallo should have been allowed to share the prize (37). But the Nobel Committee had difficulty communicating with Barré-Sinoussi since she was in Cambodia. A call on that day came from a colleague of her husband who worked at Radio France; Barré-Sinoussi recognized the number and answered the call. She feared it might be bad news since Radio France was where her husband had worked when he died.

> She asked me if I heard the news. I said, "No, I do not know what happened. Tell me." The woman was so moved that she had a problem to talk [through her tears]. At one point she said, "You got the Nobel Prize." I said, "I don't believe you." And hung up the phone.

When Barré-Sinoussi returned to the meeting in Cambodia, concerned colleagues looked at her distressed face and asked what had happened on the phone call. She told them what she was told on the call, still disbelieving. Everyone began to search on their phones. Barré-Sinoussi's phone rang again, only this time it was someone from Pasteur Institute who confirmed the news. She had won the Nobel Prize in Medicine! Colleagues in Cambodia celebrated; calls came from all over the world, including from the director of the Pasteur Institute and the director of the National Agency for AIDS Research in France.

> French Embassy [officials] in Cambodia, at the last minute, organized something at the French Embassy inviting the Minister of Health [and her colleagues] in Cambodia. So, it was really moving.

Upon her return to France, her laboratory colleagues prepared a celebration, complete with a banner, commemorating her honor (Fig. 15.2). Craig McClure, executive director of the International AIDS Society, said at the time, "She is one of

FIGURE 15.2 *Banner for Françoise Barré-Sinoussi outside her laboratory at the Pasteur Institute, Paris, congratulating her after winning the Nobel Prize in Physiology and Medicine in 2008.*

the most compassionate, most humble scientists I have ever met. It is such a thrill to see someone who was the person behind the scenes, who did all the sleuth work, being recognized for her role, which really was the fundamental role in discovering HIV" (38).

Few people have experienced the lavish ceremony associated with the receipt of the Nobel Prize in Stockholm, Sweden. As Barré-Sinoussi described it:

> Still today, I wonder if it was a reality or a dream. You are in Stockholm. Everything is so well organized. . . . You spend a week in Stockholm and it's like you are in a cloud. Every day, it's like a dream.

Barré-Sinoussi went so far as to ask the officials in Stockholm if she could return one day for the Nobel ceremonies when she was not the recipient so she could better experience the elegance and meaning of the award, a request that was granted in 2019.

Winning the Nobel Prize has been described as a life-changing experience. When asked how her life changed, she said:

> You receive solicitations from everywhere in the world. Still today, there is not a day where I do not receive an invitation to participate in an opening or closing ceremony of many conferences, . . . advisory boards, . . . research organizations . . . and from the media. And the first year after the Nobel Prize, it's like a nightmare, really. You have to learn how to make a selection on all the demands you receive [for your time]. . . . I needed a scientific assistant to help me. . . . Nobody can imagine the pressure that we had on our shoulder.

She went on to say that the pressure was not merely scientific but political. She was asked to speak to political leaders in countries that were repressive toward people with HIV. She felt compelled to do so, especially since the doors to these leaders opened more easily after the Nobel Prize.

COLD SPRING HARBOR, 2019

In 2019, a most unusual meeting occurred in Cold Spring Harbor. One hundred twenty-five luminaries in the HIV/AIDS field met there for a 3-day get-together to recount the 38+-year history of the HIV epidemic. The speakers included Nobel laureate Harold Varmus, who formerly headed the National Cancer Institute and its parent, the NIH; David Baltimore; John Coffin, whose mentor was the late Howard Temin; James Curran; Anthony Fauci; and Robert Gallo. When Barré-Sinoussi presented near the end of the meeting, she noted that she was the first woman speaker to take the podium. This drew loud applause. She joked, "Things have changed over the years, but still, you can see that males are always the first."

Neither Gallo nor Barré-Sinoussi mentioned the dispute from the 1980s. But its effects were on Barré-Sinoussi's mind.

> We're a bunch of people here who are starting to be quite old, including myself, and we have to make sure the next generation in particular will have the history in mind for HIV and other diseases. The next generation should not make the same mistakes. (39)

AWARDS AND HONORS

Over the years after the initial discovery, Barré-Sinoussi continued to focus on the virology of HIV and helping build research capacity in Africa and Asia. She has received many other honors, including Grand Officer in the National Order of the Legion of Honour in France, the highest level of honor. In keeping with her humility, she said of these honors,

> It is not the most important part of my life, personally. The things that make me most happy are when I go in a developing country, meeting with people living with HIV who are doing well because they are on antiretroviral treatment. That is the best thing for me . . . to see them alive.

The namesake for the Pasteur Institute, Louis Pasteur, took his scientific discoveries from the laboratory and converted them into improvements for society. For Barré-Sinoussi, the good fortune of working at the Pasteur Institute only became evident after her discovery of HIV.

> The Pasteur vision was not so clear to me until I started to work on HIV/AIDS. The concrete aspect of research: the relationship between the scientist and trying to develop tools for the benefit of humanity. I heard about it [his vision] when I was working at Pasteur [Institute], but it did not mean so much to me until I began work on HIV/AIDS.

After her first trip to Africa in 1985, Barré-Sinoussi committed to cultivating research and developing technology transfers on-site in developing countries.

[Pasteur] had the vision that we cannot make research in a rich country, but we are also to make research for the benefit for humanity everywhere.

The efforts of Barré-Sinoussi, along with many others, helped establish branches of the Pasteur Institute in developing countries—the first in Ho Chi Minh City in Vietnam, then in Cambodia and Laos in Southeast Asia and Senegal in Africa, among others throughout the world, realizing the vision to make research available throughout the world. Louis Pasteur would have been proud.

REFERENCES

1. **Centers for Disease Control (CDC).** 1981. Pneumocystis pneumonia—Los Angeles. *MMWR Morb Mortal Wkly Rep* **30:**250–252.
2. **World Health Organization.** Malaria testing and diagnosis. https://www.who.int/data/gho/data/themes/topics/topic-details/GHO/deaths. Accessed 26 August 2022.
3. **Centers for Disease Control (CDC).** 1981. Kaposi's sarcoma and *Pneumocystis* pneumonia among homosexual men—New York City and California. *MMWR Morb Mortal Wkly Rep* **30:**305–308.
4. **Johnson D.** 9 April 1990. Ryan White dies of AIDS at 18; his struggle helped pierce myths, p D00010. *New York Times*, New York, NY.
5. **Centers for Disease Control (CDC).** 1982. Possible transfusion-associated acquired immune deficiency syndrome (AIDS)—California. *MMWR Morb Mortal Wkly Rep* **31:**652–654.
6. **Francis DP, Curran JW, Essex M.** 1983. Epidemic acquired immune deficiency syndrome: epidemiologic evidence for a transmissible agent. *J Natl Cancer Inst* **71:**1–4.
7. **Gottlieb MS, Schroff R, Schanker HM, Weisman JD, Fan PT, Wolf RA, Saxon A.** 1981. *Pneumocystis carinii* pneumonia and mucosal candidiasis in previously healthy homosexual men: evidence of a new acquired cellular immunodeficiency. *N Engl J Med* **305:**1425–1431. http://dx.doi.org/10.1056/NEJM198112103052401.
8. **Coffin JM.** 2021. 50th anniversary of the discovery of reverse transcriptase. *Mol Biol Cell* **32:**91–97. http://dx.doi.org/10.1091/mbc.E20-09-0612.
9. **Vahlne A.** 2009. A historical reflection on the discovery of human retroviruses. *Retrovirology* **6:**40–49. http://dx.doi.org/10.1186/1742-4690-6-40.
10. **Sarngadharan MG, Sarin PS, Reitz MS, Gallo RC, Gallo RC.** 1972. Reverse transcriptase activity of human acute leukaemic cells: purification of the enzyme, response to AMV 70S RNA, and characterization of the DNA product. *Nat New Biol* **240:**67–72. http://dx.doi.org/10.1038/newbio240067a0.
11. **Morgan DA, Ruscetti FW, Gallo R.** 1976. Selective in vitro growth of T lymphocytes from normal human bone marrows. *Science* **193:**1007–1008. http://dx.doi.org/10.1126/science.181845.
12. **Poiesz BJ, Ruscetti FW, Gazdar AF, Bunn PA, Minna JD, Gallo RC.** 1980. Detection and isolation of type C retrovirus particles from fresh and cultured lymphocytes of a patient with cutaneous T-cell lymphoma. *Proc Natl Acad Sci U S A* **77:**7415–7419. http://dx.doi.org/10.1073/pnas.77.12.7415.
13. **Robert-Guroff M, Nakao Y, Notake K, Ito Y, Sliski A, Gallo RC.** 1982. Natural antibodies to human retrovirus HTLV in a cluster of Japanese patients with adult T cell leukemia. *Science* **215:**975–978. http://dx.doi.org/10.1126/science.6760397.
14. **Yoshida M, Miyoshi I, Hinuma Y.** 1982. Isolation and characterization of retrovirus from cell lines of human adult T-cell leukemia and its implication in the disease. *Proc Natl Acad Sci U S A* **79:**2031–2035. http://dx.doi.org/10.1073/pnas.79.6.2031.

15. **Popovic M, Reitz MS,** Jr, **Sarngadharan MG, Robert-Guroff M, Kalyanaraman VS, Nakao Y, Miyoshi I, Minowada J, Yoshida M, Ito Y, Gallo RC.** 1982. The virus of Japanese adult T-cell leukaemia is a member of the human T-cell leukaemia virus group. *Nature* **300:**63–66. http://dx.doi.org/10.1038/300063a0.

16. **Gallo RC.** 1991. Human retroviruses: a decade of discovery and link with human disease. *J Infect Dis* **164:**235–243. http://dx.doi.org/10.1093/infdis/164.2.235.

17. **de Royer S.** 25 April 2021. Françoise Barré-Sinoussi: La recherche, c'est un peu comme entrer au Carmel. *Le Monde*, Paris, France. https://www.lemonde.fr/sciences/article/2021/04/25/francoise-barre-sinoussi-la-recherche-c-est-un-peu-comme-entrer-au-carmel_6077957_1650684.html.

18. **The Nobel Foundation.** Women who changed science: Françoise Barré-Sinoussi. https://www.nobelprize.org/womenwhochangedscience/stories/francoise-barre-sinoussi. Accessed 26 August 2022.

19. **Barré-Sinoussi F, Chermann JC, Rey F, Nugeyre MT, Chamaret S, Gruest J, Dauguet C, Axler-Blin C, Vézinet-Brun F, Rouzioux C, Rozenbaum W, Montagnier L.** 1983. Isolation of a T-lymphotropic retrovirus from a patient at risk for acquired immune deficiency syndrome (AIDS). *Science* **220:**868–871. http://dx.doi.org/10.1126/science.6189183.

20. **Rawling A.** 22 September 1990. Forum: Montagnier, Gallo or both?—On who the scientific community believes discovered HIV. *New Scientist.* https://www.newscientist.com/article/mg12717355-200-forum-montagnier-gallo-or-both-on-who-the-scientific-community-believes-discovered-hiv/.

21. **Essex M, McLane MF, Lee TH, Falk L, Howe CW, Mullins JI, Cabradilla C, Francis DP.** 1983. Antibodies to cell membrane antigens associated with human T-cell leukemia virus in patients with AIDS. *Science* **220:**859–862. http://dx.doi.org/10.1126/science.6342136.

22. **Gelmann EP, Popovic M, Blayney D, Masur H, Sidhu G, Stahl RE, Gallo RC.** 1983. Proviral DNA of a retrovirus, human T-cell leukemia virus, in two patients with AIDS. *Science* **220:**862–865. http://dx.doi.org/10.1126/science.6601822.

23. **Popovic M, Sarngadharan MG, Read E, Gallo RC.** 1984. Detection, isolation, and continuous production of cytopathic retroviruses (HTLV-III) from patients with AIDS and pre-AIDS. *Science* **224:**497–500. http://dx.doi.org/10.1126/science.6200935.

24. **Altman LK.** 1 April 1987. US and France end rift on AIDS, p A00001. *New York Times,* New York, NY.

25. **Levy JA, Shimabukuro J, McHugh T, Casavant C, Stites D, Oshiro L.** 1985. AIDS-associated retroviruses (ARV) can productively infect other cells besides human T helper cells. *Virology* **147:**441–448. http://dx.doi.org/10.1016/0042-6822(85)90146-1.

26. **Case K.** 1986. Nomenclature: human immunodeficiency virus. *Ann Intern Med* **105:**133. http://dx.doi.org/10.7326/0003-4819-105-1-133.

27. **Montagnier L.** 2002. Historical essay. A history of HIV discovery. *Science* **298:**1727–1728. http://dx.doi.org/10.1126/science.1079027.

28. **Hilts P.** 13 November 1993. U.S. drops misconduct case against an AIDS researcher, p 1001. *New York Times,* New York, NY.

29. **Montagnier L.** 8 December 2008. 25 years after HIV discovery: prospects for cure and vaccine. Nobel Lecture. https://www.nobelprize.org/uploads/2018/06/montagnier_lecture.pdf.

30. **Alizon M, Sonigo P, Barré-Sinoussi F, Chermann JC, Tiollais P, Montagnier L, Wain-Hobson S.** 1984. Molecular cloning of lymphadenopathy-associated virus. *Nature* **312:**757–760. http://dx.doi.org/10.1038/312757a0.

31. **Wain-Hobson S, Sonigo P, Danos O, Cole S, Alizon M.** 1985. Nucleotide sequence of the AIDS virus, LAV. *Cell* **40:**9–17. http://dx.doi.org/10.1016/0092-8674(85)90303-4.

32. **Fischl MA, Richman DD, Grieco MH, Gottlieb MS, Volberding PA, Laskin OL, Leedom JM, Groopman JE, Mildvan D, Schooley RT, Jackson GG, Durack DT, King D.** 1987. The efficacy of azidothymidine (AZT) in the treatment of patients with AIDS and AIDS-related complex. A double-blind, placebo-controlled trial. *N Engl J Med* **317:**185–191. http://dx.doi.org/10.1056/NEJM198707233170401.

33. **Connor EM, Sperling RS, Gelber R, Kiselev P, Scott G, O'Sullivan MJ, VanDyke R, Bey M, Shearer W, Jacobson RL, Jimenez E, O'Neill E, Bazin B, Delfraissy J-F, Culnane M, Coombs R, Elkins M, Moye J, Stratton P, Balsley J.** 1994. Reduction of maternal-infant transmission of human immunodeficiency virus type 1 with zidovudine treatment. Pediatric AIDS Clinical Trials Group Protocol 076 Study Group. *N Engl J Med* **331:**1173–1180. http://dx.doi.org/10.1056/NEJM199411033311801.

34. **Lunardi-Iskandar Y, Nugeyre MT, Georgoulias V, Barré-Sinoussi F, Jasmin C, Chermann JC.** 1989. Replication of the human immunodeficiency virus 1 and impaired differentiation of T cells after in vitro infection of bone marrow immature T cells. *J Clin Invest* **83:**610–615. http://dx.doi.org/10.1172/JCI113924.

35. **Barré-Sinoussi F.** 2003. The early years of HIV research: integrating clinical and basic research. *Nat Med* **9:**844–846. http://dx.doi.org/10.1038/nm0703-844.

36. **Gallo RC, Montagnier L.** 2003. The discovery of HIV as the cause of AIDS. *N Engl J Med* **349:**2283–2285. http://dx.doi.org/10.1056/NEJMp038194.

37. **Pincock S.** 2008. HIV discoverers awarded Nobel Prize for medicine. *Lancet* **372:**1373. http://dx.doi.org/10.1016/S0140-6736(08)61571-8.

38. **Pincock S.** 2008. Françoise Barré-Sinoussi: shares Nobel Prize for discovery of HIV. *Lancet* **372:**1377. http://dx.doi.org/10.1016/S0140-6736(08)61575-5.

39. **Cohen J.** 26 October 2016. At gathering of HIV/AIDS pioneers, raw memories mix with current conflicts. *Science.* https://www.sciencemag.org/news/2016/10/gathering-hivaids-pioneers-raw-memories-mix-current-conflicts.

16 Barry Marshall and *Helicobacter pylori* in Peptic Ulcer Disease

With the work of Pasteur, Koch, Lister, and others, the role of microorganisms in causing human disease was unequivocally established by the 1880s. This understanding led to new concepts of the nature of disease and the role of the laboratory in defining disease at the basic science level. Medical science began to view human illnesses as belonging to two separate categories: infectious diseases or chronic diseases.

A century later, the line between infectious diseases and chronic diseases began to blur. The role of microorganisms in triggering these so-called chronic diseases began with an observation of an Australian pathologist and an Australian physician. These two individuals would eventually convince a skeptical medical world that what physicians thought they knew about peptic ulcer disease was wrong. The account of *Helicobacter pylori* and its role in peptic ulcer disease presents a golden opportunity to appreciate one of the fundamental motivations for studying the history of medicine: How does change occur in medicine?

CAUSES OF PEPTIC ULCER DISEASE, CIRCA 1980

For much of the 20th century, peptic ulcer disease was described by an authoritative medical textbook in the following manner:

> Duodenal ulcer is a chronic and recurrent disease. . . . Acid secretion by the stomach is required for production of a duodenal ulcer, but the factors which render the acid-secreting subject susceptible to duodenal ulceration are not completely understood. . . .

Germ Theory: Medical Pioneers in Infectious Diseases, Second Edition. Robert P. Gaynes.
© 2023 American Society for Microbiology.

Genetic factors appear to be important. Duodenal ulcers are approximately three times as common in first-degree relatives of duodenal ulcer patients. . . . Cigarette smoking has been associated with increased duodenal ulcer frequency. . . . Chronic anxiety and psychologic stress may be factors in exacerbation of ulcer activity (1).

Gastric [stomach] ulcers occur in the sixth decade, approximately 10 years later than for duodenal ulcers. . . . Ten to twenty percent of patients with gastric ulcer also have duodenal ulcer disease. In many instances the duodenal ulcer precedes development of the gastric ulcer. . . . Most evidence supports the importance of primary defects in gastric mucosal resistance and/or direct gastric mucosal injury as the most important elements [in the pathogenesis of gastric ulcer] (2).

COMPLICATIONS OF PEPTIC ULCER DISEASE

Ulcers in the first portion of the small intestine, the duodenum, were part of the world of chronic diseases and affected between 6 and 15% of the population. People with these ulcers had recurrent abdominal pain, which could be relieved by eating or by antacids or acid-secretion-controlling medications unless the ulcer progressed and perforated the abdominal cavity, a complication that required surgery. Bleeding from these ulcers was an additional, sometimes life-threatening complication. Obstruction of the stomach outlet due to the ulcer and failure to respond to medical treatment required consideration of surgery as well. Medical treatment and close monitoring were essential since recurrence of the ulcer was extremely common. Peptic ulcer disease was costly; resulting medical care was estimated to total as much as $6 billion annually in the United States in the 1980s when hospitalization, surgery, medication, loss of productivity, and disability were considered (3).

By the 1980s, the medical establishment and the pharmaceutical industry had created an impressive system to diagnose and treat a chronic and recurring condition that would be difficult, even daunting, to challenge. The concept that microorganisms could be involved in peptic ulcer disease was made even more improbable by the prevailing view that the presence of acid made the lining of the stomach too hostile for microbial growth.

CURVED BACTERIA ON GASTRIC BIOPSIES

A simple observation of an oddly shaped bacterium on gastric biopsy specimens (Fig. 16.1) by an Australian pathologist, Robin Warren, began the change in thinking in 1983.

The stomach must not be viewed as a sterile organ with no permanent flora. Bacteria in numbers sufficient to see by light microscopy are closely associated with an active form of gastritis a cause of considerable morbidity [peptic disease]. These organisms should be recognized, and their significance investigated (4).

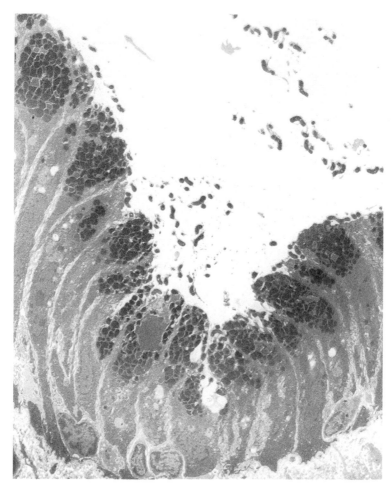

FIGURE 16.1 *One of the early images taken from the work of Drs. Robin Warren and Barry Marshall of a gastric biopsy showing curved bacteria later identified as* Helicobacter pylori. *Image courtesy of Barry Marshall.*

Warren kept a list of patients with curved bacteria present on their stomach biopsies but needed someone to follow the patients to see what clinical diseases they had. Along with Warren, the person who would challenge the medical establishment's view of peptic ulcer was an Australian physician named Barry Marshall.

EARLY INFLUENCES

Barry Marshall was born in Western Australia in 1951. He was the oldest of four children and moved around a great deal in Australia in his youth. His father was a tradesman; his mother had been a nurse until she had Barry.

He was a bright student but often did not work hard enough to earn the top position in his class. He had a boundless curiosity, which was encouraged during his childhood (5).

Marshall chose medical school and graduated from the University of Western Australia with an M.B.B.S. (bachelor of medicine, bachelor of surgery) degree in 1975. He performed an internship and residencies in internal medicine at the Queen Elizabeth II Medical Centre (Sir Charles Gairdner Hospital). He was more interested in an academic career combining research with clinical medicine in a university hospital environment and trained as a specialist physician. In 1979, he moved to Royal Perth Hospital to become more familiar with cardiology and open-heart surgery, which was only performed at that hospital in Perth.

As his training continued, he met Robin Warren in 1981 while doing a rotation in gastroenterology. As part of his training, Marshall was encouraged to perform a clinical research project each year. Warren gave him the list of patients with curved bacteria present on their stomach biopsies to see what clinical diseases they had. Marshall noted that one of the people on Warren's list was a woman he had seen on the hospital ward, who had severe stomach pain but no diagnosis. In desperation they had referred her to a psychiatrist, who prescribed an antidepressant medication for want of a better treatment. The only abnormal finding to explain the abdominal pain was some inflammation in the stomach and strangely shaped bacteria on the stomach biopsy.

Marshall took the role of these bacteria in ulcer disease more seriously than his contemporaries in gastroenterology—those doctors who cared for peptic ulcer patients. He explained his thinking:

> I was undifferentiated in that I wasn't coming from a background in gastroenterology so that my knowledge and ideas were founded in general medical basic science rather than the dogma one was required to learn in specialist medicine. As a trainee general physician with broader training, I was comfortable with the notion of infectious disease and antibiotic therapies. I am told by others that I have a lateral thinking broad approach to problems (5).

After researching the medical literature, Marshall considered that these bacteria bore a resemblance to *Campylobacter jejuni*, a newly discovered bacterium that was a frequent cause of foodborne gastroenteritis and colitis. Over the next 6 months, deeper searches of the medical literature made Marshall's interest grow.

> I found that spiral gastric bacteria had been reported again and again but passed over. I could see an interesting paper being produced, perhaps in an obscure microbiological journal, but had no idea at the time of what we were really about to discover (5).

FIRST STUDY ON PEPTIC ULCERS AND GASTRIC BACTERIA

In 1982, Marshall was in the midst of his training and raising a growing family with the help of his wife, Adrienne. He found time to perform the first study of these bacteria and their role in ulcer disease (6). Of the first 100 patients with gastrointestinal symptoms studied, 65 had gastritis when their stomachs were biopsied; the spiral bacteria were found frequently, but the percentage varied depending upon other findings. For example, 100% of patients with an ulcer in the duodenum had the bacteria, but only 80% of those patients with gastric ulcers had the bacteria. The patients with gastric ulcers where there were no bacteria had an independent reason for the ulcers—the use of medications termed nonsteroidal anti-inflammatory drugs (NSAIDs), such as ibuprofen. Thus, this unidentified spiral bacterium was present in all patients who had ulcers unless they were on NSAIDs. While this finding was suggestive of causation, the bacteria could be seen in more than 50% of people who had other diagnoses or even a normal appearance.

GROWTH CHARACTERISTICS OF THE UNIDENTIFIED CURVED BACILLI FROM THE STOMACH OF PATIENTS

A 1984 *Lancet* article described the growth characteristics of the bacteria for the first time. Notably, the bacteria were slow-growing, often requiring 3 to 7 days to grow. In stomach biopsy cultures of the first 34 patients of the study, the microbiology laboratory reported that nothing grew. When a lab technician returned from a vacation and realized that the culture plate for biopsy #35 had not been removed from the incubator, he saw the presence of minute clear colonies that were identified as like those seen on the histologic slide preparations of the gastric biopsies.

> The lab techs had been throwing the cultures out after two days because with strep [streptococcus], on the first day we may see something, but by the second day it's covered with contamination, and you might as well throw it in the bin. That was the mentality of the lab: Anything that didn't grow in two days didn't exist (7).

The spiral bacteria of the human gastric lining had never been cultured before nor had their association with active gastritis been described. Marshall and his colleagues proposed a new species closely resembling *Campylobacter* species morphologically, in growth requirements, and in DNA base composition. However, they described some differences in flagellar morphology, i.e., a filamentous projection from the cell wall. Campylobacters have a single unsheathed flagellum at one or both ends of the cell, whereas the new organism had four sheathed flagella at one end. The bacteria were initially called pyloric *Campylobacter* (pyloric referring to the region of the stomach where the bacteria were found). Soon after, the bacteria became known as *Campylobacter pylori*. Further studies explained the bacteria's unique ability to grow in the acidic stomach environment, i.e., massive urease production. The enzyme urease helped to counter the acidic environment of the stomach by breaking down urea and creating an alkaline microenvironment around the bacteria (8). In 1989, the bacterium, due to its helical shape, became known as *Helicobacter pylori* when detailed microbiologic analyses showed that these organisms were a new genus (9). While the bacteria were called *Campylobacter pylori* for a few years in the 1980s, I will refer to them as *Helicobacter pylori* hereafter.

FIRST PRESENTATION OF THE ASSOCIATION OF *HELICOBACTER PYLORI* AND PEPTIC ULCER DISEASE

Marshall submitted the work for the first time at the annual meeting of the Royal Australasian College of Physicians in Perth in 1983.

> That was my first experience of people being totally skeptical. To gastroenterologists, the concept of a germ causing ulcers was like saying that the Earth is flat. After that I realized my paper was going to have difficulty being accepted. You think, "It's science; it's got to be accepted." But it's not an absolute given. The idea was too weird (7).

In June 2022, I had the opportunity to interview Dr. Barry Marshall (you can listen to excerpts from the interview in this "Meet the Microbiologist" podcast: https://asm.org/Podcasts/MTM/Episodes/The-Self-Experimentation-of-Barry-Marshall-MTM-144). Many of the following quotations are from the interview. I asked him about this experience.

> Nobody believed it. I did a presentation, and the gastroenterologists didn't believe a bit of it . . . but I had done so much literature searching. I knew that they were wrong. It was difficult to tell your senior bosses trained at the Mayo Clinic that they were wrong.

Marshall would spend the better part of a decade trying to convince the medical world of this association. By 1983, Marshall had moved. After completing his training, he was offered a position at Fremantle Hospital as a senior registrar. Fremantle is the third and smallest of the teaching hospitals in Perth. During the next 2 years at Fremantle, Marshall worked with an enthusiastic group of people while the hospital picked up all the costs of the research. Marshall and his colleagues made two of the most important discoveries surrounding the association of *H. pylori* and peptic ulcer disease. Proving that *H. pylori* was a true pathogen, not merely colonizing the stomach, was made more plausible, first, by showing that treatment of the bacteria would alter peptic ulcer disease, and second, by demonstrating that the human body made specific antibodies to *H. pylori*, suggesting that the body was reacting to the bacteria's presence as a pathogen. This finding would also lead to the development of a serologic test for diagnosis, rather than requiring a slow-growing microbiologic culture from a stomach biopsy.

A long-recognized therapy for peptic ulcer symptoms led indirectly to successful treatment of *H. pylori*. For 200 years bismuth salts such as bismuth citrate and bismuth subsalicylate (BSS; Pepto-Bismol) had been used as a treatment for stomach disorders; many documented as cases of biopsy-proven gastritis—like the cases where Robin Warren first observed curved bacteria. When medicines such as antacids or medications that suppress the production of stomach acid, e.g., cimetidine, were used, the patient's symptoms immediately recurred when they were stopped. However, bismuth often inexplicably eradicated the symptoms permanently, even after the medication was stopped. The conundrum of why bismuth had been an effective stomach treatment was explained when investigators found that bismuth has a powerful antibacterial effect against *H. pylori* (10). In this experiment, Marshall and his colleagues grew cultures of *H. pylori*. To see if a bacterium is susceptible to an antibiotic, microbiologists place a disc impregnated with the antibiotic directly on a culture plate with a lawn of bacteria. If the antibiotic inhibits the bacterial growth, a clear zone appears around the disc; the larger the zone, the greater the effect of the drug. In his experiment, Marshall placed bismuth on the disc. Marshall explained,

> In March '83, I was busy in a clinical job, internal medicine, and right at the end of the day, before I went home, I went up to the micro lab, and with the lab tech there, we opened the incubator and got that petri dish out. And there was this [huge] zone about a 15- to 20-millimeter radius around the bismuth disc. I knew that if you drink Pepto-Bismol, you must have thousands of micrograms per mil. That's pretty, pretty effective, probably. You have a hypothesis, . . . you should be able to test it by doing this experiment and you've predicted it. Nobody ever thought of even doing that experiment at that point.

As early as 1983, clinical trials showed that gastritis healed when the bacteria disappeared (11). Investigators quickly learned that relapse of infection was common after bismuth treatment alone, eventually leading to the concept of using multiple antimicrobial agents for *H. pylori* treatment. The antimicrobial effect of bismuth in often eradicating *H. pylori* provided additional evidence of a causal role for the bacteria and launched a novel therapeutic approach to treating peptic ulcer disease.

Marshall used immunofluorescent microscopy to demonstrate the presence of antibodies in the serum of patients with *H. pylori*, leading to the development of a serologic blood test called a hemagglutination test. Ironically, the development of serology for *H. pylori* made Marshall's job of demonstrating causation of *H. pylori* in peptic ulcer disease more difficult. Serologic surveys showed that most adults had been exposed to the bacteria in developing countries (12). The role of this organism in peptic ulcer disease remained controversial.

ATTEMPT TO FULFILL KOCH'S POSTULATES

Proving the etiologic role of *H. pylori* in peptic ulcer disease would be advanced by rigorous adherence to Koch's postulates (see chapter 10):

1. The putative organism must be constantly present in diseased tissue.
2. The organism must be isolated in pure culture.
3. The pure culture must induce disease when injected into experimental animals.
4. The same organism must be isolated from these diseased animals.

Evidence for the first two postulates was clear. There were problems with numbers 3 and 4, however. There was no animal model for *H. pylori*. Marshall and his colleagues attempted to infect mice, rats, and pigs with *H. pylori*, with no success. Frustrated with his inability to convince his colleagues of the importance of *H. pylori* and with the failure to produce an animal model, Marshall undertook a brave step: a self-experiment. Marshall described what happened:

> I got the gastroenterologist to do the endoscopy on me one morning. He says, "Barry, I don't know why you're asking me to do this. And I don't want you to tell me." Those were his exact words. I had a biopsy that I was normal, no bugs or anything.
>
> The following week in July, I had prepared a mixture of *H. pylori* [with a high concentration of bacteria of 10^9 CFU/ml] and drank it. And then I didn't think anything was going to happen for a few years. I thought, "It's a chronic infection gastritis; you might get ulcers."
>
> But I had this acute illness. At the time, I was going crazy with a project; I was a bit overtired. I wasn't sleeping well. Then my wife says, "You've got a bad breath. You don't look well." I was getting up and vomiting in the mornings. I thought, "I wonder if this is something to do with the bacteria." So, on day 10, I had another endoscopy and biopsy. I had *H. pylori* bacteria everywhere in there. I was pretty excited.

The culture was positive for *H. pylori* (13). After the development of his symptoms, Marshall told his wife what he had done. Marshall described her reaction.

> I went home and told my wife about it. And, of course, she was like, "Oh my God, you're going to infect the whole family. That's why you've got such a bad breath. This is a disaster."

She became concerned that the entire family would get *H. pylori* and develop gastritis or ulcers. Marshall knew he could treat himself with bismuth but added an antibiotic, metronidazole. He was cured. (Marshall said that he believes his acute infection may have already been eradicated by his innate immune system by the time he started his antibiotics since this sometimes happens [B. Marshall, personal communication].) Koch's postulates had been proven, albeit in a single individual.

PHYSICIAN SELF-EXPERIMENTATION

Marshall was, of course, not the first physician to experiment on himself. Like Marshall's, the overwhelming majority of medical self-experiments have been performed by single individuals in relative isolation (14, 15). In cataloging the instances of self-experimentation, Weisse found 465 examples of this practice in the last 2 centuries. Infectious diseases, including vaccine administration, were the most frequent area of interest, followed by anesthesiology (general and local), physiology, pharmacology, radiology (including X rays and other radiation sources), and oncology. A total of 12 Nobel Prizes were awarded to individuals who performed self-experimentation, although 5 of them were for work unrelated to the self-experiment. Among those who received the Nobel Prize for infectious disease-related self-experimentation were Charles Nicolle (1928, for discovery of the cause of typhus) and Gerhard Domagk (1939, for discovery of sulfa drugs). Both Jonas Salk and Albert Sabin were said to have administered their polio vaccines to themselves (and family members) before large-scale trials were performed (14). While several great discoveries have been made with self-experimentation, seven deaths due to infectious disease-related self-experimentation have been documented (14). Still, Weisse stated,

> My own conclusion is that, despite some unwise decisions in the past to indulge in this activity, many self-experiments have proved invaluable to the medical community and to the patients we are seeking to help (14).

NATURAL HISTORY OF *H. PYLORI* DISEASE

While Marshall's self-experiment was risky, it was revealing. The self-experiment, along with other studies, suggested the natural history of *H. pylori*.

> The acute form of the disease should be most common in previously unexposed individuals, that is, children or young adults. It will be seen in families with a history of peptic ulceration or chronic dyspepsia. After ingestion of the infective agent, the patient will have no symptoms for about one week. Subsequently, an acute gastrointestinal disturbance, characterized by epigastric discomfort, nausea, and vomiting, will develop in 50% of cases, the remainder being asymptomatic.

If the organism is not cleared by natural immune processes, a mild gastrointestinal disturbance may persist, but the patient will have achlorhydria [a lack of stomach acid], and so will not suffer from acidity to any degree. The achlorhydric phase will last three to 12 months during which time the histological pattern of chronic gastritis will develop.

As immunity to the infection increases, the inflammation in the body of the stomach will regress and acid secretion will return. The most severely affected mucosa will then be in the antrum and pyloric canal where bacterial growth will be less inhibited by acid secretion. In this final, "chronic" stage of the infection, chronic inflammation of the gastroduodenal mucosa will persist, but acid secretion will return to baseline levels. The patient will then have the potential for the development of peptic ulceration (13).

Marshall and his colleagues proposed that *H. pylori* was the common link between achlorhydric gastritis (of childhood), chronic gastritis, and peptic ulcer disease. This progression explained the observation that the serology for *H. pylori* was common, nearly universally positive, in adults in developing countries. Most children in those countries were exposed to the bacteria in the first few years of life; children either developed a few days of vomiting, often before they could talk to describe their symptoms, or were asymptomatic. Since adult exposure to *H. pylori* was less common in developed countries—usually 20 to 50% of adults had an antibody response—this hypothesis of disease also explained why peptic ulcer disease tended to cluster in families in developed countries, where close family contact would be needed to provide exposure to the bacteria. Convincing the medical community of this link would be long and laborious.

A DIFFICULT YEAR, 1984

Marshall traveled in 1983 to sway opinion leaders in the United Kingdom in favor of his hypothesis. Physicians in Worcester, England, were able to culture *H. pylori* from ulcer patients, showing that isolation of the bacteria was not limited to Australia. Marshall said,

> But 1984 was a difficult year. I was unsuccessfully attempting to infect an animal model. There was interest and support from a few but most of my work was rejected for publication and even accepted papers were significantly delayed. I was met with constant criticism that my conclusions were premature and not well supported. When the work was presented, my results were disputed and disbelieved, not on the basis of science but because they simply could not be true. It was often said that no one was able to replicate my results. This was untrue but became part of the folklore of the period. I was told that the bacteria were either contaminants or harmless commensals (5).

Marshall began to understand the forces in the medical establishment with which he was contending when he tried to get funding from the drug companies that had a stake in treating peptic ulcer disease.

> They all wrote back saying how difficult times were and they didn't have any research money. But they were making a billion dollars a year for the antacid drug Zantac and another billion for Tagamet. You could make a patient feel better by removing the acid. Treated, most patients didn't die from their ulcer and didn't need surgery, so it was worth $100 a month per patient, a hell of a lot of money in those days. In America in the 1980s, 2 to 4 percent of the population had Tagamet tablets in their pocket. There was no incentive to find a cure (7).

A BOOST FROM AN UNEXPECTED SOURCE

Marshall's efforts to convince the world of the importance of *H. pylori* in peptic ulcer disease were boosted from an unexpected source. Even though Marshall was working at Fremantle Hospital in 1984, he was still in regular communication with Robin Warren. Marshall told Warren of his experiences after his self-experimentation. Warren was interviewed by a journalist from the U.S. and let the journalist know of Marshall's self-experiment and subsequent illness and treatment. The tabloid, the *Star*, ran the story, headlined "Guinea-pig doctor discovers new cause for ulcers . . . and the cure"; the story was picked up by other news outlets, including the *New York Times* (16).

After publication of the story, patients with ulcers from as far away as the U.S. quickly contacted Marshall, desperate for treatment.

> I had no support, no funding for a secretary or a research fellow. It's just me. And so, then the *Star* came out and Robin Warren and I were getting letters from America. The word kind of got out in America. We could cure ulcers.
>
> [At that time] I was successfully experimentally treating patients who had suffered with life threatening ulcer disease for years. Some of my patients had postponed surgery which became unnecessary after a simple 2-week course of antibiotics and bismuth (5).

At the end of 1984, the Australian Medical Research Council funded a study to conduct a prospective double-blind trial to see if antibiotics could cure duodenal ulcers. The study required enrollment of a large number of patients, so Marshall moved back to Royal Perth Hospital, where the patient load was far higher (17). Eventually, this study showed that eradication of the bacteria was strongly associated with disappearance of a duodenal ulcer and decreased the likelihood of a recurrence. Marshall described the significance of the double-blind study:

I had to do the double-blind study. . . . No matter how much data I've got, until I do the double-blind study, it's worth nothing. So, I then spent the next 3 years doing that double-blind study. Everyone thinks you've died because you don't publish. It's all secret and blinded. At the end of the 3 years, when you get a good publication in the *Lancet*, everyone says you are the conquering hero.

But further work would be needed to determine the optimal therapy for *H. pylori*. As his study at Royal Perth Hospital was enrolling patients, investigators in Portugal, France, Germany, Italy, the Netherlands, the U.K., and other countries in Western Europe began making contributions to the medical literature on the role of *H. pylori* in gastrointestinal disease (18, 19). Several key experts in the United States were initially more difficult to convince but eventually supported Marshall (20, 21). A trip to Asia by Marshall in 1985 set off a burst of interest in *H. pylori* (22).

The *Star* article had an additional benefit. The story encouraged Michael Manhart, a microbiologist at Procter & Gamble (P&G), to examine Marshall's publications. P&G was the maker of the bismuth compound Pepto-Bismol. Marshall commented,

That made an easy life for me because the people selling Pepto-Bismol in the United States were Procter & Gamble. They heard about it and their head microbiologist in Cincinnati, Mike Manhart, called me at midday in Fremantle. So that was midnight in Cincinnati. And I had this long discussion. A few days later I was on a plane to Cincinnati. So, it changed my life.

Realizing the economic potential in Marshall's work, P&G eventually established business relationships with him and helped Marshall gain patents for diagnostic tests for *H. pylori*. P&G also funded a fellowship and a laboratory in the U.S. for Marshall to further his research. In 1985, Marshall left Australia for the University of Virginia. While he expected to be in the U.S. for 2 to 3 years, his stay in America would last 10 years, but it was there that he would finally convince the world that his hypothesis about the role of *H. pylori* in gastrointestinal disease was correct.

THE MOVE TO THE UNITED STATES: RESEARCH AT THE UNIVERSITY OF VIRGINIA

The research at University of Virginia centered around diagnostic tests and treatment of *H. pylori*. In the 1980s, patients with suspected ulcer disease would have to undergo endoscopy with a fiber-optic endoscope and have a biopsy and culture to look for *H. pylori*. A Gram stain could be performed immediately, but confirmation with a microbiologic culture would take days. Marshall and others recognized that only *H. pylori* produced urease in the stomach. They developed a rapid urease test to detect changes in the pH of gastric specimens that could be performed in a few minutes simply and without special equipment. However, endoscopy would still need to be performed. The solution was to develop a rapid carbon-14-labeled breath test. This test involved a patient swallowing material with urea containing a minute amount of radioactive carbon (^{14}C). If *H. pylori* was in the stomach, this urea would be broken down by the bacteria's urease enzyme, producing radioactive CO_2 which could be detected in the patient's exhaled breath within minutes, eliminating the need for endoscopy (23). The breath test allowed investigators to embark on large-scale trials of *H. pylori* therapy using noninvasive methods for diagnosis and follow-up.

THE TIDE TURNS

In 1991, the first convincing study of the cure of peptic ulcers in the United States was published, stating:

> Combined therapy with anti-*H. pylori* agents and ranitidine was superior to ranitidine alone for duodenal ulcer healing. Our results indicate that *H. pylori* plays a role in duodenal ulcer disease (24).

The tide was finally turning. The medical establishment was beginning to accept a role for *H. pylori* in peptic ulcer disease. Challenges remained, however. Marshall's self-experiment suggested a role of the bacteria in acute gastritis. Treating with antimicrobial agents helped heal duodenal ulcers. But if a large number of adults had already been exposed to *H. pylori*, particularly in developing countries, how did the bacteria cause duodenal ulcers only in certain people (1% of adults who were infected per year) (25)? The key factor associated with development of a peptic ulcer appeared to be a toxin produced by *H. pylori*—a cytotoxin. This soluble element stimulated the production of the body's defenses, including interleukin-8, a substance that calls in neutrophils, a type of white blood cell. Persons who harbored toxin-producing *H. pylori* had greater amounts of inflammation from the neutrophils on the lining of the stomach or duodenum. The body's own inflammatory response seemed to be responsible for producing the ulcers. People with this toxin-producing *H. pylori* also appeared to have greater numbers of the bacteria in the lining of the stomach.

For development of peptic ulcers, however, an additional factor appeared to be involved: hypersecretion of stomach acid. The production of stomach acid is normally stimulated by the production of a hormone known as gastrin. When food enters the stomach, certain cells in the stomach release gastrin, which sends signals to other cells in the stomach to release stomach acid. Motility of the stomach and the first portion of the small intestine is increased by gastrin as well. When *H. pylori* affects the stomach, causing chronic gastritis, affected patients tend to develop high gastrin levels and high basal acid secretion due to hypertrophy (overgrowth) of the gastric acid-secreting mucosal cells. While the acid secretion could be suppressed with medicines such as cimetidine, the high levels of acid would immediately return if the medicine was stopped; often the medicine could not fully suppress the hypersecretion of acid. Treating the *H. pylori* seemed to reverse the hypersecretion. The optimal treatment of *H. pylori* continues to be debated, but evidence has shown that treatment to eradicate *H. pylori* can greatly reduce the risks of peptic ulcer disease and nonulcer dyspepsia (indigestion), although the data are less consistent regarding nonulcer dyspepsia (26).

A WATERSHED YEAR, 1994

As his time at the University of Virginia progressed in the early 1990s, Marshall began to see more praise than criticism for his hypothesis. I asked Dr. Marshall at what point he personally believed that the medical establishment was starting to accept his theory about *H. pylori*.

> At the American Gastroenterology Association meeting in New Orleans in 1993. In that meeting, there were rows and rows of *H. pylori* posters, but I guarantee there was like 250 *H. pylori* posters. [Members of the BBC, filming a documentary on my work,] just followed me. They put a microphone on me. And then, I just walked around and looked at the posters. And this guy said, "Oh, this is great, Dr. Marshall, you know, I really think you should win the Nobel Prize." And this is on air. I said, "Never mention the Nobel Prize; that would jinx me."

Patients who were successfully treated, ending years of ulcer symptoms and eliminating the need for surgery, would send donations to a foundation that Marshall had set up. By 1994, everything changed.

> 1994 was a watershed year for us. In February 1994 the NIH held a consensus meeting in Washington DC which ended after 2 days with the statement to the effect that the key to treatment of duodenal and gastric ulcer was detection and eradication of *Helicobacter pylori* (27).
>
> I had been waiting for ten years for this day and I felt a combination of relief and satisfaction that I had achieved what I set out to do. Years before, I had developed the hypothesis, tested it, proved it and now it had reached official acceptance (5).

Marshall's efforts defending his theory to the world had taken a toll, with constant traveling to present his findings. While he was beginning to receive awards such as the Lasker Prize (28), he was increasingly dissatisfied and took a leave from his university research. In 1996, the family moved back to Australia. His research continues with monies from businesses devoted to *H. pylori* diagnostic tests.

H. PYLORI AND GASTRIC CANCER

Marshall's initial hypothesis, an association of this bacterium with peptic ulcer disease, was sent in an entirely unanticipated direction—into cancer and its treatment. Patients with chronic gastritis often had follicles or sacs of cells called lymphocytes in the lining of the stomach seen on biopsies. In some patients, these lymphoid follicles develop into a low-grade cancer called mucosa-associated lymphoid tissue (MALT) lymphoma. In 1993, investigators found a close association between gastric MALT lymphoma and *H. pylori*. After treatment of the bacteria in patients with MALT lymphoma, in all cases *H. pylori* was eradicated, and in five of six patients repeated biopsies showed no evidence of lymphoma.

> These results suggest that eradication of *H. pylori* causes regression of low-grade B-cell gastric MALT lymphoma, and that anti-H-pylori treatment should be given for this lymphoma (29).

A more recent observational study in Hong Kong showed a significantly lower incidence of gastric cancer among patients older than 60 years of age who had received treatment to eradicate *H. pylori* infection (30). An even more convincing randomized, double-blind, placebo-controlled trial in South Korea showed that patients with early gastric cancer who received treatment for *H. pylori* infection had lower rates of metachronous (a cancer diagnosed more than 6 months after the index cancer) gastric cancer after a median of 5.9 years than those who received placebo (31). More information is needed to fully assess the effect of *H. pylori* on the development, treatment, and prevention of gastric cancer, but a recent consensus conference recommended expanding *H. pylori* testing to include persons with a family history of gastric cancer (32).

INFECTIOUS AGENTS AND OTHER CHRONIC DISEASES

The association between a microorganism, *H. pylori*, and a chronic disease, peptic ulcer disease, was a startling and revolutionary concept that took nearly a decade to become accepted by the medical establishment. However, this association prompted a flurry of interest in other microorganisms triggering chronic diseases, including numerous cancers, arthritis, and type 1 diabetes, and may eventually extend to diseases such as atherosclerosis, stroke, Alzheimer's disease, multiple sclerosis, and schizophrenia (33–39). The associations may be more complex than a single microorganism triggering a chronic disease, but rather alterations of a community of microorganisms in the body known as the microbiome. Obtaining a comprehensive view of the human microbiome has become possible with advances in culture-independent techniques using next-generation DNA sequencing methods. Several studies have suggested a correlation between our microbiome and various diseases, including metabolic disorders and cancers (40, 41).

AWARDS AND HONORS

After a decade of frustration, Marshall finally began to receive credit for his dogged efforts to convince the medical community of his hypothesis. Marshall received the Warren Alpert Prize in 1994; the Australian Medical Association Award and the Albert Lasker Award for Clinical Medical Research in 1995; the Gairdner Foundation International Award in 1996; the Paul Ehrlich and Ludwig Darmstaedter Prize in 1997; and the Golden Plate Award of the American Academy of Achievement, the Dr. A.H. Heineken Prize for Medicine, the Florey Medal, and the Buchanan Medal of the Royal Society in 1998.

THE NOBEL PRIZE

In 2005, Barry Marshall and Robin Warren were corecipients of the Nobel Prize in Physiology or Medicine "for their discovery of the bacterium *Helicobacter pylori* and its role in gastritis and peptic ulcer disease." In his Nobel address, Marshall said,

> One of the truly great things about winning the Nobel Prize in 2005 was that I was living and working back home. I got to share it and celebrate with those who had been involved in the initial work at Royal Perth and Fremantle Hospital.

Marshall's experience differed from Françoise Barré-Sinoussi's hectic life in her post-Nobel time. Marshall said,

> It is fantastic because you get this great publicity, and it's great publicity for your hospital and university. . . . That sort of thing happens all of a sudden. You are right up there, and everybody is getting a bit of the Nobel charisma. . . . You are getting all this media attention. People then appreciate you and value you. So, that's great. All of a sudden, you've got an easy life as far as getting things done.

CHALLENGES REMAINING

Uniformly successful strategies to find regimens to eradicate *H. pylori* and improve adherence to the multidrug regimens remain elusive, requiring guidelines that need continual updating (42). Better information is needed to complement observational trials of treatment for *H. pylori* infection in regions in which there is a high prevalence of infection and an increased incidence of gastric cancer. *H. pylori* antibiotic resistance is threatening effective treatment of the infection in various areas of the United States. Efforts to develop a vaccine against *H. pylori* have thus far been unsuccessful (43).

CONCLUSIONS

In 1983, when Marshall and Warren first published in the *Lancet*, they embarked on a quest to convince the medical world of a paradigm-shifting hypothesis. Marshall commented on the process:

> We couldn't knock down our own hypothesis. That's the thing: be critical of your own data. And then go with the data. And once you've got a pathway and you are getting facts, then it doesn't matter how many people are out there who don't believe you. . . . Science is not a democracy.

Medical history is replete with stories of contributions where the discoverers suffered in their efforts to convince the world of their correctness, including Vesalius, Harvey, and Lister. But Marshall and Warren toiled in the modern age, when physicians had full access to clear, carefully performed studies. Still, a full decade was required before full acceptance, owing, in part, not to medical literature but exposure via a tabloid. This story is a lesson to everyone in medicine to be open to ideas of innovation. As the Royal Society stated when it awarded Marshall the Buchanan Medal in 1998, "The work of Marshall has produced one of the most radical and important changes in medical perception in the last 50 years."

REFERENCES

1. **Isselbacher KJ, et al (ed).** 1980. p 1373–1374. *In Harrison's Principles of Internal Medicine.* McGraw-Hill Book Company, New York, NY.
2. **Isselbacher KJ, et al (ed).** 1980. p 1377–1378. *In Harrison's Principles of Internal Medicine.* McGraw-Hill Book Company, New York, NY.
3. **Kurata JH, Haile BM.** 1984. Epidemiology of peptic ulcer disease. *Clin Gastroenterol* **13:**289–307. http://dx.doi.org/10.1016/S0300-5089(21)00614-3.
4. **Warren JR, Marshall B.** 1983. Unidentified curved bacilli on gastric epithelium in active chronic gastritis. *Lancet* **1:**1273–1275.

5. **The Nobel Prize.** 2005. Barry J. Marshall. https://www.nobelprize.org/prizes/medicine/2005/marshall/facts/. Accessed 26 August 2022.

6. **Marshall BJ, Warren JR.** 1984. Unidentified curved bacilli in the stomach of patients with gastritis and peptic ulceration. *Lancet* **1:**1311–1315. http://dx.doi.org/10.1016/S0140-6736(84)91816-6.

7. **Weintraub P.** 8 April 2010. The doctor who drank infectious broth, gave himself an ulcer, and solved a medical mystery. *Discover*. https://www.discovermagazine.com/health/the-doctor-who-drank-infectious-broth-gave-himself-an-ulcer-and-solved-a-medical-mystery. Accessed 26 August 2022.

8. **Marshall BJ, Barrett L, Guerrant RL.** 1988. Protection of *Campylobacter pylori* but not *C. jejuni* against acid susceptibility by urea, p 402–403. *In* Kaiser B, Falsen E (ed), Campylobacter. *IV. Proceedings of the Fourth International Workshop on* Campylobacter *Infections*. Goterna, Göteborg, Sweden.

9. **Goodwin CS, Armstrong JA, Chilvers T, Peters M, Collins MD, Sly L, McConnell W, Harper WES.** 1989. Transfer of *Campylobacter pylori* and *Campylobacter mustelae* to *Helicobacter* gen. nov. as *Helicobacter pylori* comb. nov. and *Helicobacter mustelae* comb. nov., respectively. *Int J Syst Evol Microbiol* **39:**397–405.

10. **Marshall BJ, Armstrong JA, Francis GJ, Nokes NT, Wee SH.** 1987. Antibacterial action of bismuth in relation to *Campylobacter pyloridis* colonization and gastritis. *Digestion* **37**(Suppl 2):16–30. http://dx.doi.org/10.1159/000199555.

11. **Marshall BJ, McGechie DB, Rogers PAR, Glancy RJ.** 1985. Pyloric *Campylobacter* infection and gastroduodenal disease. *Med J Aust* **142:**439–444. http://dx.doi.org/10.5694/j.1326-5377.1985.tb113444.x.

12. **Mégraud F, Brassens-Rabbé MP, Denis F, Belbouri A, Hoa DQ.** 1989. Seroepidemiology of *Campylobacter pylori* infection in various populations. *J Clin Microbiol* **27:**1870–1873. http://dx.doi.org/10.1128/jcm.27.8.1870-1873.1989.

13. **Marshall BJ, Armstrong JA, McGechie DB, Glancy RJ.** 1985. Attempt to fulfil Koch's postulates for pyloric *Campylobacter*. *Med J Aust* **142:**436–439. http://dx.doi.org/10.5694/j.1326-5377.1985.tb113443.x.

14. **Weisse AB.** 2012. Self-experimentation and its role in medical research. *Tex Heart Inst J* **39:**51–54.

15. **Altman LK.** 1987. *Who Goes First? The Story of Self-Experimentation in Medicine*. Random House, New York, NY.

16. **Doctor Yak.** 2019 July 13. Challenging science's status-quo: the tale of Barry Marshall. https://medium.com/doctoryak/challenging-sciences-status-quo-the-tale-of-barry-marshall-c80a873412a6. Accessed 26 August 2022.

17. **Marshall BJ, Goodwin CS, Warren JR, Murray R, Blincow ED, Blackbourn SJ, Phillips M, Waters TE, Sanderson CR.** 1988. Prospective double-blind trial of duodenal ulcer relapse after eradication of *Campylobacter pylori*. *Lancet* **2:**1437–1442. http://dx.doi.org/10.1016/S0140-6736(88)90929-4.

18. **Tytgat GN, Rauws EA.** 1987. Significance of *Campylobacter pylori*. *Aliment Pharmacol Ther* **1**(Suppl 1): 527S–539S. http://dx.doi.org/10.1111/j.1365-2036.1987.tb00662.x.

19. **Brunner H, Mittermayer H, Regele H.** 1987. Die Campylobacter-pylori-Besiedelung der Antrumschleimhaut bei Patienten mit chronischen Gastritiden und peptischem Ulkus. [*Campylobacter pylori* colonization of the antrum mucosa in patients with chronic gastritis and peptic ulcer.] *Z Gastroenterol* **25**(Suppl 4):20–23.

20. **Blaser MJ.** 1987. Gastric *Campylobacter*-like organisms, gastritis, and peptic ulcer disease. *Gastroenterology* **93:**371–383. http://dx.doi.org/10.1016/0016-5085(87)91028-6.

21. **Graham DY, Klein PD.** 1987. *Campylobacter pyloridis* gastritis: the past, the present, and speculations about the future. *Am J Gastroenterol* **82:**283–286.

22. **Matsui K.** 1987. [*Campylobacter pylori* in gastric diseases]. *Nippon Shokakibyo Gakkai Zasshi* **84:**1864. In Japanese.

23. **Marshall BJ, Surveyor I.** 1988. Carbon-14 urea breath test for the diagnosis of *Campylobacter pylori* associated gastritis. *J Nucl Med* **29:**11–16.

24. **Graham DY, Lew GM, Evans DG, Evans DJ Jr, Klein PD.** 1991. Effect of triple therapy (antibiotics plus bismuth) on duodenal ulcer healing. A randomized controlled trial. *Ann Intern Med* **115:**266–269. http://dx.doi.org/10.7326/0003-4819-115-4-266.

25. **Cullen DJ, Collins BJ, Christiansen KJ, Epis J, Warren JR, Surveyor I, Cullen KJ.** 1993. When is *Helicobacter pylori* infection acquired? *Gut* **34:**1681–1682. http://dx.doi.org/10.1136/gut.34.12.1681.

26. **Crowe SE.** 2019. *Helicobacter pylori* infection. *N Engl J Med* **380:**1158–1165. http://dx.doi.org/10.1056/NEJMcp1710945.

27. **NIH Consensus Conference.** 1994. *Helicobacter pylori* in peptic ulcer disease. NIH Consensus Development Panel on *Helicobacter pylori* in Peptic Ulcer Disease. *JAMA* **272:**65–69. http://dx.doi.org/10.1001/jama.272.1.65.

28. **Marshall BJ.** 1995. The 1995 Albert Lasker Medical Research Award. *Helicobacter pylori.* The etiologic agent for peptic ulcer. *JAMA* **274:**1064–1066. http://dx.doi.org/10.1001/jama.1995.03530130070032.

29. **Wotherspoon AC, Doglioni C, Diss TC, Pan L, Moschini A, de Boni M, Isaacson PG.** 1993. Regression of primary low-grade B-cell gastric lymphoma of mucosa-associated lymphoid tissue type after eradication of *Helicobacter pylori*. *Lancet* **342:**575–577. http://dx.doi.org/10.1016/0140-6736(93)91409-F.

30. **Leung WK, Wong IOL, Cheung KS, Yeung KF, Chan EW, Wong AYS, Chen L, Wong ICK, Graham DY.** 2018. Effects of *Helicobacter pylori* treatment on incidence of gastric cancer in older individuals. *Gastroenterology* **155:**67–75. http://dx.doi.org/10.1053/j.gastro.2018.03.028.

31. **Choi IJ, Kook MC, Kim YI, Cho SJ, Lee JY, Kim CG, Park B, Nam BH.** 2018. *Helicobacter pylori* therapy for the prevention of metachronous gastric cancer. *N Engl J Med* **378:**1085–1095. http://dx.doi.org/10.1056/NEJMoa1708423.

32. **El-Serag HB, Kao JY, Kanwal F, Gilger M, LoVecchio F, Moss SF, Crowe SE, Elfant A, Haas T, Hapke RJ, Graham DY.** 2018. Houston Consensus Conference on testing for *Helicobacter pylori* infection in the United States. *Clin Gastroenterol Hepatol* **16:**992–1002.e6. http://dx.doi.org/10.1016/j.cgh.2018.03.013.

33. **Walboomers JM, Jacobs MV, Manos MM, Bosch FX, Kummer JA, Shah KV, Snijders PJ, Peto J, Meijer CJ, Muñoz N.** 1999. Human papillomavirus is a necessary cause of invasive cervical cancer worldwide. *J Pathol* **189:**12–19. http://dx.doi.org/10.1002/(SICI)1096-9896(199909)189:1<12::AID-PATH431>3.0.CO;2-F.

34. **Parkin DM.** 2006. The global health burden of infection-associated cancers in the year 2002. *Int J Cancer* **118:**3030–3044. http://dx.doi.org/10.1002/ijc.21731.

35. **Hyrich KL, Inman RD.** 2001. Infectious agents in chronic rheumatic diseases. *Curr Opin Rheumatol* **13:**300–304. http://dx.doi.org/10.1097/00002281-200107000-00010.

36. **Principi N, Berioli MG, Bianchini S, Esposito S.** 2017. Type 1 diabetes and viral infections: what is the relationship? *J Clin Virol* **96:**26–31. http://dx.doi.org/10.1016/j.jcv.2017.09.003.

37. **Leinonen M, Saikku P.** 2002. Evidence for infectious agents in cardiovascular disease and atherosclerosis. *Lancet Infect Dis* **2:**11–17. http://dx.doi.org/10.1016/S1473-3099(01)00168-2.

38. **Sochocka M, Zwolińska K, Leszek J.** 2017. The infectious etiology of Alzheimer's disease. *Curr Neuropharmacol* **15:**996–1009.

39. **Lipkin WI.** 1997. The search for infectious agents in neuropsychiatric disorders: lessons from multiple sclerosis. *Mol Psychiatry* **2:**437–438. http://dx.doi.org/10.1038/sj.mp.4000338.

40. **Young VB.** 2017. The role of the microbiome in human health and disease: an introduction for clinicians. *BMJ* **356:**j831. http://dx.doi.org/10.1136/bmj.j831.

41. **Sepich-Poore GD, Zitvogel L, Straussman R, Hasty J, Wargo JA, Knight R.** 2021. The microbiome and human cancer. *Science* **371:**eabc4552. http://dx.doi.org/10.1126/science.abc4552.

42. **Chey WD, Leontiadis GI, Howden CW, Moss SF.** 2017. ACG clinical guideline: treatment of *Helicobacter pylori* infection. *Am J Gastroenterol* **112:**212–239. http://dx.doi.org/10.1038/ajg.2016.563.

43. **Stubljar D, Jukic T, Ihan A.** 2018. How far are we from vaccination against *Helicobacter pylori* infection? *Expert Rev Vaccines* **17:**935–945. http://dx.doi.org/10.1080/14760584.2018.1526680.

17 Anthony Fauci: America's Top Infectious Disease Doctor

Despite predictions in the early 1980s that infectious diseases had been conquered, new or reemerging microorganisms have continued to wreak havoc on humanity since then. From HIV, anthrax, West Nile virus, SARS, and H1N1 influenza to Zika virus and Ebola, humanity has been repeatedly subjected to these new dangers, culminating in the COVID-19 pandemic beginning in 2020. As Director of the National Institute of Allergy and Infectious Diseases (NIAID), Dr. Anthony Fauci has helped set the American research agenda in infectious diseases and the governmental response to these dangers under seven different U.S. Presidents. Dr. Fauci's prominent role as the "voice of science" during the COVID-19 pandemic is well known to the U.S. public. However, his other contributions have been less recognized and bear scrutiny to understand how the United States and the world have benefited from his achievements, knowledge, and experience.

EARLY INFLUENCES

Anthony Fauci was born in Brooklyn, NY, a second-generation American whose grandparents immigrated from Italy. His father was a pharmacist in Brooklyn, where Tony grew up above the family store. Fauci worked behind the counter and delivered prescriptions on his bicycle through his high school days. Fauci described his father's influence:

> My father was very involved as a pharmacist in being a public servant. Back then, they called him "Doc"; he would be the psychiatrist, the doctor, the friend, the confidant

Germ Theory: Medical Pioneers in Infectious Diseases, Second Edition. Robert P. Gaynes.
© 2023 American Society for Microbiology.

of people in the neighborhood who did not have access [to other healthcare professionals] . . . That was kind of ingrained in me at the time I was a young boy. That has permeated the things that I have done throughout my school years, then in my deciding to become a physician, even up to the things that I am doing today (1).

Fauci went to a Jesuit high school in Manhattan, commuting over an hour each way from Brooklyn. He was influenced by the Jesuit spirit of "service for others," which became an important component of his life. Fauci also attributes the Jesuit schooling for instilling an academic rigor and adherence to scientific principles. The combination of the Jesuit training in classics and humanism and his strong abilities in science in high school led Fauci to pursue the idea of becoming a physician. After completing undergraduate work at the College of the Holy Cross in Worcester, MA, Fauci attended Cornell Medical School in Manhattan, NY, graduating first in his class.

As medical school training was nearing completion, Fauci had two major decisions to make. The first was what specialty in medicine to enter. In June 2022, I had the opportunity to interview Dr. Fauci (you can listen to excerpts from the interview in this "Meet the Microbiologist" podcast: https://asm.org/Podcasts/MTM/Episodes/The-Career-of-Tony-Fauci-MTM-143.) When I asked him why he chose internal medicine as a specialty, he said,

I liked the intellectual challenge of internal medicine, where you have the entire body that you're dealing with. And someone comes in with a constellation of symptoms and you really have to be the medical detective for an individual. Then when I got into internal medicine, it became clear that there are multiple subspecialties. And I happened to be attracted to one that at the time was a growing subspecialty, which was the interface between the emerging field of human immunology and infectious diseases. So, when they both came together, that's when I took a combined fellowship in infectious diseases and clinical immunology.

But then came Fauci's second important decision in medical school.

I wanted to do a fellowship. But we were all drafted automatically [into one of the uniformed services]. I had no hesitation of going into any of the armed services. During my senior year, a recruiting officer, I believe it was a United States Marine major, came into our auditorium in Cornell Medical School and said, "You know, at the end of this year, everybody, except the few women in the class, is going to be either in the army, the navy, the air force, or the public health service. So put down your choices and we'll see what comes out."

FAUCI AND THE NATIONAL INSTITUTES OF HEALTH

What "came out" for Dr. Fauci was an acceptance into the U.S. Public Health Service (USPHS), which meant taking a position at the National Institutes of

Health (NIH), the Centers for Disease Control (CDC), or, possibly, other site assignments in the USPHS. Based upon a medical school elective in immunology and infectious disease, Fauci was interested in developing a research background in these fields. He chose to go to the NIH.

> One of my mentors at Cornell was a physician named Marvin Sleisenger, who was the director of the GI [gastrointestinal] program. Marvin had a good friend named Sheldon Wolff, who was the head of the infectious diseases section at the NIH. So, when I got down to the NIH, Marvin told me, "You really should take a serious look at Shelly Wolff." . . . I interviewed with a whole bunch of people at the NIH but the chemistry between Shelly Wolff and I hit right off the bat. We just took to each other. I put him down as my first choice at the NIH and he picked me; it was a matching program. I got there; I was in his lab for three years and then he made me a permanent senior investigator. He, ultimately, years later, became the best man at my wedding.

CHIEF RESIDENCY

After 3 years at the NIH, mostly in the research laboratory, Fauci went back to Cornell Hospital to be the chief resident in internal medicine. I asked him what he got out of that year.

> That was one of the best years of my life because I knew I was going to go into both basic and clinical research, but I felt I needed to be the best clinical doctor . . . that was my goal. I thought that this would be a great opportunity. I was on every other night and every other weekend. And for one full year, I was a practicing physician for the sickest patients in the hospital. The clinical experience that I got during that chief residency year was something that was invaluable to me. And it helped me to appreciate the interface between clinical research and pure clinical medicine.

BREAKTHROUGH IN VASCULITIS TREATMENT

Fauci returned to the NIH and was immediately assigned to work on patients with vasculitis, severe inflammation of blood vessels, which had a variety of categories/ diseases, many with very high mortality. One disease, known as Wegener's granulomatosis (now known as granulomatosis with polyangiitis), is characterized by a necrotizing vasculitis (inflammation of blood vessels that leads to death of the cells comprising the vessels) and inflammation of cells in the upper and lower respiratory tracts, and sometimes the kidneys. The cause remains unknown. When Fauci began studying these patients, mortality was nearly 100%. Fauci explained how he and his colleagues made a breakthrough in treatment.

> My mentor, Shelly Wolff, said, "Why don't we figure out a way to be bold about this?" And we made a decision. When I was seeing infectious diseases consults on the cancer ward at the NIH, the cancer patients were being treated with

immunosuppressive drugs, including cyclophosphamide, in very high doses. And very often it suppressed the bone marrow so much that the patient died of an opportunistic infection. We knew that vasculitis was an aberrant immunological response. We got the idea that if we titrated the cyclophosphamide in a very careful way to suppress the immune system enough to shut off the vasculitis, but not enough to completely suppress the bone marrow, we had to give it in a different way than for cancer [where you give a high dose intravenously]. For Wegener's, you give moderately small doses on a daily basis, chronically over time. And you follow the white count. When the white count goes down, you pull back the dose a little; when the white count goes back up, you give them the dose so that you keep the counts within a safe range. We found out, to our great gratification, that we had taken a disease that had a 99% mortality, we gave it a 93% remission rate. We essentially had a major, major breakthrough in rheumatology by introducing the concept that you could suppress a nonmalignant disease with low doses of drugs that you treat malignancies with. And that became the beginning of the immune suppression of autoimmune diseases (2).

Fauci took on other diseases whose causes were unknown but considered autoimmune; i.e., the human body's immunologic response was aberrantly directed against itself (3–7). In a period of 9 years following his chief residency, Fauci had made major breakthroughs in the treatment of highly fatal diseases, writing over 100 scientific publications and book chapters. At an early stage in his career, Fauci received great recognition for his work on these unusual diseases that required careful therapeutic approaches to fine-tuning the body's pathologic immunologic reaction. Fauci never gave up his interest in clinical medicine during this period, performing infectious disease consultations on patients on the Clinical Service of the NIH. He did not realize it at the time, but his interest in the human body's immune system and in diagnosing and treating infectious diseases would place him in the perfect position to study one of the most important microorganisms of the 20th century—human immunodeficiency virus (HIV).

FAUCI AND AIDS

In June 1981, the CDC's *Morbidity and Mortality Weekly Report* (*MMWR*) first reported a cluster of unusual infections among five homosexual men in Los Angeles, California, the first account of a syndrome that would ultimately become known as acquired immunodeficiency syndrome (AIDS) (8). Initially, Fauci thought perhaps it was just a fluke. When, a few months later, an account appeared of a group of men in New York and California with an unusually aggressive cancer called Kaposi's sarcoma (9), Fauci took notice.

I remember sitting back and saying to myself, "Holy cow, this could be a brand-new disease. And if this is sexually transmitted, which it looked like it was, it's got to be a new infection." Because, remember, for those 9 years, in addition to my research, I was the infectious disease consultant for the NIH. And I never saw anything like this. [I asked myself] why all young gay men? And why a very specific immunological defect, a diminution dramatically in their CD4 T cells? [CD4 T cells are a type of cell called a T lymphocyte that are central to coordinating the body's immune function.] I said, "Whoa, we have a likely infection that's destroying the immune system." And I just looked up and thought, if ever my training was directly geared into this—I'm board certified in infectious disease, I'm board certified in immunology, and I'm a practicing immunosuppressive guy. It was like I was created for this disease.

In 1981, Fauci had two positions at the NIAID at the NIH. He was Chief of the Laboratory of Immunoregulation and the Deputy Clinical Director of NIAID. He decided to devote his entire time to this puzzling new disease. He hired others to continue the work on immunosuppression of vasculitis diseases. He began working on what this new disease/microorganism was doing to the human immune system, knowing in the back of his mind that there were virologists who were looking for and eventually identified the virus (chapter 15). Fauci tried to stay out of the 1984 controversy over who first isolated HIV, involving a colleague at the NIH, Robert Gallo, and the French scientists. Fauci used his laboratory to examine the irregularities in the immune function of patients with AIDS (10, 11). He also cared for AIDS patients and wrote articles and book chapters on the diagnosis and treatment of these patients (12–14).

FAUCI NAMED DIRECTOR OF THE NIAID

The success that Fauci had in his first decade at the NIH led one of his mentors, Dr. Sheldon Wolff, to urge him to put his name in as a candidate for the Director of the NIAID when the position became open in 1984. Fauci did not aspire to be the head of one of NIH's institutes but reconsidered when he was encouraged by Dr. Wolff.

I thought that we, as a nation and certainly the NIH, were not paying as much attention to nor were they emphasizing enough the importance of emerging infectious diseases as exemplified by what I saw as a lack of a robust response from a research standpoint with HIV. . . . The challenge of what I predicted in 1981 would turn out to be a global catastrophe. Yet the amount of money that was allocated and the amount of interest in the scientific community about getting involved in studying this was a bit below what I thought [it should be]. Maybe if I took this job or even put my name in, I could do something about it, which would have really broken the mold of Institute directors because Institute directors almost invariably don't lobby for their own cause. They just accept what's given to them.

Fauci placed his name in for the position but with certain conditions.

> I told [the search committee] that my interest would be a little bit different than any other Institute director. Because I would not give up seeing patients and I would not give up my lab, which would be unprecedented.

In 1984, Fauci was named the Director of the National Institute of Allergy and Infectious Diseases, a position he held until the end of 2022, an extraordinary duration, nearly 40 years, for such a prominent federal post.

> When I got into the job, I realized the potential of the bully pulpit of the director of the infectious disease institute in the United States. . . . I could lobby for money from the administration and from the Congress, even though that was unprecedented. I made my case publicly and privately [as to] why we needed to get more people involved. I made dozens and dozens of trips throughout the country, stoking up interest in getting people who were not involved in HIV to get involved in HIV. It had an effect because we had an upwell of young scientists who said, "Maybe this would be a good career choice to get involved in this new disease."
>
> I [also] became acutely aware of so many other diseases of global health impor-tance, like malaria and tuberculosis. . . . That's when I started to get involved, not only in science and clinical medicine, but also in global health and emerging infections. I was in the unique position that no other Institute director ever was in. I became . . . a spokesperson and an advocate for broadening our impact globally in infectious dis-eases. I came to be the advisor to multiple Presidents, one after the other. It was breaking the mold of what an Institute director had been up to that point and for decades and decades up to that point.

As Director of NIAID, Fauci would be repeatedly called upon to testify in front of Congress and advise the U.S. President on matters involving infectious diseases. He would help set the U.S. research agenda in infectious diseases at a critical time with the onset of the HIV epidemic. The role also propelled him into the public eye.

OPENING THE DOOR OF THE AIDS ACTIVIST COMMUNITY

In those early days of AIDS, there was no effective treatment. Once it was diag-nosed, the mortality from AIDS was nearly 100%. When a blood test became avail-able for HIV, a positive HIV test was one of the strongest impetuses for suicide, which was seen at a rate 60 times higher than in the general population (15). People who tested positive were desperate. One of the most visible channels of that desperation was advocacy groups, and AIDS patients and those at risk for AIDS formed one of the most vocal: AIDS Coalition to Unleash Power (ACT

UP). ACT UP was formed on 12 March 1987 at the Lesbian and Gay Community Services Center in New York City. Its founder was Larry Kramer, an American playwright, author, film producer, public health advocate, and gay rights activist. His aim was gaining more public action to fight the AIDS crisis, and he did so by making sensational and bombastic comments in the media. Members of ACT UP would engage in civil disobedience that would result in arrests; Kramer was arrested dozens of times. One of ACT UP's early targets was the Food and Drug Administration (FDA), which Kramer accused of neglecting the development and approval of badly needed medications for HIV-infected Americans.

In a 1987 open letter to the *San Francisco Examiner*, Kramer targeted Fauci.

Anthony Fauci, you are a murderer and should not be the guest of honor at any event that reflects on the past decade of the AIDS crisis. Your refusal to hear the screams of AIDS activists early in the crisis resulted in the deaths of thousands. . . . Anthony Fauci, you are a murderer because you oversee government sponsored clinical trials that test and retest combinations of immunosuppressive, toxic therapies that kill people with HIV (16).

That letter was the first time Fauci became aware of Larry Kramer. Kramer viewed Fauci as the face of a failing federal government. Kramer repeatedly lashed out directly at him. At the 1989 International AIDS Conference in Montreal, the two finally met. Fauci recalled,

I ran into him on the streets of Montreal—like two generals in a war on opposites coming together for a little bit of a truce and saying, you know, "We're both soldiers and maybe we're on the same side and we're not on opposite sides." He was being totally opposite from the fist-raising, in-your-face Larry, . . . and we found out at the end of that that even though we still disagreed on how to do things, we didn't disagree on what needed to be done (17).

Instead of ignoring the activists like many scientists, Fauci reached out and began a dialogue with members of ACT UP. Fauci explained his thinking.

One of the best things I did was, when they were out there disrupting to get attention, I said, "Let me do something. Let me dial down the rhetoric, the theatrics, the attacks on me personally, and listen to what they were saying." Because nobody was listening to what they were saying. And when you listen to what they were saying, they made absolutely perfect sense. I did what I often do, even with a patient, put myself in their shoes and say, "What would I do if I were a young gay man that was having a disease that we barely understood, and I was seeing all my friends die?" I would do exactly the same as what they did. So that's when I said, "I have to listen to these people." And once I did that, we began a years-long process of building up trust. It wasn't overnight. . . . Finally, over a period of time, we really trusted each other. They knew I was a person of my word. I would never go back on my word, even though there were things we disagreed on, even at the best of our relationship.

That dialogue did not stop ACT UP from turning on the NIH. On 21 May 1990, more than 1,000 ACT UP protestors stormed the campus in Bethesda, MD, to demand that the NIH accelerate the pace of AIDS research, broaden its investigations, and include AIDS activists and community members in the committees that oversaw AIDS research.

Fauci assured activists he would include them on those committees, even though he discovered scientific colleagues who bitterly opposed the idea.

I promised them [AIDS activists] that I would do that. Several of my staff members pushed back very, very strongly against it and even undermined it. I had to do something that also was painful. I had to let go some of my own staff who were not amenable to embracing the activists. . . . It made very clear that well-informed and well-intentioned activists have a major, major, positive impact on the scientific agenda, its development, and its implementation. It allows you to see what actually works in the trenches. It takes away that separatism between the ivory tower and what's going on in the community. It brings them together. Instead of making proclamations of what you can do with the community, you link up with the community and say, what is feasible? . . . To me it just blew open the relationship between the scientific community, the regulatory community, and the activist community. . . . If they have a memorial for me someday, I would love them to say that . . . one of the best things that I've ever done was to open up the door of the activist community.

The effect of Fauci's insistence on the role of the activists completely changed the practices at the FDA and NIH, and not just in HIV-related protocols. As one journalist wrote,

It is difficult to overstate the impact of ACT UP. The average approval time for some critical drugs fell from a decade to a year, and the character of placebo-controlled trials was altered for good. The National Institutes of Health even recognized ACT UP's role in getting drugs to more people earlier in the process of testing; soon changes in the way AIDS drugs were approved were adopted for other diseases, ranging from breast cancer to Alzheimer's Disease (18).

Fauci developed a close friendship with Larry Kramer over the years. He was his physician for a time and helped Kramer connect with appropriate people to receive a liver transplant in 2001. Kramer died of pneumonia in 2020. Fauci said, "Although Larry could be confrontative and combative, I considered him a dear friend and a hero of the AIDS activist movement" (18).

ADVISING U.S. PRESIDENTS ON INFECTIOUS DISEASES

As Director of the NIAID, Fauci advised the U.S. President on matters pertaining to infectious diseases. He recalled some of those interactions:

> Reagan: I liked the guy. He was a gentleman and a man, I believe, of considerable integrity, quite conservative. But I felt really badly that he did not use the bully pulpit of the presidency to call attention to the emerging AIDS outbreak. I was never friends with him, but it was cordial. And I respected him.
>
> Clinton: Clinton was very intellectually curious. I had more interactions with him about HIV than West Nile, particularly because he fully embraced the AIDS community that's what I had been involved with. So that brought me close, not only to him, but to Hillary.

However, the most noteworthy relationships that he had with U.S. Presidents were with George H. W. Bush and eventually his son, George W. Bush. Fauci explained,

> George H. W. Bush allowed me to realize that you can have a relationship with the President that is really meaningful, more than just the courtesy of "Thank you, Mr. President, have a nice day." Somebody who cares about things. I saw for the first time the complexity of a conservative administration run by a person who really cared about people. He began to fully open up the doors of better budget, even though the [AIDS] activist community absolutely jumped all over him, saying he didn't do enough. He did start to open up the situation. Then Clinton took over and he fully embraced [the AIDS community.]

Fauci's relationship with Bush's son, George W. Bush, was to change the landscape of infectious diseases and bring about one of the most significant contributions to the field that the U.S. government has ever made.

> My interaction [with George W. Bush] was really positive. I knew him before when he was in the White House with his dad. We just took to each other. He relied on my judgment about how we could best protect the country against the bioterrorism attacks, which after 9/11 was a real possibility. . . . He gave me $1.5 billion in my Institute [NIAID] to build up a biodefense program. . . . I had the opportunity to convince him of something that he had an open mind for. Instead of using all the resources to prepare for the unlikely threat of a deliberate bioterrorism event, why don't we broaden it and protect the country against what I told him was the worst bioterrorist, which was nature. Could we use the money, not only to worry about smallpox and anthrax, but for pandemic flu and coronaviruses and to worry about other things? And he said, "That's a good idea . . . do it!" . . . That's when I really got to know him.

HIGHLY ACTIVE ANTIRETROVIRAL THERAPY FOR HIV

The 1990s were a time of amazing transformation for HIV-positive patients in the United States. Highly active antiretroviral therapy (HAART) provided a second chance at life for these patients, turning a fatal disease into a treatable disease; HAART-treated HIV-infected patients had a life span nearly equivalent to people who did not have HIV (19). Unfortunately, the majority of patients who did not receive HAART died within 2 years of the onset of AIDS.

CREATION OF PRESIDENT'S EMERGENCY PLAN FOR AIDS RELIEF (PEPFAR)

The success of HAART in the United States in the mid-1990s did not translate to success everywhere in the world. At the turn of the 21st century, enormous discrepancies existed in the numbers of people receiving antiretroviral drugs between the developed countries of North America and Europe and the poor countries of Africa and Asia. George W. Bush and his wife, Laura, developed a serious interest in improving the outcomes of people in Africa with AIDS after visiting Gambia in 1990. Bush supported the Global Fund to Fight AIDS, Tuberculosis and Malaria, conceived in 2000. The Millennium Challenge Account (MCA) followed in 2002 as a testing ground for a new approach to foreign aid, emphasizing partnership over paternalism (20). President Bush sent Fauci and Department of Health and Human Services Secretary Tommy Thompson to several countries in sub-Saharan Africa to develop an initiative related to the recent proof in a clinical trial run by NIAID of the efficacy of nevirapine in preventing mother-to-child transmission of HIV. Fauci made a proposal to the President for $500 million to implement a nevirapine program in sub-Saharan Africa. The President immediately agreed to the plan. On 19 June 2002, Bush announced his International Mother and Child HIV Prevention Initiative, pledging $500 million in aid. But then Bush went on to ask White House Chief of Staff Joshua Bolton to have Fauci put together a much larger and transforming program that would go beyond the mother-to-child transmission initiative and that would have a much broader impact. Fauci then asked his assistant, Dr. Mark Dybul (who later would become PEPFAR Director), and his budget officer, Ralph Tate, an NIAID executive with experience in operations management, to help put together a much broader program for the President's consideration.

Fauci and his NIAID colleagues put together lists of countries based on the prevalence of HIV and AIDS and on prevailing social, economic, and political conditions that could be included in a wide-ranging program to combat AIDS. They estimated the costs of prevention measures, the costs of antiretroviral drugs, and the costs of other aspects of care such as treatment of secondary infections and support for orphan children of parents who died from AIDS.

The cost of AIDS drugs was a critical issue for the plan. Unless they could be brought down, treatment of HIV in Africa would not be possible. The U.S. government position at the turn of the 21st century was to protect patents of medications, including HIV medications, which effectively blocked their distribution to African countries. It would be necessary to break those patents for the program to be successful, an unexpected position for the Bush administration to take. Fauci explained how it happened:

> That came about with sitting in the room with me, the President, Josh Bolton, Kristen Silverberg [White House Staff], and a bunch of other people, including Gary Edson [White House Staff], Margaret Spellings [Director of the Domestic Policy Council], all the unsung heroes in that administration to say, "The only way we're going to get this done is to get the drugs down to a price that's so low that we can do it. So, do you want the program to succeed or not?" George W. Bush said, "Let's do it."

The reason Fauci referred to those people as unsung heroes was because all the planning for PEPFAR was done in strict secrecy, to avoid the interagency discord that Bush and others feared would result from a more open process. During the mere 7 months when the PEPFAR proposal was formed, skeptical officials in the White House and, in particular, the Office of Management and Budget were convinced of its likely success; Fauci helped shepherd the process along by providing PEPFAR with a set of metrics that allowed accountability.

On 28 January 2003, President Bush announced the PEPFAR proposal in his State of the Union speech, to the surprise of many in Congress and elsewhere. The Bush administration moved quickly to allay concerns and secured funding of $15 billion over 5 years thanks to bipartisan support in Congress (20).

PEPFAR'S SUCCESSES

PEPFAR became the largest foreign aid program ever dedicated to a particular disease. After 10 years, the Institute of Medicine was asked to evaluate the program.

> PEPFAR has played a transformative role . . . [in] . . . the global response to HIV. . . . The pride, gratitude and appreciation expressed by partner country governments, implementing partners, providers, and others [has made PEPFAR] a lifeline that has restored hope. PEPFAR has achieved—and in some cases surpassed—its initial ambitious aims (21).

After 15 years of PEPFAR, Fauci wrote in a 2018 *New England Journal of Medicine* article:

> PEPFAR-funded programs have provided antiretroviral therapy to 13.3 million people, averted nearly 2.2 million perinatal HIV infections, and provided care for more than 6.4 million orphans and vulnerable children.

With regard to international public relations, PEPFAR has done as much as or more than any other program in enhancing the humanitarian image of the United States and has firmly established it as a key player in the response to a historic global public health crisis (22).

In our interview, Fauci reflected on his role in PEPFAR:

We would not have been in a position to create and develop PEPFAR the way we did had we not had the trust of President Bush . . . with critical help from certain White House staff, particularly Josh Bolton and Gary Edson, among others, and the invaluable help of Mark Dybul. We put together a program that . . . when you look at the big picture of life, is the most important thing that I have ever done. I would not have been able to do it without the President and his staff totally supporting me. I worked for [7] months on that. When I came back from Africa, I did as much work on PEPFAR as I did at my job as the NIAID Director. . . . Finally, the President made the decision to accept my proposal, which was a $15 billion program to treat 2 million people, to save 7 million lives . . . that snowballed now into something that has saved at least 18 million lives. For me personally, it's one of the crowning achievements of my career.

A SERIES OF NEW INFECTIOUS DISEASE THREATS IN THE 21ST CENTURY

In the years following PEPFAR, several new infectious disease threats loomed over the world. Each time Fauci would be called upon to lend expertise, brief the U.S. Congress and the President to forge responses, and direct research at his Institute.

SARS

In 2003, an outbreak of severe acute respiratory syndrome (SARS) began in Asia. It was caused by a coronavirus, named for the protein spikes that project from the viral surface, which in electron micrographs are reminiscent of a crown or corona. Human coronaviruses have been known since the 1960s, producing respiratory infections that are generally mild. However, the coronavirus that caused this epidemic, known as SARS-CoV, emerged in southern China in late 2002, spread rapidly to other Chinese cities, and then spread to 28 other countries in 2003. More than 8,000 people had severe infection; 775 died (23, 24). Chinese officials initially covered up a SARS outbreak for weeks before a growing death toll and leaks from the scientific community forced the government to reveal the epidemic, apologize, and pledge openness in future outbreaks (25). That pledge would be tested 16 years later.

2009 Influenza Pandemic

In 2007, work on bioterrorism, AIDS, and other emerging infectious disease threats prompted Fauci and an NIH colleague, David Morens, to examine the 1918 influenza pandemic for insights (26). That pandemic, which killed an estimated 50 million to 100 million people worldwide, was the greatest single demographic shock mankind has ever experienced (27–29). Of the three types of influenza virus (A, B, and C), type A carries the most significant public health risk and is subtyped by the major antigens on the viral surface—hemagglutinin (H) and neuraminidase (N) proteins; 16 H and 9 N subtypes have been identified so far. The authors examined the 1918 influenza pandemic, caused by an H1N1 virus. That virus was maintained in humans until reassortment events led to the emergence of an H2N2 pandemic virus in 1957, which circulated until it was replaced by an H3N2 virus in 1968. Morens and Fauci determined that while influenza pandemics tend to occur in cycles, there was no predictable periodicity or pattern of major influenza pandemics. From a medical treatment point of view, comparing approaches from 1918 to "now" was largely encouraging with regard to diagnosis, treatment, and prevention. However, Morens and Fauci made a prophetic comment, that if another pandemic occurred,

> The most difficult challenge would probably not be to increase medical knowledge about treatment and prevention but to increase medical capacity and resource availability [e.g., hospital beds, medical personnel, drugs, and supplies] and public-health and community-crisis responses to an event in which 25–50% of the population could fall ill during a few weeks' time. Health-care systems could be rapidly overwhelmed by the sheer volume of cases; ensuring production and delivery of sufficient quantities of antivirals, vaccines, and antibiotics, as well as providing widespread access to medications and medical care, particularly in impoverished regions, would be a sobering challenge. And the just-in-time nature of our supply chain of necessary medications and equipment for medical care could easily be disrupted by such a global public-health catastrophe (26).

In 2009, influenza did indeed become a global problem due to an unusual strain of the virus. The World Health Organization (WHO) declared a pandemic in June 2009 from a new influenza virus (H1N1) that spread rapidly to a total of 74 countries and territories. Unlike typical seasonal flu patterns, the new virus caused high levels of summer infections in the Northern Hemisphere, and then even higher levels of activity during cooler months. The new virus also led to patterns of death and illness not normally seen in influenza infections.

The majority of these influenza cases occurred in younger age groups, those age 12 to 17 years. Most individuals had a mild, self-limiting, upper respiratory tract illness, but 2 to 5% of cases required hospitalization. Notably, pregnant women were found to be at higher risk for severe disease. The overall case-fatality rate was 0.15 to 0.25%, which was lower than in other influenza pandemics. Death from influenza typically is more common in the very young or very old. In the 2009 pandemic, deaths mostly occurred in middle-aged adults (median age, 40 to 50 years). Deaths amongst hospitalized patients mostly occurred in those with underlying health conditions, but nearly one-third who died were previously healthy (30).

The CDC and NIH made progress in a variety of areas to prepare for a pandemic after 2009 (31). The CDC emphasized surveillance and prevention of influenza; the NIH, specifically the NIAID under Fauci's direction, stressed research into diagnosis, treatment, and especially vaccines (32). The National Strategy for Pandemic Influenza and National Strategy for Pandemic Influenza Implementation Plan, first prepared in 2005 and 2006, were updated in 2009 and 2017 (33).

MERS-CoV

In September 2012, another new coronavirus was identified, initially called Novel Coronavirus 2012, and now officially named Middle East respiratory syndrome coronavirus (MERS-CoV) (34). MERS-CoV can cause infection that ranges from no symptoms (asymptomatic) or mild respiratory symptoms to severe acute respiratory disease and death. MERS had a higher mortality (nearly 40%) than most coronavirus infections. However, this coronavirus did not spread easily from human to human (35).

Zika Virus Outbreak

In April 2015, a different virus, called Zika, caused an outbreak that began in Brazil and spread to other countries in South America, Central America, North America, and the Caribbean. With Zika virus infection, humans can have clinical illness that ranges from asymptomatic infection to mild febrile illness, sometimes with a rash. However, in pregnant women, Zika infection was associated with a cluster of infants born with microcephaly, a birth defect where a baby's head is smaller than expected, causing a range of neurologic problems, which was first reported in Brazil. It was estimated that 1.5 million people were infected by Zika in Brazil, with more than 3,500 cases of microcephaly reported between October 2015 and January 2016. As of August 2017, the number of new Zika virus cases in the Americas had fallen dramatically.

Ebola Virus Epidemic

On 23 March 2014, the WHO reported cases of Ebola virus disease in southeastern Guinea, marking the beginning of the West Africa Ebola epidemic, the largest in history. Two and a half years after the first case was discovered, the outbreak ended with more than 28,600 cases and 11,325 deaths. Overall, 11 people were treated for Ebola in the United States during the 2014–2016 epidemic.

With each of these emerging infections, Fauci was called upon to aid in the U.S. governmental response, even though their impact on the United States was limited. Fauci recalled the Ebola response:

> As the Director of NIAID, the research on Ebola came under my auspices. During the Obama administration, we put together a working group to fashion what the response would be to this outbreak in West Africa, which spilled over in a very slight way, with a couple of cases here. . . . We created policy in collaboration with the CDC and with other agencies of the federal government. . . . I had people like my colleague, Dr. Cliff Lane, develop very effective clinical trials that tested drugs and vaccines in West Africa. I oversaw the overall research component of the United States' response.

Fauci did more than simply fashion a response to Ebola with members of the Obama administration. He got personally involved in the care of Americans who contracted Ebola.

> When we started to get cases of Americans who were volunteering over there, who were getting infected, our unit at the NIH became one of the three qualified centers in the country to care for people with Ebola [along with Emory University Hospital and University of Nebraska Hospital]. . . . Many healthcare providers in Africa were dying because they were getting infected when they were taking care of Ebola patients. When I made the decision to accept Ebola patients in our special clinical studies unit, in the NIH Clinical Center, I decided that I would personally take care of those patients because I did not want my staff to feel they had to do something that I wasn't willing myself to do.

Each of these outbreaks—SARS, 2009 influenza, MERS, Zika, and Ebola— was a warning that infectious diseases presented ongoing threats to the United States and the world, requiring public health preparation. However, no one was adequately prepared for what was about to come in late 2019.

COVID-19 PANDEMIC

In late 2019, reports began to circulate about a severe type of pneumonia occurring in Wuhan, China. In late December, I asked a friend who was a relatively high-placed official at the CDC what was happening. He responded, "I don't know; they [the Chinese] won't tell us." Reports remained murky into early January 2020.

On 11 and 12 January 2020, the WHO reported detailed information about the outbreak, strongly suggesting that it was associated with exposures in one seafood market in Wuhan. The market closed on 1 January 2020. At this stage, there were

41 cases; there was no infection among health care workers, and the Chinese reported limited evidence of human-to-human transmission. The Chinese authorities continued their intensive surveillance and epidemiological investigations. China shared the genetic sequence of the novel coronavirus on 12 January 2020, which aided other countries in developing specific diagnostic kits and ultimately vaccines (36).

The official name of the virus became SARS-CoV-2, and the disease that the virus caused was referred to as COVID-19. A 2 January 2020 online article reported on these first 41 cases admitted to hospital. Most of the infected patients were men (73%); less than half had underlying diseases (32%). The median age was 49 years; 66% of the 41 patients had been exposed to the Wuhan seafood market. Common symptoms at onset of illness were fever (98%), cough (76%), and muscle pain or fatigue (44%). Significant shortness of breath, or dyspnea, developed in half of the patients. All 41 patients had pneumonia with abnormal findings on chest computed tomography (CT) scans (37). Within 3 weeks the number of cases had ballooned to over 800, which now included health care workers, signifying human-to-human transmission. Mortality was initially estimated to be 10%.

By mid-January 2020 there were isolated reports of cases outside of China, in both the United States and Europe. China implemented a lockdown of Wuhan on 23 January. The CDC issued a Level 3 travel warning for China, recommending against nonessential travel 4 days later. On 30 January, the WHO declared COVID-19 a public health emergency of international concern, the same day that the CDC confirmed the first domestic transmission of SARS-CoV-2. On 31 January, the United States implemented travel restrictions for mainland China. The rapidity of events caused heightened concern everywhere, but the trajectory of the outbreak remained difficult to predict.

I asked Dr. Fauci when he was aware of the looming threat.

> We were not allowed in there to see what was going on. And then after a couple of weeks, when we saw the Chinese building these 10,000-bed hospitals over a 10-day period, we said, "Wait a minute. Why would they be building 10,000-bed hospitals with a construction program to put up buildings in literally 2 weeks unless something really bad was going on there?" That's when I knew they had a problem. The real question was, could it be contained?

The question about containing COVID-19 to China was quickly answered. By the end of February 2020, the WHO reported more than 82,000 cases worldwide, which included at least 1 case on all continents except Antarctica; South Korea overtook China in the number of new daily cases reported (38).

FAUCI AND THE WHITE HOUSE CORONAVIRUS TASK FORCE

The White House Coronavirus Task Force was established on 29 January 2020 to coordinate the administration's efforts to monitor, prevent, and mitigate the spread of the coronavirus. Initially, Secretary of Health and Human Services Alex Azar was the chair. Dr. Fauci was one of the original members of the task force.

On 5 February 2020, the FDA issued an Emergency Use Authorization for a CDC coronavirus test, clearing the way for it to be used in state labs. However, problems with the test kits forced federal officials to suspend for a month the launch of a nationwide detection program for the coronavirus.

On 25 February, the CDC held a press conference about the steps Americans might need to take to protect themselves. Leading that briefing was Dr. Nancy Messonnier, Director of the CDC's Division of Immunization and Respiratory Diseases, who told Americans that they needed to prepare for a significant disruption of their lives. The next day, in response to mounting coronavirus concerns including the testing difficulties, U.S. Vice President Mike Pence was named to chair the White House Task Force; Dr. Deborah Birx, who had been U.S. Global AIDS Coordinator as part of PEPFAR since 2014, was named the response coordinator. In early March 2020, the White House Task Force deployed a team to try to cope with test kit shortages across the country. The CDC lost credibility as the nation's leading public health agency; the country lost ground as COVID-19 was able to spread undetected within its borders (39).

Both the CDC and the WHO monitored symptomatic, laboratory-confirmed cases as a measure of the magnitude of COVID-19. In the spring of 2020, reports began to circulate that COVID-19 could be spread by asymptomatic persons (40). Fauci commented,

> The thing we didn't know, which was absolutely critical, but we ultimately found out after one and a half to two months or so that at least half of the transmission came from asymptomatic persons. . . . If you use the flu model and you only concentrate on people with symptoms, [that] was completely incorrect. . . . It was being spread very effectively by people who had no symptoms, either those who would remain asymptomatic or those who were in the presymptomatic phase of their infection. Once that became clear, I knew we were in deep trouble.

DR. FAUCI AND PRESIDENT TRUMP

On 11 March 2020, the WHO declared COVID-19 a pandemic, with more than 118,000 cases in 114 countries and 4,291 deaths. The White House Coronavirus Task Force held daily press briefings until 25 April 2020. The task force reviewed all coronavirus-related actions by federal agencies. Struggles occurred when the task force overruled the CDC recommendations on the use of masks on public transportation and the CDC's order to keep cruise ships docked. Fauci found himself disagreeing with President Trump and administration officials on several matters, including Trump's comparing illness with COVID-19 to that of the common flu (41). Fauci felt compelled to disagree publicly with Trump's overly rosy forecast of COVID-19 and when Trump and others promoted an unproven potential treatment, hydroxychloroquine, which could cause heart problems and other harms (42). Fauci recounted his interaction with Trump:

> On a personal level, we got along really well. I think it was the sort of subtle camaraderie we felt because we were both born and raised in New York City. . . . He took to me and liked me. It was only when he started saying things that were clearly not true . . . when he started saying the virus was going to disappear and hydroxychloroquine was [going to] end it all, I had to do something that was very uncomfortable for me. I had to publicly, when asked, disagree with the President of the United States. Interestingly, he got less upset about that than his staff did.

After the CDC's coronavirus testing debacle, the Trump administration exploited events to take control of the agency's messaging, blocking CDC scientists from press briefings, changing CDC recommendations, eventually causing damage to the trust in the agency (43). With the CDC effectively sidelined, Dr. Fauci became the major scientific voice to the U.S. public for COVID-19.

In the spring of 2020, Americans seemed to rely on Fauci for COVID-19 information. In one national poll, released in April 2020, 78% of participants approved of Fauci's efforts in press briefings; only 7% disapproved (44). As David Relman, a prominent microbiologist at Stanford University, stated,

> Tony has essentially become the embodiment of the biomedical and public-health research enterprise in the United States. Nobody is a more tireless champion of the truth and the facts. I am not entirely sure what we would do without him (42).

Not everyone agreed. Right-wing social media and talk radio denigrated Fauci as someone who was exaggerating the effect of COVID-19 and was hurting the economy. Trump supporters spread misinformation about the virus, purporting that the experts, especially Fauci, were not trustworthy, with agendas that were not in line with their beliefs. Fauci, who was no stranger to criticisms after ACT UP, began to receive serious threats to his life. Fauci observed,

> There are people right now in jail for threatening my life because they were more than credible threats. There were people who were intercepted, armed with guns, getting ready to kill me. So, they were really serious. It wasn't like just the usual

harassment against me and my wife and my children: obscene phone calls, harassing my kids where they work, where they travel, where they live in three different cities. It was an organized "Get Fauci and Fauci family campaign." But tucked within those were credible, serious threats to my life that were intercepted.

During March 2020 the Trump administration gave mixed messaging on mask wearing and other ways to reduce the risk of COVID-19 infection. However, the administration's most significant move was to put the burden on states—an approach that turned hospital systems and state governments into rivals (45). This predicament served to confirm Fauci's forewarning comment that "the just-in-time nature of our supply chain of necessary medications and equipment for medical care could easily be disrupted by a global public-health catastrophe" (26). On 18 March 2020, to respond to the spread of COVID-19 in the United States, President Trump issued an executive order on allocating health and medical resources, citing the Defense Production Act of 1950, although shortages of medical supplies continued for months.

On 16 March, Trump announced new national guidelines to be implemented for only 15 days that encouraged the practice of social distancing, including starting homeschooling, if possible; avoiding crowds of over 10 people; avoiding bars and restaurants, and avoiding unessential travel. Fauci said,

> The only trouble was that when we tried to extend that to shutting down in the United States, he half-heartedly went into it and he said, "Okay, let's do 15 days, flatten the curve, then let's extend it for 30 days." He was ambivalent. He allowed Deb Birx and I to do it, but then no sooner did he allow it that he got up publicly and said, "Liberate Michigan, liberate Virginia," which means "Don't listen to these people who are telling you to shut down."

By the end of March 2020, the United States had surpassed Italy as the country with the highest number of COVID-19 cases in the world, a position that continues to the time of this writing in 2022. By October 2020, COVID-19 was the leading cause of death in the United States, with many deaths deemed preventable if the country had followed public health measures more consistently (46).

OPERATION WARP SPEED

Social distancing and the use of personal protective equipment such as masks were the major tools to prevent spread of the COVID-19 virus for most of 2020, the first year of the pandemic. Despite controversy, masking was effective, saving an estimated 200,000 American lives (47). The development of a vaccine for controlling the outbreak was an urgent priority (48). Traditional vaccines stimulate the body's immune response by injecting into the body one of several elements—a portion of a microorganism (antigen), an attenuated (weakened) microorganism, or an inactivated (dead) microorganism. More recent vaccine technology has used a recombinant antigen-encoding viral vector (harmless carrier virus with an antigen's gene). These antigens or attenuated microorganisms are produced outside the body and then injected. In contrast, another technology, mRNA (messenger RNA) vaccines, introduce a short-lived, synthetically created fragment of the RNA sequence of a virus into the individual being vaccinated (49). These mRNA fragments are taken up in cells and use the cell's internal machinery to produce the viral protein antigens that the mRNA encodes. The body quickly degrades the mRNA fragments within a few days, but the viral antigen is expressed on the cell's surface and induces a powerful immune response. All these technologies were available in 2020, but the time needed to produce a vaccine was typically measured in years, too long to help control the virus in the middle of a pandemic. A "fast track" approach to vaccine development was needed.

Operation Warp Speed (OWS), as it came to be known, was tasked with facilitating the development of multiple vaccines against COVID-19. The U.S. government agreed to place advance orders of millions of vaccine doses, costing billions of dollars, for a vaccine that could be shown to be safe and effective. OWS greatly accelerated the timeline for the development of vaccines through clinical trials, FDA review, and mass distribution to eventually produce the most rapid vaccine rollout in history (50). When the search for a COVID-19 vaccine began, the public-private collaboration had 26 vaccine candidates in clinical trials, not knowing if any would be effective.

By October 2020, six vaccines were in advanced clinical trials: two recombinant protein (antigen)-based vaccines (Novavax and Sanofi/GlaxoSmithKline); two recombinant antigen-encoding viral vector vaccines (AstraZeneca and Johnson & Johnson/Janssen); and two mRNA vaccines (Pfizer and Moderna). A "successful" vaccine was deemed to be one with at least 60% vaccine efficacy, i.e., one in which there was at least a 60% decrease in the number of COVID-19 infection cases among a group of vaccinated people compared with a group in which nobody was vaccinated.

The mRNA vaccines were reported to have approximately 95% efficacy initially (51, 52) and were the first to receive the FDA's Emergency Use Authorization. Fauci, who had been encouraging OWS, said,

> The idea of Operation Warp Speed was brilliant. . . . That was very much the brain-child of a combination of Peter Marks from the FDA and Alex Azar . . . that occurred during the Trump administration. The fact that he [Trump] allowed it to happen—he deserves all the credit in the world, but it [Operation Warp Speed] came from the rank and file.

[The NIH Vaccine Center personnel] collaborated with the mRNA people. We created a highly effective vaccine in absolute record time. . . . That would not have been available to the American people for a long, long time had it not been for the decision via Operation Warp Speed to make major, major investments, to the tune of hundreds of millions, if not billions of dollars, to actually prepurchase vaccine and manufacture it even before you knew that it was effective. That is never ever, ever done in private industry.

COVID-19 VACCINE HESITANCY

Despite the high vaccine efficacy, vaccine administration presented problems. The first was inadequate supply to meet initial demand, requiring prioritization for who would first receive the vaccine. Distribution was challenging but was largely coordinated in the United States when supply became adequate. Vaccine hesitancy became a notable problem, with many sources for it (53, 54). In a review of vaccine hesitancy, Drs. G. Troiano and A. Nardi wrote,

One of the most interesting aspects of our review is the influence of political ideology on vaccine acceptance or refusal: people who declared Democratic political partisanship were significantly more likely to choose to receive vaccination; those who felt close to radical parties or those who did not vote/did not feel close to any party were significantly more likely to refuse the vaccine; those who voted for a far left or far right candidate in the last elections were more likely to refuse vaccination. This kind of analysis has already been conducted by Kennedy et al. with a focus on populist party: they observed that the support for populist parties could be used as a proxy for vaccine hesitancy (53).

POLITICS OF COVID-19

While Fauci has tried to stay out of politics, he has recognized the intrusion of politics in scientific matters, particularly in crises. However, in the COVID-19 pandemic he saw something different.

There are always political intonations in any response to a crisis. Usually, it's at a noninterfering level. There are people who want to make political points here versus political points there. But never in my experience, and certainly in my knowledge of history, have we ever faced a crisis as historic as a pandemic, the likes of which we haven't seen in a hundred-plus years, where the divisiveness in the country was palpable. It occurred in the last year of an administration, which was an election year. The divisiveness along party lines was as deep and profound as I've ever seen it under any circumstances. Instead of taking the unified approach of putting ideological differences aside and realizing the common enemy is the virus, [saying] "Let's all pull together against this virus," . . . decisions about masking, about vaccine, about shutting down, about congregating, about anything were shockingly divided along ideologic lines. This is the worst possible thing that you can do when you're trying to get your arms around and control a deadly outbreak.

DEVELOPMENT OF COVID-19 VARIANTS

Unfortunately, nature had an additional vexing scheme—the development of COVID-19 variants (55). As with the influenza virus, successive variants of COVID-19 have emerged as the virus mutates. Based on the epidemiological update by the WHO, as of 18 July 2022, five SARS-CoV-2 variants of concern have been identified since the beginning of the pandemic:

- Alpha (B.1.1.7): first variant of concern, described in the United Kingdom in late December 2020
- Beta (B.1.351): first reported in South Africa in December 2020
- Gamma (P.1): first reported in Brazil in early January 2021
- Delta (B.1.617.2): first reported in India in December 2020
- Omicron (B.1.1.529): first reported in South Africa in November 2021; now includes BA.1, BA.2, BA.3, BA.4, BA.5, and descendent lineages

With the emergence of each of these variants, peaks in cases, hospitalizations, and deaths occurred. COVID-19 variants can increase the transmissibility of the virus; increase its virulence or worsen the clinical disease presentation; and decrease the effectiveness of public health and social measures or available diagnostic tests, vaccines, and therapeutics (56). The more COVID-19 circulates around the globe, the more opportunities it has to change. More variants will undoubtably emerge until COVID-19 infection decreases to low levels everywhere.

COVID-19 AND HERD IMMUNITY

During the initial period of the pandemic, there was talk of herd immunity, which occurs when a sufficient percentage of a population has become immune to an infection, whether through previous infections or vaccination, and reduces the likelihood of infection for individuals who lack immunity. With COVID-19, there are significant obstacles to achieving herd immunity. Variants have circumvented natural and vaccine-induced immunity; asymptomatic transmission has made public health measures less effective; neither infection nor vaccination appears to induce prolonged protection against SARS-CoV-2 in most people; and the public health community has encountered substantial resistance to efforts to control the spread of SARS-CoV-2 by vaccination, mask wearing, and other interventions. We will have to learn to live with COVID-19. As Fauci and his coauthors wrote in 2022,

> Living with COVID-19 is best considered not as reaching a numerical threshold of immunity, but as optimizing population protection without prohibitive restrictions on our daily lives. Effective tools for prevention and control of COVID-19 [vaccines, prevention measures] are available; if utilized, the road back to normality is achievable even without achieving classical herd immunity (57).

POST-COVID CONDITION (LONG COVID)

The whole story of the COVID-19 pandemic is not yet written and may not be apparent for decades. COVID-19's end or at least low-level existence in our society is not yet in view. We do not entirely understand the long-term impact of COVID-19. The post-COVID condition, or "long COVID," may leave people who are infected with this virus and survive the acute phase of the illness at risk of developing longer-term consequences including cutaneous, respiratory, cardiovascular, musculoskeletal, mental health, neurologic, and renal involvement. Given the global scale of this pandemic, the health care needs of patients with long-term consequences of COVID-19 will continue to increase for the foreseeable future (58).

LESSONS FROM THE COVID-19 PANDEMIC

Despite the incomplete story, there are lessons from this pandemic that can already be discerned. I asked Dr. Fauci about them.

> The first is scientific preparedness and response. And the other is public health preparedness and response. I think the example of the investment made in biomedical research that we put in for decades prior to this outbreak was the reason why we were able to do what would be considered the seemingly impossible—to get vaccines and drugs, particularly vaccines that have already saved millions of lives, within a time frame that was unprecedented. We were not so successful with the public health response; the reason is that we've let our public health infrastructure really deteriorate over years, probably because we're victims of our own success with vaccines and drugs. . . . Our ability to respond to a threatening outbreak like this was proven to be subpar at best—the lack of the connectivity with the local [circumstances and] people, the [inflexible bond] to certain paradigms like syndromic approach toward respiratory diseases, [which hindered] the realization of asymptomatic spread [and] the lack of transparency [on the nature and extent of the outbreak] early on.
>
> We have to get the rest of the world to remember that no one is to blame when there is an outbreak. No one should feel that they have to cover up or downplay

something. They need to be as open and honest as possible. We've got to make sure that we have the capability of responding as a global community, not as individual nations. While we need to continue to support the basic and clinical biomedical research [such as a universal COVID-19 vaccine (59)], we've got to do much better in our public health security plans and our capabilities of being transparent, interactive, cooperative, and collaborative.

AWARDS AND HONORS

Dr. Anthony Fauci has received numerous awards, too many to list, but including the Maxwell Finland Award for Scientific Achievement presented by the National Foundation for Infectious Diseases, Thomas H. Ham-Louis R. Wasserman Award of the American Society of Hematology, John Phillips Memorial Award for Achievement in Internal Medicine from the American College of Physicians, 1999 Bristol Award of the Infectious Diseases Society of America, Lifetime Achievement Award of the American Association of Immunologists, Mary Woodard Lasker Award for Public Service, National Medal of Science, 2013 Robert Koch Gold Medal from the Robert Koch Foundation (Berlin, Germany), Lifetime Achievement Award of the Institute of Human Virology, Eleanor Roosevelt Prize for Global Human Rights Advancement from the American Bar Association Center for Human Rights, Joseph F. Boyle Award for Distinguished Service from the American College of Physicians, and *People* magazine 2020 Person of the Year. In Fauci's mind, one award stands above the others: the 2008 Presidential Medal of Freedom, awarded by President George W. Bush at the White House (Fig. 17.1). Fauci explained why the award stood out as so significant:

> Because of the reason why it was given to me [his work developing PEPFAR], I'd have to say the Presidential Medal of Freedom, because it was given to me by the President who allowed me to go to Africa to create a program that has actually saved 15, 18, 20 million lives. To me, that award not only for what the award is, but for the President who actually bestowed it on me, to me is the most meaningful.

DR. FAUCI'S LEGACY

Dr. Fauci announced his retirement from the NIAID at the end of 2022. "Because of Dr. Fauci's many contributions to public health, lives here in the United States and around the world have been saved," President Joseph Biden said (60). Fauci's enormous influence is difficult to calculate but is evident in the recognition by the medical/scientific community and the public, including a rise in medical school applications, attributed, in part, to Dr. Fauci (61). For someone who has devoted his lifetime to infectious diseases, Dr. Fauci's legacy is best described by the man himself:

FIGURE 17.1 *President George W. Bush presenting Anthony Fauci with the 2008 Presidential Medal of Freedom at the White House.*

I would hope [my legacy] would be that I've devoted my entire professional career, as an individual physician taking care of patients and as a clinical researcher, designing clinical trials that have saved lives, but also as a leader of the scientific community in the field of infectious disease that have allowed us to respond from a scientific standpoint, very effectively to a number of public health threats. . . . I had the privilege of advising seven Presidents of the United States that I hope and believe had a positive influence on how our country responded to global health threats.

No better tribute can be made than what Dr. Fauci said when I asked him what he would like his epitaph to read.

I tried my very best and gave all of my energy and all of my passion to preserving the public health of people in the United States directly and, by doing that, indirectly for people throughout the world.

REFERENCES

1. **Axelrod D.** 23 July 2020. The Axe Files with David Axelrod, Episode 396. https://www.cnn.com/audio/podcasts/axe-files/episodes/d4ba4c2d-6208-423c-822b-ac010162ea90.
2. **Fauci AS, Wolff SM, Johnson JS.** 1971. Effect of cyclophosphamide upon the immune response in Wegener's granulomatosis. *N Engl J Med* **285:**1493–1496. http://dx.doi.org/10.1056/NEJM197112302852701.

3. **Fauci AS, Johnson RE, Wolff SM.** 1976. Radiation therapy of midline granuloma. *Ann Intern Med* **84:**140–147. http://dx.doi.org/10.7326/0003-4819-84-2-140.

4. **Wright DG, Wolff SM, Fauci AS, Alling DW.** 1977. Efficacy of intermittent colchicine therapy in familial Mediterranean fever. *Ann Intern Med* **86:**162–165. http://dx.doi.org/10.7326/0003-4819-86-2-162.

5. **Parrillo JE, Fauci AS, Wolff SM.** 1977. The hypereosinophilic syndrome: dramatic response to therapeutic intervention. *Trans Assoc Am Physicians* **90:**135–144.

6. **Fauci AS, Doppman JL, Wolff SM.** 1978. Cyclophosphamide-induced remissions in advanced polyarteritis nodosa. *Am J Med* **64:**890–894. http://dx.doi.org/10.1016/0002-9343(78)90533-8.

7. **Fauci AS.** 1979. Granulomatous hepatitis, p 1070–1075. *In* Mandell GL, Douglas RG Jr, Bennett JE (ed), *Principles and Practice of Infectious Disease.* John Wiley and Sons, New York, NY.

8. **Centers for Disease Control (CDC).** 1981. *Pneumocystis* pneumonia—Los Angeles. *MMWR Morb Mortal Wkly Rep* **30:**250–252.

9. **Centers for Disease Control (CDC).** 1981. Kaposi's sarcoma and *Pneumocystis* pneumonia among homosexual men—New York City and California. *MMWR Morb Mortal Wkly Rep* **30:**305–308.

10. **Lane HC, Masur H, Edgar LC, Whalen G, Rook AH, Fauci AS.** 1983. Abnormalities of B-cell activation and immunoregulation in patients with the acquired immunodeficiency syndrome. *N Engl J Med* **309:**453–458. http://dx.doi.org/10.1056/NEJM198308253090803.

11. **Rook AH, Masur H, Lane HC, Frederick W, Kasahara T, Macher AM, Djeu JY, Manischewitz JF, Jackson L, Fauci AS, Quinnan GV Jr.** 1983. Interleukin-2 enhances the depressed natural killer and cytomegalovirus-specific cytotoxic activities of lymphocytes from patients with the acquired immune deficiency syndrome. *J Clin Invest* **72:**398–403. http://dx.doi.org/10.1172/JCI110981.

12. **Fauci AS.** 1983. The acquired immune deficiency syndrome. The ever-broadening clinical spectrum. *JAMA* **249:**2375–2376. http://dx.doi.org/10.1001/jama.1983.03330410061029.

13. **Macher AM, Masur H, Lane HC, Fauci AS.** 1983. *AIDS Diagnosis and Management.* Burroughs Wellcome Co, Research Triangle Park, NC.

14. **Fauci AS, Macher AM, Longo DL, Lane HC, Rook AH, Masur H, Gelmann EP.** 1984. NIH conference. Acquired immunodeficiency syndrome: epidemiologic, clinical, immunologic, and therapeutic considerations. *Ann Intern Med* **100:**92–106. http://dx.doi.org/10.7326/0003-4819-100-1-92.

15. **Marzuk PM, Tierney H, Tardiff K, Gross EM, Morgan EB, Hsu MA, Mann JJ.** 1988. Increased risk of suicide in persons with AIDS. *JAMA* **259:**1333–1337. http://dx.doi.org/10.1001/jama.1988.03720090023028.

16. **Pasquarelli D.** 12 May 1995. ACT UP zaps Tony Fauci. *Queer Resources Directory.* http://www.qrd.org/qrd/aids/orgs/ACTUP/sf/1995/fauci.zap.flyer-05.06.95.

17. **Carlson A.** 2 June 2020. Looking back at Dr. Fauci's enduring bond with AIDS activist Larry Kramer—and their final phone call. *People.* https://people.com/health/dr-anthony-fauci-friendship-larry-kramer/.

18. **Specter M.** 13 May 2002. Larry Kramer, public nuisance. *The New Yorker.* https://www.newyorker.com/magazine/2002/05/13/public-nuisance.

19. **Poorolajal J, Hooshmand E, Mahjub H, Esmailnasab N, Jenabi E.** 2016. Survival rate of AIDS disease and mortality in HIV-infected patients: a meta-analysis. *Public Health* **139:**3–12. http://dx.doi.org/10.1016/j.puhe.2016.05.004.

20. **Varmus H.** 1 December 2013. Making PEPFAR: a triumph of medical diplomacy. *Science & Diplomacy.* https://www.sciencediplomacy.org/article/2013/making-pepfar.

21. **Institute of Medicine.** 2013. *Evaluation of PEPFAR.* National Academies Press, Washington, DC.

22. **Fauci AS, Eisinger RW.** 2018. PEPFAR—15 years and counting the lives saved. *N Engl J Med* **378:**314–316. http://dx.doi.org/10.1056/NEJMp1714773.

23. **Ksiazek TG, Erdman D, Goldsmith CS, Zaki SR, Peret T, Emery S, Tong S, Urbani C, Comer JA, Lim W, Rollin PE, Dowell SF, Ling AE, Humphrey CD, Shieh WJ, Guarner J, Paddock CD, Rota P, Fields B, DeRisi J, Yang JY, Cox N, Hughes JM, LeDuc JW, Bellini WJ, Anderson LJ, SARS Working Group.** 2003. A novel coronavirus associated with severe acute respiratory syndrome. *N Engl J Med* **348:**1953–1966. http://dx.doi.org/10.1056/NEJMoa030781.

24. **Drosten C, Günther S, Preiser W, van der Werf S, Brodt HR, Becker S, Rabenau H, Panning M, Kolesnikova L, Fouchier RA, Berger A, Burguière AM, Cinatl J, Eickmann M, Escriou N, Grywna K, Kramme S, Manuguerra JC, Müller S, Rickerts V, Stürmer M, Vieth S, Klenk HD, Osterhaus AD, Schmitz H, Doerr HW.** 2003. Identification of a novel coronavirus in patients with severe acute respiratory syndrome. *N Engl J Med* **348:**1967–1976. http://dx.doi.org/10.1056/NEJMoa030747.

25. **Reuters Staff.** 31 December 2019. Chinese officials investigate cause of pneumonia outbreak in Wuhan. Reuters. https://www.reuters.com/article/us-china-health-pneumonia/chinese-officials-investigate-cause-of-pneumonia-outbreak-in-wuhan-idUSKBN1YZ0GP.

26. **Morens DM, Fauci AS.** 2007. The 1918 influenza pandemic: insights for the 21st century. *J Infect Dis* **195:**1018–1028. http://dx.doi.org/10.1086/511989.

27. **Porter R.** 1997. *The Greatest Benefit to Mankind: a Medical History of Humanity from Antiquity to the Present.* HarperCollins, London, United Kingdom.

28. **Johnson NP, Mueller J.** 2002. Updating the accounts: global mortality of the 1918-1920 "Spanish" influenza pandemic. *Bull Hist Med* **76:**105–115. http://dx.doi.org/10.1353/bhm.2002.0022.

29. **Barry JM.** 2004. *The Great Influenza: the Epic Story of the Greatest Plague in History.* Viking Penguin, New York, NY.

30. **Broadbent AJ, Subbarao K.** 2011. Influenza virus vaccines: lessons from the 2009 H1N1 pandemic. *Curr Opin Virol* **1:**254–262. http://dx.doi.org/10.1016/j.coviro.2011.08.002.

31. **Centers for Disease Control and Prevention.** 6 June 2019. Summary of progress since 2009. https://www.cdc.gov/flu/pandemic-resources/h1n1-summary.htm.

32. **National Institute of Allergy and Infectious Diseases.** 8 June 2017. Influenza vaccines. https://www.niaid.nih.gov/diseases-conditions/influenza-vaccines.

33. **U.S. Department of Health and Human Services.** 2017. *Pandemic Influenza Plan: 2017 Update.* https://www.cdc.gov/flu/pandemic-resources/pdf/pan-flu-report-2017v2.pdf.

34. **World Health Organization.** 18 December 2019. Middle East respiratory syndrome coronavirus (MERS-CoV)—Saudi Arabia. https://www.who.int/emergencies/disease-outbreak-news/item/2019-DON222.

35. **Zumla A, Hui DS, Perlman S.** 2015. Middle East respiratory syndrome. *Lancet* **386:**995–1007. http://dx.doi.org/10.1016/S0140-6736(15)60454-8.

36. **World Health Organization.** 12 January 2020. COVID-19—China. https://www.who.int/emergencies/disease-outbreak-news/item/2020-DON233.

37. **Huang C, Wang Y, Li X, Ren L, Zhao J, Hu Y, Zhang L, Fan G, Xu J, Gu X, Cheng Z, Yu T, Xia J, Wei Y, Wu W, Xie X, Yin W, Li H, Liu M, Xiao Y, Gao H, Guo L, Xie J, Wang G, Jiang R, Gao Z, Jin Q, Wang J, Cao B.** 2020. Clinical features of patients infected with 2019 novel coronavirus in Wuhan, China. *Lancet* **395:**497–506. http://dx.doi.org/10.1016/S0140-6736(20)30183-5.

38. **World Health Organization.** 27 February 2020. Coronavirus disease 2019 (COVID-19): Situation Report–38. https://www.who.int/docs/default-source/coronaviruse/situation-reports/20200227-sitrep-38-covid-19.pdf.

39. **Kaplan S.** 18 April 2020. C.D.C. labs were contaminated, delaying coronavirus testing, officials say. *New York Times.* https://www.nytimes.com/2020/04/18/health/cdc-coronavirus-lab-contamination-testing.html.

40. **Cheng HY, Jian SW, Liu DP, Ng TC, Huang WT, Lin HH, Taiwan COVID-19 Outbreak Investigation Team.** 2020. Contact tracing assessment of COVID-19 transmission dynamics in Taiwan and risk at different exposure periods before and after symptom onset. *JAMA Intern Med* **180:**1156–1163. http://dx.doi.org/10.1001/jamainternmed.2020.2020.

41. **Paules CI, Marston HD, Fauci AS.** 2020. Coronavirus infections—more than just the common cold. *JAMA* **323:**707–708. http://dx.doi.org/10.1001/jama.2020.0757.

42. **Bump P.** 31 March 2022. Ivermectin is the signature example of politics trumping health. *Washington Post.* https://www.washingtonpost.com/politics/2022/03/31/ivermectin-is-signature-example-politics-trumping-health/.

43. **Bandler J, Callahan P, Rotella S, Berg K.** 15 October 2020. Inside the fall of the CDC. ProPublica. https://www.propublica.org/article/inside-the-fall-of-the-cdc.

44. **Specter M.** 10 April 2002. How Anthony Fauci became America's doctor. *The New Yorker.* https://www.newyorker.com/magazine/2020/04/20/how-anthony-fauci-became-americas-doctor.

45. **Bender MC, Ballhaus R.** 31 August 2020. How Trump sowed Covid supply chaos. 'Try getting it yourselves.' *Wall Street Journal.* https://www.wsj.com/articles/how-trump-sowed-covid-supply-chaos-try-getting-it-yourselves-11598893051.

46. **Woolf SH, Chapman DA, Lee JH.** 2021. COVID-19 as the leading cause of death in the United States. *JAMA* **325:**123–124. http://dx.doi.org/10.1001/jama.2020.24865.

47. **Lerner AM, Folkers GK, Fauci AS.** 2020. Preventing the spread of SARS-CoV-2 with masks and other "low-tech" interventions. *JAMA* **324:**1935–1936. http://dx.doi.org/10.1001/jama.2020.21946.

48. **U.S. Department of Health and Human Services.** 15 May 2020. Trump Administration announces framework and leadership for 'Operation Warp Speed.' https://www.hhs.gov/about/news/2020/05/15/trump-administration-announces-framework-and-leadership-for-operation-warp-speed.html.

49. **Dolgin E.** 2021. The tangled history of mRNA vaccines. *Nature* **597:**318–324. http://dx.doi.org/10.1038/d41586-021-02483-w.

50. **Caldwell A.** 11 February 2021. The most rapid vaccine rollout in history: how researchers developed COVID-19 vaccines so quickly. SciTechDaily. https://scitechdaily.com/the-most-rapid-vaccine-rollout-in-history-how-researchers-developed-covid-19-vaccines-so-quickly/.

51. **Skowronski DM, De Serres G.** 2021. Safety and efficacy of the BNT162b2 mRNA Covid-19 vaccine. *N Engl J Med* **384:**1576–1577. http://dx.doi.org/10.1056/NEJMc2036242.

52. **Baden LR, El Sahly HM, Essink B, Kotloff K, Frey S, Novak R, Diemert D, Spector SA, Rouphael N, Creech CB, McGettigan J, Khetan S, Segall N, Solis J, Brosz A, Fierro C, Schwartz H, Neuzil K, Corey L, Gilbert P, Janes H, Follmann D, Marovich M, Mascola J, Polakowski L, Ledgerwood J, Graham BS, Bennett H, Pajon R, Knightly C, Leav B, Deng W, Zhou H, Han S, Ivarsson M, Miller J, Zaks T, COVE Study Group.** 2021. Efficacy and safety of the mRNA-1273 SARS-CoV-2 vaccine. *N Engl J Med* **384:**403–416. http://dx.doi.org/10.1056/NEJMoa2035389.

53. **Troiano G, Nardi A.** 2021. Vaccine hesitancy in the era of COVID-19. *Public Health* **194:**245–251. http://dx.doi.org/10.1016/j.puhe.2021.02.025.

54. **Kennedy J.** 2019. Populist politics and vaccine hesitancy in Western Europe: an analysis of national-level data. *Eur J Public Health* **29:**512–516. http://dx.doi.org/10.1093/eurpub/ckz004.

55. **Ciotti M, Ciccozzi M, Pieri M, Bernardini S.** 2022. The COVID-19 pandemic: viral variants and vaccine efficacy. *Crit Rev Clin Lab Sci* **59:**66–75. http://dx.doi.org/10.1080/10408363.2021.1979462.

56. **World Health Organization.** Tracking SARS-CoV-2 variants. https://www.who.int/activities/tracking-SARS-CoV-2-variants.

57. **Morens DM, Folkers GK, Fauci AS.** 2022. The concept of classical herd immunity may not apply to COVID-19. *J Infect Dis* **226:**195–198. http://dx.doi.org/10.1093/infdis/jiac109.

58. **Nalbandian A, Sehgal K, Gupta A, Madhavan MV, McGroder C, Stevens JS, Cook JR, Nordvig AS, Shalev D, Sehrawat TS, Ahluwalia N, Bikdeli B, Dietz D, Der-Nigoghossian C, Liyanage-Don N, Rosner GF, Bernstein EJ, Mohan S, Beckley AA, Seres DS, Choueiri TK, Uriel N, Ausiello JC, Accili D, Freedberg DE, Baldwin M, Schwartz A, Brodie D, Garcia CK, Elkind MSV, Connors JM, Bilezikian JP, Landry DW, Wan EY.** 2021. Post-acute COVID-19 syndrome. *Nat Med* **27:**601–615. http://dx.doi.org/10.1038/s41591-021-01283-z.

59. **Morens DM, Taubenberger JK, Fauci AS.** 2022. Universal coronavirus vaccines—an urgent need. *N Engl J Med* **386:**297–299. http://dx.doi.org/10.1056/NEJMp2118468.

60. **Doherty E.** 22 August 2022. Fauci stepping down in December. Axios. https://www.axios.com/2022/08/22/fauci-stepping-down-in-december.

61. **Budryk Z.** 7 December 2020. Medical schools call 18 percent increase in applications 'Fauci effect.' The Hill. https://thehill-com.cdn.ampproject.org/c/s/thehill.com/policy/healthcare/529045-medical-schools-call-18-percent-increase-in-applications-fauci-effect?amp.

18 Conclusions

History can provide a base from which we can approach the future challenges from infectious diseases in a variety of ways. For example, the study of history can teach us about change and how people and society adapt to change. Each of the individuals profiled in this book was responsible, directly or indirectly, for a paradigm shift in medical thought. The process by which they came to their discovery or discoveries, and the reaction of their societies, can reveal that change does not always occur from one momentous discovery. Major contributions to infectious diseases were often the result of lifelong studies and efforts—such as those by Pasteur, Koch, Wald, Ehrlich, and Fauci—not single, isolated discoveries. Breakthroughs are often touted in the media, but in medical science they are actually quite rare. Major shifts in thinking often are products not just of the discovery but also of the age and culture. The innovations of Avicenna, Fracastoro, and Leeuwenhoek could have advanced the field of infectious diseases centuries earlier than they did. However, a model in medicine, the humoral theory, had to be torn away and the innovations had to be, essentially, rediscovered. Sadly, innovators often pay a high price for their innovations—a problem in all scientific disciplines, including medicine. Significant contributions to medicine have not always been well received by contemporaries. The contributions of Jenner, Semmelweis, and Lister, some of the most monumental in the history of science, were harshly received. The negative reaction of contemporaries to their discoveries befuddled and angered their discoverers for years. Even in the 20th century, Barry Marshall paid a price when he

Germ Theory: Medical Pioneers in Infectious Diseases, Second Edition. Robert P. Gaynes.
© 2023 American Society for Microbiology.

and his colleague, Robin Warren, tried to convince the modern health care world of a paradigm shift in thinking about peptic ulcer disease.

Despite infectious diseases claiming countless lives from prehistoric times, the theory of "contagion" due to living entities, i.e., the germ theory, is a relatively recent one in the annals of Western medicine. Yet progress in just over a century has been dramatic. However, serious challenges exist in the practice of infectious diseases medicine. The difficulties with antibiotic resistance are threatening not only progress in the management of infectious diseases but all medical advances. New medical procedures, surgery, and transplantation cannot be successful if a patient develops an untreatable infection afterwards. We have been repeatedly reminded in the 21st century of threats of infections with new or reemerging pathogens, including severe acute respiratory syndrome (SARS), Middle East respiratory syndrome (MERS), Zika virus, and Ebola virus, culminating in a worldwide pandemic with COVID-19. Finally, human immunodeficiency virus (HIV) will remain an enormous challenge for decades, both for treating those individuals already infected with HIV and preventing HIV in new generations with an HIV vaccine.

THE DEVELOPMENT OF ANTIBIOTIC RESISTANCE

When medicine's greatest contributions to humanity are tallied, antibiotics are invariably mentioned. In general, antibiotics are remarkably free of adverse effects, eradicating the microorganism and usually leaving the host unaffected. They truly are Ehrlich's vision of "magic bullets." However, antibiotics are also virtually the only class of drugs that lose their effectiveness once they are in widespread use. For example, one of our oldest drugs, aspirin, is just as effective an analgesic today as the first time it was used. The same cannot be said of penicillin. In the 1950s, the use and overuse of penicillin resulted in penicillin resistance. In the 1960s and 1970s, the use and overuse of streptomycin resulted in streptomycin resistance. With each new class of antibiotics, the observation that "resistance follows use" has been repeatedly made. Because antibiotics are typically well tolerated and effective, they are often used when the diagnosis of a bacterial infection is unclear. This liberal use of antibiotics has contributed to antibiotic resistance—a problem that has become a crisis. According to the World Health Organization, antimicrobial resistance is considered one of the three greatest threats to human health (1). A more recent analysis suggests that antimicrobial-resistant infections claimed nearly 5 million lives worldwide in 2019 alone (2).

With ever-increasing levels of antibiotic resistance, some bacterial infections are now virtually untreatable, prompting the need for continued searches for new antibiotics. Yet many drug companies have completely abandoned antibiotic discovery, finding that it is no longer economically viable. From the beginning of

2008 until the end of 2010, the U.S. Food and Drug Administration approved only one new antibacterial antibiotic, telavancin (3). The dry pipeline for antibiotics prompted the Infectious Diseases Society of America (IDSA) to call for a global commitment to develop new antibacterial drugs (4). While there were some promising developments in the first few years after IDSA's plea, there were no new antibacterial antibiotics approved by the Food and Drug Administration in the United States from 2019–2022. Changes in economics may be needed for progress (5). It will be many years, if not decades, until the production of new antibiotics overcomes the need for new drugs due to antibiotic resistance. We have squandered these wonder drugs and need to find ways to conserve the effectiveness of the antibiotics that we currently have. The rarity of the strain that produced penicillin in Fleming's laboratory and the difficulty of purifying it, especially in large quantities, should have demonstrated to us the treasured nature of these drugs. Climbing out of this crisis will require painstaking efforts to develop new therapies and use them with care. Ehrlich and Fleming taught us that antibiotics are difficult to find, a lesson we should have learned.

THE NEED FOR A NEW PARADIGM IN ANTIMICROBIAL TREATMENT

Antibiotic-resistant pathogens and the lack of new, effective antibiotics threaten the current practice of infectious diseases medicine. History may be able to suggest a path to follow. Prior to the introduction of sulfonamides, Salvarsan, quinine, and serum therapies were the only treatments for infectious diseases. These treatments required pathogen-specific diagnoses. When modern antibacterial antibiotic therapy began with sulfa drugs and penicillin, the need for specific pathogen diagnosis became less important, since sulfa drugs treated a variety of bacteria. As our antibiotics became broader in their spectra of antibacterial activity, the need for specific pathogen diagnosis became even less crucial. Today, despite years of advances in laboratory techniques, we find specific bacterial diagnoses in less than one-third of all hospitalized patients who are treated with antibiotics. Most patients are treated empirically, without knowledge of the specific pathogen causing the infection. This practice of empiricism has led to problems with antibiotic selection among providers (6). The indiscriminate, broad-spectrum antibiotic use has greatly contributed to the development of antibiotic resistance.

Initially, we believed that the breathtaking success of antibiotics was wholly the result of the drugs' lethal action on the microorganisms. The early studies with antibiotics were on patients whose immune system was entirely normal. We did not fully appreciate the role of the host's immunity in fighting infection until we began to manipulate it. We have successfully altered the human body's immune system to prolong the survival of transplant recipients. However, we have also discovered the essential role that these intentionally weakened host defenses of the body play in fighting infectious diseases. Infections can overwhelm a patient with such a weakened immune system even when we use seemingly effective antibiotic therapy. One investigator has proposed a new paradigm that is inspired by some of the earliest of all therapies in infectious diseases—serum therapy (7). He proposed a return to the days of pathogen-specific serum therapy such as von Behring's efforts in diphtheria, but with an important technological change. Human monoclonal antibodies have been developed against specific microorganisms, such as SARS-CoV-2, that avoid the toxicity of the old serum therapies (8). However, the evolution of that virus rendered some of the monoclonal antibodies ineffective. The overall concept would require improved microbiological diagnostics to determine the specific pathogen causing a patient's infection. With the limited availability of antibacterial antibiotics, this proposed direction for treatment might be our only path if we are without any effective antibiotics to treat resistant pathogens.

THREATS FROM NEW OR REEMERGING PATHOGENS

HIV/AIDS, SARS, influenza H1N1, MERS, Ebola virus, and COVID-19 share a common trait. All of these diseases are animal-related, i.e., zoonotic diseases—those that can be transmitted between animals and humans. Zoonotic diseases represent approximately 75% of the newly emerging diseases currently affecting people (9).

The history of the COVID-19 pandemic remains incomplete; it may be decades before we have a full understanding of its impact. On the heels of COVID-19, monkeypox has reemerged; its impact is unclear at the time of this writing. Unknown pathogens will continue to be found, some with structures and mechanisms totally unrelated to our currently described pathogens and carrying the risk of causing a pandemic. Demonstrating a relationship between newly discovered microorganisms and human disease will remain an enduring challenge. We owe a great debt to trailblazing scientists such as Pasteur, Koch, Barré-Sinoussi, Marshall, and their colleagues, who showed openness to the discovery of previously undescribed microorganisms and proved their relationship to disease to a doubting scientific community. More importantly, these scientists adhered to rigorous methods to determine the exact nature of the relationship of the microorganism to a disease. We may not be able to strictly follow Koch's postulates to prove the relationship between a new pathogen and human disease, since we may not be able to cultivate newly discovered microorganisms in the laboratory as simply and easily as Koch did. New technology that can determine the presence of a microorganism by its DNA fingerprint will uncover potential pathogens whose role in human disease may be difficult to prove. Microorganisms have been inaccurately linked to many clinical entities; e.g., the agent of chronic fatigue syndrome has been variably ascribed to Epstein-Barr virus, *Candida albicans*, *Borrelia burgdorferi*, enterovirus, and cytomegalovirus, among others (10). We would do well to remember the rigor and meticulousness with which Pasteur, Koch, Barré-Sinoussi, Marshall, and others worked out the methods to prove the causal role of microorganisms in human disease. The path that they forged can be an inspiration, although their feuds remind us of human frailty, even among the most gifted minds.

The alarming spread of SARS-CoV-2, monkeypox, methicillin-resistant *Staphylococcus aureus* (MRSA), and other antibiotic-resistant pathogens through hospitals and communities is forcing a reevaluation of measures to control the spread of pathogens. Hand hygiene has never been more important, as increasingly resistant strains are spread in the health care environment. Despite the Joint Commission requirement to implement Centers for Disease Control and Prevention (CDC) guidelines on hand hygiene, those individuals charged with the task of trying to ensure good hand hygiene compliance among health care workers find frustration. Rates of hand washing or hand hygiene rarely exceed 60%. Physicians nearly always have the worst adherence (11). With all the research efforts over the last 150 years, studies on hand hygiene have never equaled Semmelweis's work in power or clarity. Yet health care workers need to be continually reminded about Semmelweis's lessons on hand hygiene (12).

STRENGTHENING PUBLIC HEALTH

Weaknesses in public health responsiveness and infrastructure were exposed with the COVID-19 pandemic. Whether these weaknesses will be remedied once the public health crises abate remains unclear. History suggests that they may not be fully addressed. In 2009, the U.S. Agency for International Development (USAID) launched the Emerging Pandemic Threats program, a 5-year program targeting "the early detection of new disease threats and enhanced 'national-level' preparedness and response capacities for their effective control." The PREDICT project established a network of partnerships in 36 countries and strengthened the virus detection capacity of 67 collaborating labs, which detected a new Ebola virus, Bombali virus, in free-tailed bats in Sierra Leone and Guinea; this discovery marked the first detection of an Ebola virus before disease has been reported in humans or animals. PREDICT found animal reservoirs of other viruses such as coronavirus (bats in Uganda), Ebola virus (bats in Liberia), and Marburg virus (bats in Sierra Leone). The PREDICT funding was cut in September 2019, just months before the COVID-19 pandemic began (13). SARS-CoV-2 will likely not be the last microorganism to launch a pandemic. We must learn lessons from the

COVID-19 pandemic, especially with regard to public health. During periods when public health is not actively threatened, we should not let down our guard. Rather, it is the time to strengthen public health measures in preparation.

THE CHALLENGE OF VACCINES

The unprecedented speed of the preparation, approval, and administration of COVID-19 vaccines represented a major success, but implementation of wide-scale vaccination and the evolution of SARS-CoV-2 remain serious challenges for control of the pandemic. For the HIV vaccine, new concepts and approaches to our fundamental understanding of immunity and vaccine formation are needed. Despite our considerable experience with vaccines over the last century, the choice of vaccine antigens for HIV-1 has been largely empirical, usually including HIV-1 structural genes (14). Perhaps we have not advanced far from Jenner's innovative empirical trials with cowpox. Even though Jenner's vaccination led to major gains for humanity, he spent much of his life combating the antivaccinationists, a struggle that continues today (15). History shows us how the medical community and society dealt with Jenner's tireless efforts to demonstrate that vaccination protects the human constitution from smallpox. His efforts took a great personal toll on him. But we can also learn from and be inspired by his labors to prove to the world the value of vaccination. His opponents sounded many of the same unscientific alarms that can be heard today over the use of vaccines. But an HIV vaccine will require a better understanding of aspects of the immune response to HIV infection, especially at the entry point of HIV in most individuals, the mucous membranes. The complex task of trying new formulations for an HIV vaccine will hopefully build on the modest success of the Thai trial (16). In 2010, when I asked Dr. Mark Mulligan, then Executive Director of The Hope Clinic of the Emory Vaccine Center, how long he thought it would take for a successful HIV vaccine, he said,

> We are still talking about a time frame that is longer than any of us would like, perhaps 15 years, minimum.

In 2022, we are no closer to an HIV vaccine than we were in 2010 (17). HIV vaccine efforts may require more innovative or courageous approaches than those currently planned. With the stakes so high with HIV infection worldwide, we must entertain thoughtful but bold approaches and can look back to Jenner's daring for inspiration.

Finally, we often focus only on the scientific issues related to infectious diseases and need to be reminded that social and economic problems will continue to lead to countless infectious disease problems. Global warming is affecting not only human habitats but those of microorganisms. Because of thermal adaptation, certain fungi, e.g., *Candida auris*, appear to have responded to global climate change (18). As we struggle with social, environmental, and economic problems, the strength, passion, and single-mindedness of people like Lillian Wald and Anthony Fauci should serve as inspiring examples of creative ways to fund the path for progress in combating not just the microorganisms but societal ills.

REFERENCES

1. **World Health Organization.** 2001. *WHO Global Strategy for Containment of Antimicrobial Resistance.* World Health Organization, Geneva, Switzerland.
2. **Collaborators AR, Antimicrobial Resistance Collaborators.** 2022. Global burden of bacterial antimicrobial resistance in 2019: a systematic analysis. *Lancet* **399:**629–655. http://dx.doi.org/10.1016/S0140-6736(21)02724-0.
3. **Gaynes RP.** 2010. Preserving the effectiveness of antibiotics. *JAMA* **303:** 2293–2294. http://dx.doi.org/10.1001/jama.2010.766.
4. **Infectious Diseases Society of America.** 2010. The 10 x 20 Initiative: pursuing a global commitment to develop 10 new antibacterial drugs by 2020. *Clin Infect Dis* **50:**1081–1083. http://dx.doi.org/10.1086/652237.
5. **Hutchings MI, Truman AW, Wilkinson B.** 2019. Antibiotics: past, present and future. *Curr Opin Microbiol* **51:**72–80. http://dx.doi.org/10.1016/j.mib.2019.10.008.
6. **Kollef MH.** 1994. Antibiotic use and antibiotic resistance in the intensive care unit: are we curing or creating disease? *Heart Lung* **23.**363–367.
7. **Casadevall A.** 1996. Crisis in infectious diseases: time for a new paradigm? *Clin Infect Dis* **23:** 790–794. http://dx.doi.org/10.1093/clinids/23.4.790.
8. **van de Veerdonk FL, Giamarellos-Bourboulis E, Pickkers P, Derde L, Leavis H, van Crevel R, Engel JJ, Wiersinga WJ, Vlaar APJ, Shankar-Hari M, van der Poll T, Bonten M, Angus DC, van der Meer JWM, Netea MG**. 2022. A guide to immunotherapy for COVID-19. *Nat Med* **28:**39–50. http://dx.doi.org/10.1038/s41591-021-01643-9.
9. **EcoHealth Alliance.** PREDICT. https://www.ecohealthalliance.org/program/predict.
10. **Engleberg N.** 2000. Chronic fatigue syndrome, p 1529–1534. *In* Mandell GL, Bennett JE, Dolin R (ed), *Principles and Practice of Infectious Diseases*, 5th ed. Churchill Livingstone, Philadelphia, PA.
11. **Erasmus V, Daha TJ, Brug H, Richardus JH, Behrendt MD, Vos MC, van Beeck EF.** 2010. Systematic review of studies on compliance with hand hygiene guidelines in hospital care. *Infect Control Hosp Epidemiol* **31:**283–294. http://dx.doi.org/10.1086/650451.
12. **Haas JP, Larsen EL.** 2008. Compliance with hand hygiene guidelines: where are we in 2008? *Am J Nurs* **108:**40–44.
13. **U.S. Agency for International Development** 12 July 2021. Emerging Pandemic Threats program. https://www.usaid.gov/ept2.
14. **McElrath MJ**. 2010. Immune responses to HIV vaccines and potential impact on control of acute HIV-1 infection. *J Infect Dis* **202**(Suppl 2):S323–S326. http://dx.doi.org/10.1086/655658.
15. **Poland GA, Jacobson RM.** 2011. The age-old struggle against the antivaccinationists. *N Engl J Med* **364:**97–99. http://dx.doi.org/10.1056/NEJMp1010594.
16. **Rerks-Ngarm S, Pitisuttithum P, Nitayaphan S, Kaewkungwal J, Chiu J, Paris R, Premsri N, Namwat C, de Souza M, Adams E, Benenson M, Gurunathan S, Tartaglia J, McNeil JG, Francis DP, Stablein D, Birx DL, Chunsuttiwat S, Khamboonruang C, Thongcharoen P, Robb ML, Michael NL, Kunasol P, Kim JH; MOPH-TAVEG Investigators.** 2009. Vaccination with ALVAC and AIDSVAX to prevent HIV-1 infection in Thailand. *N Engl J Med* **361(23):**2209–2220.
17. **Kim J, Vasan S, Kim JH, Ake JA.** 2021. Current approaches to HIV vaccine development: a narrative review. *J Int AIDS Soc* **24**(Suppl 7):e25793.
18. **Casadevall A, Kontoyiannis DP, Robert V.** 2021. Environmental *Candida auris* and the global warming emergence hypothesis. *MBio* **12:**e00360–e21. http://dx.doi.org/10.1128/mBio.00360-21.

Index